D1765827

Texts in Computational Science and Engineering 10

For further volumes:
http://www.springer.com/series/5151

Mats G. Larson • Fredrik Bengzon

The Finite Element Method: Theory, Implementation, and Applications

Mats G. Larson
Fredrik Bengzon
Department of Mathematics
Umeå University
Umeå
Sweden

ISSN 1611-0994
ISBN 978-3-642-33286-9 ISBN 978-3-642-33287-6 (eBook)
DOI 10.1007/978-3-642-33287-6
Springer Heidelberg New York Dordrecht London

Library of Congress Control Number: 2012954548

Mathematics Subject Classification (2010): 65N30, 65N15, 65M60, 65M15

Printed on acid-free paper

Springer is part of Springer Science+Business Media (www.springer.com)

Preface

This book gives an introduction to the finite element method as a general computational method for solving partial differential equations approximately. Our approach is mathematical in nature with a strong focus on the underlying mathematical principles, such as approximation properties of piecewise polynomial spaces, and variational formulations of partial differential equations, but with a minimum level of advanced mathematical machinery from functional analysis and partial differential equations.

In principle, the material should be accessible to students with only knowledge of calculus of several variables, basic partial differential equations, and linear algebra, as the necessary concepts from more advanced analysis are introduced when needed.

Throughout the text we emphasize implementation of the involved algorithms, and have therefore mixed mathematical theory with concrete computer code using the numerical software MATLAB[1] and its PDE-Toolbox. A basic knowledge of the MATLAB language is therefore necessary. The PDE-Toolbox is used for pre and post processing (i.e., meshing and plotting).

We have also had the ambition to cover some of the most important applications of finite elements and the basic finite element methods developed for those applications, including diffusion and transport phenomena, solid and fluid mechanics, and also electromagnetics.

The book is loosely divided into two parts Chaps. 1–6 which provides basic material and Chaps. 7–14 which covers more advanced material and applications. In the first part Chaps. 1–4 gives an introduction to the finite element method for stationary second order elliptic problems. Here we consider the one dimensional case in Chaps. 1 and 2 and then extend to two dimensions in Chaps. 3 and 4. In Chap. 5 we consider time dependent problems and in Chap. 6 we give an introduction to numerical linear algebra for sparse linear systems of equations. In the second more advanced part we present the abstract theory in Chap. 7, various finite elements in Chap. 8, and a short introduction to nonlinear problems

[1] MATLAB is a registered trademark of The MathWorks Inc. (www.mathworks.com)

in Chap. 9. In Chaps. 10–13 we consider applications to transport problems, solid mechanics, fluid mechanics, and electromagnetics. Finally, in Chap. 14 we give a short introduction to discontinuous Galerkin methods.

The book is based on lecture notes used in various courses given by the authors and their coworkers during the last eight years at Chalmers University of Technology, Umeå University, Uppsala University, and at the University of Oslo. Several different courses in engineering and mathematics programs can be based on the material in the book. The following are some examples of possible courses that we have had in mind while developing the material:

- Short introduction to finite elements as part of a calculus or numerical analysis course. Chapters 1 and 2.
- Introduction to finite elements as part of a calculus or numerical analysis course, with focus on one dimensional stationary and time dependent problems. Chapters 1 and 2 and parts of Chapter 5.
- Introduction to finite elements as part of a calculus or numerical analysis course, with focus on stationary problems in one and two dimensions. Chapters 1–4.
- First course on finite elements. Chapters 1–6.
- Second course on finite elements and its applications. Chapters 7–14.

Umeå, Sweden

Mats G. Larson
Fredrik Bengzon

Acknowledgements

This book is based on material developed for courses given at Chalmers University of Technology, Umeå University, Uppsala University, and the University of Oslo during the last 8 years and the authors gratefully acknowledge the contributions of the teachers and students involved.

We also wish to thank Timothy J. Barth, (Nasa Ames), Axel Målqvist (Uppsala University), Hans-Petter Langtangen (Simula Research Laboratory and Oslo University), Thomas Ericsson (Chalmers University of Technology), and Håkan Jakobsson, Robert Söderlund, Karl Larsson, Per Vesterlund, and Jakob Öhrman (Umeå University), for valuable discussions and comments on the manuscript.

Contents

Chapter 1
Piecewise Polynomial Approximation in 1D

Abstract In this chapter we introduce a type of functions called piecewise polynomials that can be used to approximate other more general functions, and which are easy to implement in computer software. For computing piecewise polynomial approximations we present two techniques, interpolation and L^2-projection. We also prove estimates for the error in these approximations.

1.1 Piecewise Polynomial Spaces

1.1.1 The Space of Linear Polynomials

Let $I = [x_0, x_1]$ be an interval on the real axis and let $P_1(I)$ denote the vector space of linear functions on I, defined by

$$P_1(I) = \{v : v(x) = c_0 + c_1 x,\ x \in I,\ c_0,\ c_1 \in \mathbb{R}\} \tag{1.1}$$

In other words $P_1(I)$ contains all functions of the form $v(x) = c_0 + c_1 x$ on I.

Perhaps the most natural basis for $P_1(I)$ is the monomial basis $\{1, x\}$, since any function v in $P_1(I)$ can be written as a linear combination of 1 and x. That is, a constant c_0 times 1 plus another constant c_1 times x. In doing so, v is clearly determined by specifying c_0 and c_1, the so-called coefficients of the linear combination. Indeed, we say that v has two degrees of freedom.

However, c_0 and c_1 are not the only degrees of freedom possible for v. To see this, recall that a line, or linear function, is uniquely determined by requiring it to pass through any two given points. Now, obviously, there are many pairs of points that can specify the same line. For example, $(0, 1)$ and $(2, 3)$ can be used to specify $v = x + 1$, but so can $(-1, 0)$ and $(4, 5)$. In fact, any pair of points within the interval I will do as degrees of freedom for v. In particular, v can be uniquely determined by its values $\alpha_0 = v(x_0)$ and $\alpha_1 = v(x_1)$ at the end-points x_0 and x_1 of I.

M.G. Larson and F. Bengzon, *The Finite Element Method: Theory, Implementation, and Applications*, Texts in Computational Science and Engineering 10,
DOI 10.1007/978-3-642-33287-6_1, © Springer-Verlag Berlin Heidelberg 2013

To prove this, let us assume that the values $\alpha_0 = v(x_0)$ and $\alpha_1 = v(x_1)$ are given. Inserting $x = x_0$ and $x = x_1$ into $v(x) = c_0 + c_1 x$ we obtain the linear system

$$\begin{bmatrix} 1 & x_0 \\ 1 & x_1 \end{bmatrix} \begin{bmatrix} c_0 \\ c_1 \end{bmatrix} = \begin{bmatrix} \alpha_0 \\ \alpha_1 \end{bmatrix} \tag{1.2}$$

for c_i, $i = 1, 2$.

Computing the determinant of the system matrix we find that it equals $x_1 - x_0$, which also happens to be the length of the interval I. Hence, the determinant is positive, and therefore there exist a unique solution to (1.2) for any right hand side vector. Moreover, as a consequence, there is exactly one function v in $P_1(I)$, which has the values α_0 and α_1 at x_0 and x_1, respectively. In the following we shall refer to the points x_0 and x_1 as the nodes.

We remark that the system matrix above is called a Vandermonde matrix.

Knowing that we can completely specify any function in $P_1(I)$ by its node values α_0 and α_1 we now introduce a new basis $\{\lambda_0, \lambda_1\}$ for $P_1(I)$. This new basis is called a nodal basis, and is defined by

$$\lambda_j(x_i) = \begin{cases} 1, & \text{if } i = j \\ 0, & \text{if } i \neq j \end{cases}, \quad i, j = 0, 1 \tag{1.3}$$

From this definition we see that each basis function λ_j, $j = 0, 1$, is a linear function, which takes on the value 1 at node x_j, and 0 at the other node.

The reason for introducing the nodal basis is that it allows us to express any function v in $P_1(I)$ as a linear combination of λ_0 and λ_1 with α_0 and α_1 as coefficients. Indeed, we have

$$v(x) = \alpha_0 \lambda_0(x) + \alpha_1 \lambda_1(x) \tag{1.4}$$

This is in contrast to the monomial basis, which given the node values requires inversion of the Vandermonde matrix to determine the corresponding coefficients c_0 and c_1.

The nodal basis functions take the following explicit form on I

$$\lambda_0(x) = \frac{x_1 - x}{x_1 - x_0}, \qquad \lambda_1(x) = \frac{x - x_0}{x_1 - x_0} \tag{1.5}$$

This follows directly from the definition (1.3), or by solving the linear system (1.2) with $[1,\ 0]^T$ and $[0,\ 1]^T$ as right hand sides.

1.1.2 *The Space of Continuous Piecewise Linear Polynomials*

A natural extension of linear functions is piecewise linear functions. In constructing a piecewise linear function, v, the basic idea is to first subdivide the domain of v into smaller subintervals. On each subinterval v is simply given by a linear function.

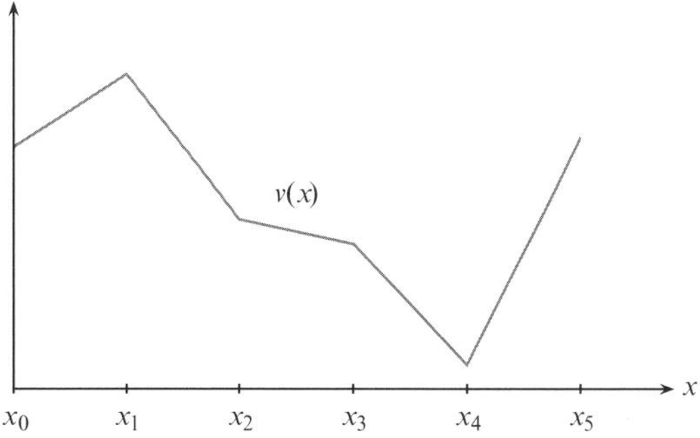

Fig. 1.1 A continuous piecewise linear function v

Continuity of v between adjacent subintervals is enforced by placing the degrees of freedom at the start- and end-points of the subintervals. We shall now formalize this more mathematically stringent.

Let $I = [0, L]$ be an interval and let the $n + 1$ node points $\{x_i\}_{i=0}^n$ define a partition

$$\mathcal{I} : 0 = x_0 < x_1 < x_2 < \ldots < x_{n-1} < x_n = L \tag{1.6}$$

of I into n subintervals $I_i = [x_{i-1}, x_i]$, $i = 1, 2 \ldots, n$, of length $h_i = x_i - x_{i-1}$. We refer to the partition $\mathcal{I}$ as to a mesh.

On the mesh $\mathcal{I}$ we define the space V_h of continuous piecewise linear functions by

$$V_h = \{v : v \in C^0(I),\ v|_{I_i} \in P_1(I_i)\} \tag{1.7}$$

where $C^0(I)$ denotes the space of continuous functions on I, and $P_1(I_i)$ denotes the space of linear functions on I_i. Thus, by construction, the functions in V_h are linear on each subinterval I_i, and continuous on the whole interval I. An example of such a function is shown in Fig. 1.1

It should be intuitively clear that any function v in V_h is uniquely determined by its nodal values

$$\{v(x_i)\}_{i=0}^n \tag{1.8}$$

and, conversely, that for any set of given nodal values $\{\alpha_i\}_{i=0}^n$ there exist a function v in V_h with these nodal values. Motivated by this observation we let the nodal values define our degrees of freedom and introduce a basis $\{\varphi_j\}_{j=0}^n$ for V_h associated with the nodes and such that

$$\varphi_j(x_i) = \begin{cases} 1, & \text{if } i = j \\ 0, & \text{if } i \neq j \end{cases}, \quad i, j = 0, 1, \ldots, n \tag{1.9}$$

The resulting basis functions are depicted in Fig. 1.2.

Because of their shape the basis functions φ_i are often called hat functions. Each hat function is continuous, piecewise linear, and takes a unit value at its own node x_i, while being zero at all other nodes. Consequently, φ_i is only non-zero on the two

Fig. 1.2 A typical hat function φ_i on a mesh. Also shown is the "half hat" φ_0

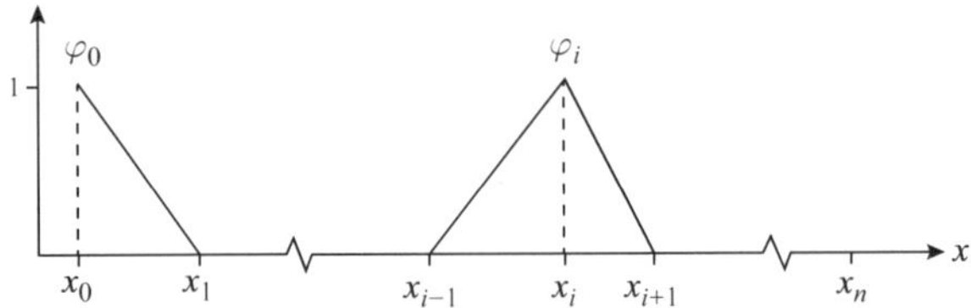

intervals I_i and I_{i+1} containing node x_i. Indeed, we say that the support of φ_i is $I_i \cup I_{i+1}$. The exception is the two "half hats" φ_0 and φ_n at the leftmost and rightmost nodes $a = x_0$ and $x_n = b$ with support only on one interval.

By construction, any function v in V_h can be written as a linear combination of hat functions $\{\varphi_i\}_{i=0}^n$ and corresponding coefficients $\{\alpha_i\}_{i=0}^n$ with $\alpha_i = v(x_i)$, $i = 0, 1, \dots, n$, the nodal values of v. That is,

$$v(x) = \sum_{i=0}^{n} \alpha_i \varphi_i(x) \tag{1.10}$$

The explicit expressions for the hat functions are given by

$$\varphi_i = \begin{cases} (x - x_{i-1})/h_i, & \text{if } x \in I_i \\ (x_{i+1} - x)/h_{i+1}, & \text{if } x \in I_{i+1} \\ 0, & \text{otherwise} \end{cases} \tag{1.11}$$

1.2 Interpolation

We shall now use the function spaces $P_1(I)$ and V_h to construct approximations, one from each space, to a given function f. The method we are going to use is very simple and only requires the evaluation of f at the node points. It is called interpolation.

1.2.1 Linear Interpolation

As before, we start on a single interval $I = [x_0, x_1]$. Given a continuous function f on I, we define the linear interpolant $\pi f \in P_1(I)$ to f by

$$\pi f(x) = f(x_0)\varphi_0 + f(x_1)\varphi_1 \tag{1.12}$$

We observe that interpolant approximates f in the sense that the values of πf and f are the same at the nodes x_0 and x_1 (i.e., $\pi f(x_0) = f(x_0)$ and $\pi f(x_1) = f(x_1)$).

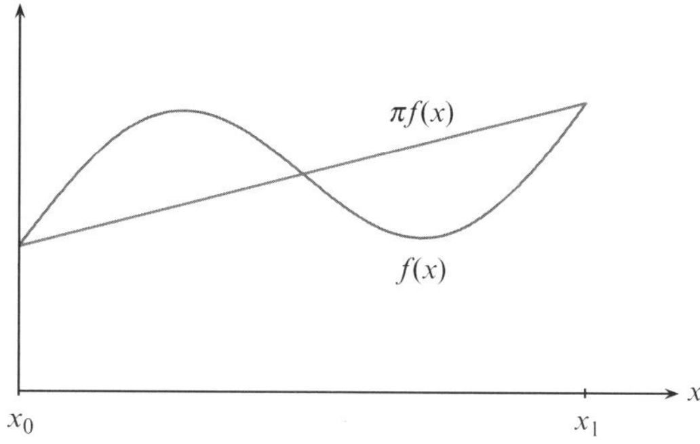

Fig. 1.3 A function f and its linear interpolant πf

In Fig. 1.3 we show a function f and its linear interpolant πf.

Unless f is linear, πf will only approximate f, and it is therefore of interest to measure the difference $f - \pi f$, which is called the interpolation error. To this end, we need a norm. Now, there are many norms and it is not easy to know which is the best. For instance, should we measure the interpolation error in the infinity norm, defined by

$$\|v\|_\infty = \max_{x \in I} |v(x)| \tag{1.13}$$

or the $L^2(I)$-norm defined, for any square integrable function v on I, by

$$\|v\|_{L^2(I)} = \left(\int_I v^2 \, dx \right)^{1/2} \tag{1.14}$$

We shall use the latter norm, since it captures the average size of a function, whereas the former only captures the pointwise maximum.

In this context we recall that the $L^2(I)$-norm, or any norm for that matter, obeys the Triangle inequality

$$\|v + w\|_{L^2(I)} \leq \|v\|_{L^2(I)} + \|w\|_{L^2(I)} \tag{1.15}$$

as well as the Cauchy-Schwarz inequality

$$\int_I vw \, dx \leq \|v\|_{L^2(I)} \|w\|_{L^2(I)} \tag{1.16}$$

for any two functions v and w in $L^2(I)$.

Then, using the L^2-norm to measure the interpolation error, we have the following results.

Proposition 1.1. *The interpolant πf satisfies the estimates*

$$\|f - \pi f\|_{L^2(I)} \leq Ch^2 \|f''\|_{L^2(I)} \tag{1.17}$$

$$\|(f - \pi f)'\|_{L^2(I)} \leq Ch \|f''\|_{L^2(I)} \tag{1.18}$$

where C is a constant, and $h = x_1 - x_0$.

Proof. Let $e = f - \pi f$ denote the interpolation error.

From the fundamental theorem of calculus we have, for any point y in I,

$$e(y) = e(x_0) + \int_{x_0}^{y} e' \, dx \tag{1.19}$$

where $e(x_0) = f(x_0) - \pi f(x_0) = 0$ due to the definition of πf.

Now, using the Cauchy-Schwarz inequality we have

$$e(y) = \int_{x_0}^{y} e' \, dx \tag{1.20}$$

$$\leq \int_{x_0}^{y} |e'| \, dx \tag{1.21}$$

$$\leq \int_I 1 \cdot |e'| \, dx \tag{1.22}$$

$$\leq \left(\int_I 1^2 \, dx \right)^{1/2} \left(\int_I e'^2 \, dx \right)^{1/2} \tag{1.23}$$

$$= h^{1/2} \left(\int_I e'^2 \, dx \right)^{1/2} \tag{1.24}$$

or, upon squaring both sides,

$$e(y)^2 \leq h \int_I e'^2 \, dx = h \|e'\|^2_{L^2(I)} \tag{1.25}$$

Integrating this inequality over I we further have

$$\|e\|^2_{L^2(I)} = \int_I e(y)^2 \, dy \leq \int_I h \|e'\|^2_{L^2(I)} \, dy = h^2 \|e'\|^2_{L^2(I)} \tag{1.26}$$

since the integrand to the right of the inequality is independent of y. Thus, we have

$$\|e\|_{L^2(I)} \leq h \|e'\|_{L^2(I)} \tag{1.27}$$

With a similar, but slightly different argument, we also have

$$\|e'\|_{L^2(I)} \leq h \|e''\|_{L^2(I)} \tag{1.28}$$

Hence, we conclude that

$$\|e\|_{L^2(I)} \leq h \|e'\|_{L^2(I)} \leq h^2 \|e''\|_{L^2(I)} \tag{1.29}$$

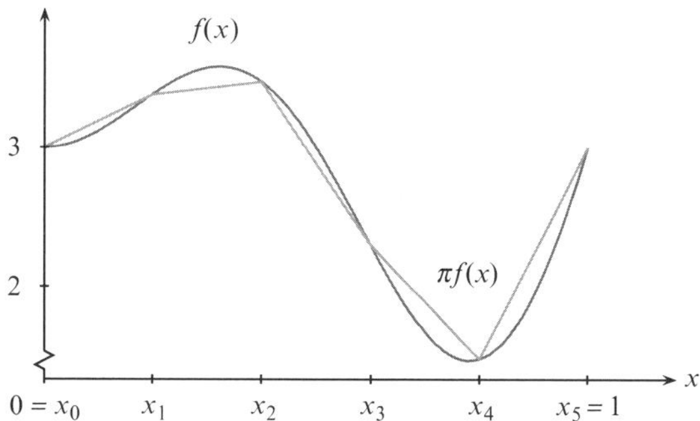

Fig. 1.4 The function $f(x) = 2x\sin(2\pi x) + 3$ and its continuous piecewise linear interpolant $\pi f(x)$ on a uniform mesh of $I = [0, 1]$ with six nodes x_i, $i = 0, 1, \dots, 5$

from which the first inequality of the proposition follows by noting that since πf is linear $e'' = f''$. The second inequality of the proposition follows similarly from (1.26)

The difference in argument between deriving (1.27) and (1.28) has to do with the fact that we can not simply replace e with e' in (1.19), since $e'(x_0) \neq 0$. However, noting that $e(x_0) = e(x_1) = 0$, there exist by Rolle's theorem a point $\bar{x}$ in I such that $e'(\bar{x}) = 0$, which means that

$$e'(y) = e'(\bar{x}) + \int_{\bar{x}}^{y} e''\,dx = \int_{\bar{x}}^{y} e''\,dx \tag{1.30}$$

Starting instead from this and proceeding as shown above (1.28) follows. □

Examining the proof of Proposition 1.1 we note that the constant C equals unity and could equally well be left out. We have, however, chosen to retain this constant, since the estimates generalize to higher spatial dimensions, where C is not unity. The important thing to understand is how the interpolation error depends on the interpolated function f, and the size of the interval h.

1.2.2 Continuous Piecewise Linear Interpolation

It is straight forward to extend the concept of linear interpolation on a single interval to continuous piecewise linear interpolation on a mesh. Indeed, given a continuous function f on the interval $I = [0, L]$, we define its continuous piecewise linear interpolant $\pi f \in V_h$ on a mesh $\mathcal{I}$ of I by

$$\pi f(x) = \sum_{i=1}^{n} f(x_i)\varphi_i(x) \tag{1.31}$$

Figure 1.4 shows the continuous piecewise linear interpolant $\pi f(x)$ to $f(x) = 2x\sin(2\pi x) + 3$ on a uniform mesh of $I = [0, 1]$ with 6 nodes.

Regarding the interpolation error $f - \pi f$ we have the following results.

Proposition 1.2. *The interpolant πf satisfies the estimates*

$$\|f - \pi f\|_{L^2(I)}^2 \leq C \sum_{i=1}^{n} h_i^4 \|f''\|_{L^2(I_i)}^2 \tag{1.32}$$

$$\|(f - \pi f)'\|_{L^2(I)}^2 \leq C \sum_{i=1}^{n} h_i^2 \|f''\|_{L^2(I_i)}^2 \tag{1.33}$$

Proof. Using the Triangle inequality and Proposition 1.1, we have

$$\|f - \pi f\|_{L^2(I)}^2 = \sum_{i=1}^{n} \|f - \pi f\|_{L^2(I_i)}^2 \tag{1.34}$$

$$\leq \sum_{i=1}^{n} C h_i^4 \|f''\|_{L^2(I_i)}^2 \tag{1.35}$$

which proves the first estimate. The second follows similarly. □

Proposition 1.2 says that the interpolation error vanish as the mesh size h_i tends to zero. This is natural, since we expect the interpolant πf to be a better approximation to f where ever the mesh is fine. The proposition also says that if the second derivative f'' of f is large then the interpolation error is also large. This is also natural, since if the graph of f bends a lot (i.e., if f'' is large) then f is hard to approximate using a piecewise linear function.

1.3 L^2-Projection

Interpolation is a simple way of approximating a continuous function, but there are, of course, other ways. In this section we shall study so-called orthogonal-, or L^2-projection. L^2-projection gives a so to speak good on average approximation, as opposed to interpolation, which is exact at the nodes. Moreover, in contrast to interpolation, L^2-projection does not require the function we seek to approximate to be continuous, or have well-defined node values.

1.3.1 Definition

Given a function $f \in L^2(I)$ the L^2-projection $P_h f \in V_h$ of f is defined by

$$\int_I (f - P_h f) v \, dx = 0, \quad \forall v \in V_h \tag{1.36}$$

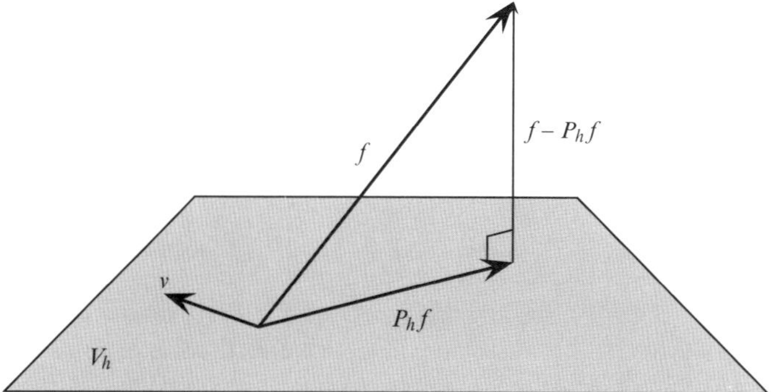

Fig. 1.5 Illustration of the function f and its L^2-projection $P_h f$ on the space V_h

Fig. 1.6 The function $f(x) = 2x\sin(2\pi x) + 3$ and its L^2-projection $P_h f$ on a uniform mesh of $I = [0, 1]$ with six nodes, x_i, $i = 1, 2, \dots, 6$

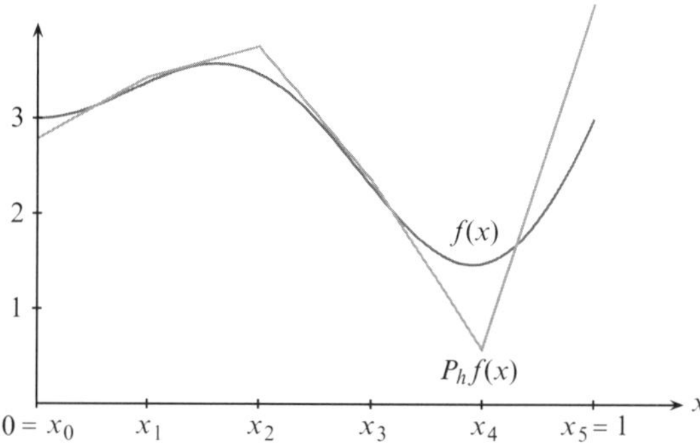

In analogy with projection onto subspaces of $\mathbb{R}^n$, (1.34) defines a projection of f onto V_h, since the difference $f - P_h f$ is required to be orthogonal to all functions v in V_h. This is illustrated in Fig. 1.5.

As we shall see later on, $P_h f$ is the minimizer of $\min_{v \in V_h} \|f - v\|_{L^2(I)}$, and therefore we say that it approximates f in a least squares sense. In fact, $P_h f$ is the best approximation to f when measuring the error $f - P_h f$ in the L^2-norm.

In Fig. 1.6 we show the L^2-projection of $f(x) = 2x\sin(2\pi x) + 3$ computed on the same mesh as was used for showing the continuous piecewise linear interpolant πf in Fig. 1.4. It is instructive to compare these two approximations because it highlights their different characteristics. The interpolant πf approximates f exactly at the nodes, while the L^2-projection $P_h f$ approximates f on average. In doing so, it is common for $P_h f$ to over and under shoot local maxima and minima of f, respectively. Also, both the interpolant and the L^2-projection have difficulty with approximating rapidly oscillating or discontinuous functions unless the node positions are adjusted appropriately.

1.3.2 Derivation of a Linear System of Equations

In order to actually compute the L^2-projection $P_h f$, we first note that the definition (1.36) is equivalent to

$$\int_I (f - P_h f)\varphi_i \, dx = 0, \quad i = 0, 1, \dots, n \tag{1.37}$$

where φ_i, $i = 0, 1, \dots, n$, are the hat functions. This is a consequence of the fact that if (1.36) is satisfied for v anyone of the hat functions, then it is also satisfied for v a linear combination of hat functions. Conversely, since any function v in V_h is precisely such a linear combination of hat functions, (1.37) implies (1.36).

Then, since $P_h f$ belongs to V_h it can be written as the linear combination

$$P_h f = \sum_{j=0}^{n} \xi_j \varphi_j \tag{1.38}$$

where ξ_j, $j = 0, 1, \dots, n$, are $n + 1$ unknown coefficients to be determined.

Inserting the ansatz (1.38) into (1.37) we get

$$\int_I f\varphi_i \, dx = \int_I \left(\sum_{j=0}^{n} \xi_j \varphi_j \right) \varphi_i \, dx \tag{1.39}$$

$$= \sum_{j=0}^{n} \xi_j \int_I \varphi_j \varphi_i \, dx, \quad i = 0, 1, \dots, n \tag{1.40}$$

Further, introducing the notation

$$M_{ij} = \int_I \varphi_j \varphi_i \, dx, \quad i, j = 0, 1, \dots, n \tag{1.41}$$

$$b_i = \int_I f\varphi_i \, dx, \quad i = 0, 1, \dots, n \tag{1.42}$$

we have

$$b_i = \sum_{j=0}^{n} M_{ij} \xi_j, \quad i = 0, 1, \dots, n \tag{1.43}$$

which is an $(n + 1) \times (n + 1)$ linear system for the $n + 1$ unknown coefficients ξ_j, $j = 0, 1, \dots, n$. In matrix form, we write this

$$M\xi = b \tag{1.44}$$

where the entries of the $(n+1) \times (n+1)$ matrix M and the $(n+1) \times 1$ vector b are defined by (1.41) and (1.42), respectively.

We, thus, conclude that the coefficients ξ_j, $j = 0, 1, \dots, n$ in the ansatz (1.38) satisfy a square linear system, which must be solved in order to obtain the L^2-projection $P_h f$.

For historical reasons we refer to M as the mass matrix and to b as the load vector.

1.3.3 Basic Algorithm to Compute the L^2-Projection

The following algorithm summarizes the basic steps for computing the L^2-projection $P_h f$:

Algorithm 1 Basic algorithm to compute the L^2-projection

1: Create a mesh with n elements on the interval I and define the corresponding space of continuous piecewise linear functions V_h.

2: Compute the $(n+1) \times (n+1)$ matrix M and the $(n+1) \times 1$ vector b, with entries

$$M_{ij} = \int_I \varphi_j \varphi_i \, dx, \qquad b_i = \int_I f \varphi_i \, dx \tag{1.45}$$

3: Solve the linear system

$$M\xi = b \tag{1.46}$$

4: Set

$$P_h f = \sum_{j=0}^{n} \xi_j \varphi_j \tag{1.47}$$

1.3.4 A Priori Error Estimate

Naturally, we are interested in knowing how good $P_h f$ is at approximating f. In particular, we wish to derive bounds for the error $f - P_h f$. The next theorem gives a key result for deriving such error estimates. It is a so-called a best approximation result.

Theorem 1.1. *The L^2-projection $P_h f$, defined by* (1.36), *satisfies the best approximation result*

$$\|f - P_h f\|_{L^2(I)} \le \|f - v\|_{L^2(I)}, \quad \forall v \in V_h \tag{1.48}$$

Proof. Using the definition of the L^2-norm and writing $f - P_h f = f - v + v - P_h f$, with v an arbitrary function in V_h, we have

$$\|f - P_h f\|^2_{L^2(I)} = \int_I (f - P_h f)(f - v + v - P_h f)\,dx \tag{1.49}$$

$$= \int_I (f - P_h f)(f - v)\,dx + \int_I (f - P_h f)(v - P_h f)\,dx \tag{1.50}$$

$$= \int_I (f - P_h f)(f - v)\,dx \tag{1.51}$$

$$\leq \|f - P_h f\|_{L^2(I)} \|f - v\|_{L^2(I)} \tag{1.52}$$

where we used the definition of the L^2-projection to conclude that

$$\int_I (f - P_h f)(v - P_h f)\,dx = 0 \tag{1.53}$$

since $v - P_h f \in V_h$. Dividing by $\|f - P_h f\|_{L^2(I)}$ concludes the proof. □

This shows that $P_h f$ is the closest of all functions in V_h to f when measuring in the L^2-norm. Hence, the name best approximation result.

We can use best approximation result together with interpolation estimates to study how the error $f - P_h f$ depends on the mesh size. In doing so, we have the following basic so-called a priori error estimate.

Theorem 1.2. *The L^2-projection $P_h f$ satisfies the estimate*

$$\|f - P_h f\|^2_{L^2(I)} \leq C \sum_{i=1}^{n} h_i^4 \|f''\|^2_{L^2(I_i)} \tag{1.54}$$

Proof. Starting from the best approximation result, choosing $v = \pi f$ the interpolant of f, and using the interpolation error estimate of Proposition 1.1, we have

$$\|f - P_h f\|^2_{L^2(I)} \leq \|f - \pi f\|^2_{L^2(I)} \tag{1.55}$$

$$\leq \sum_{i=1}^{n} \|f - \pi f\|^2_{L^2(I_i)} \tag{1.56}$$

$$\leq \sum_{i=1}^{n} C h_i^4 \|f''\|^2_{L^2(I_i)} \tag{1.57}$$

which proves the estimate. □

Defining $h = \max_{1 \leq i \leq n} h_i$ we conclude that

$$\|f - P_h f\|_{L^2(I)} \leq C h^2 \|f''\|_{L^2(I)} \tag{1.58}$$

Thus, the L^2-error $\|f - P_h f\|_{L^2(I)}$ tends to zero as the maximum mesh size h tends to zero.

1.4 Quadrature

To compute the L_2-projection we need to compute the mass matrix M whose entries are integrals involving products of hat functions. One way of doing this is to use quadrature, or, numerical integration. To this end, f be a continuous function on the interval $I = [x_0, x_1]$, and consider the problem of evaluating, approximately, the integral

$$J = \int_I f(x)\,dx \tag{1.59}$$

A quadrature rule is a formula that is used to compute integrals approximately. It it usually derived by first interpolating the integrand f by a polynomial and then integrating the interpolant. Depending on the degree of the interpolating polynomial one obtains quadrature rules of different computational complexity and accuracy. Evaluating a quadrature rule generally involves summing values of the integrand f at a set of carefully selected quadrature points within the interval I times the interval length $h = x_1 - x_0$. We shall next describe three classical quadrature rules called the Mid-point rule, the Trapezoidal rule, and Simpson's formula, which corresponds to using polynomial interpolation of degree 0, 1, and 2 on f, respectively.

1.4.1 The Mid-Point Rule

Interpolating f with the constant $f(m)$, where $m = (x_0 + x_1)/2$ is the mid-point of I, we get

$$J \approx f(m)h \tag{1.60}$$

which is the Mid-point rule. Geometrically this means that we approximate the area under the integrand f with the area of the square $f(m)h$, see Fig. 1.7. The Mid-point rule integrates linear polynomials exactly.

1.4.2 The Trapezoidal Rule

Continuing, interpolating f with the line passing through the points $(x_0, f(x_0))$ and $(x_1, f(x_1))$ we get

$$J \approx \frac{f(x_0) + f(x_1)}{2}h \tag{1.61}$$

which is the Trapezoidal rule. Geometrically this means that we approximate the area under f with the area under the trapezoidal with the four corner points $(x_0, 0)$,

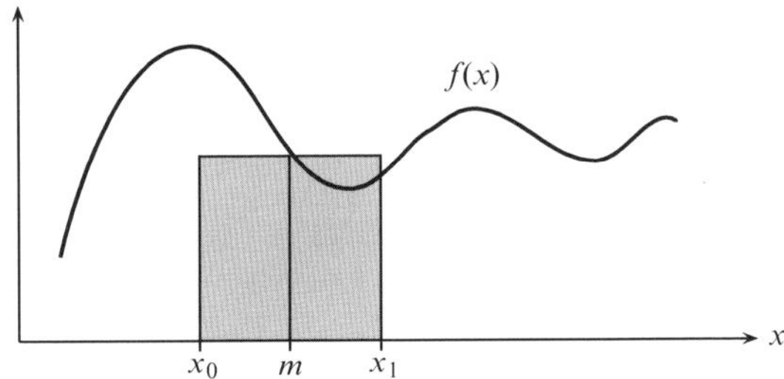

Fig. 1.7 The area of the *shaded square* approximates $J = \int_I f(x)\,dx$

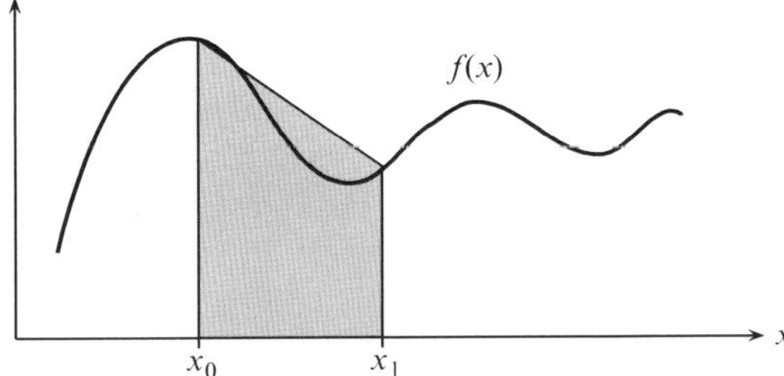

Fig. 1.8 The area of the *shaded* trapezoidal approximates $J = \int_I f(x)\,dx$

$(x_0, f(x_0))$, $(x_1, 0)$, and $(x_1, f(x_1))$, see Fig. 1.8. The Trapezoidal rule is also exact for linear polynomials.

1.4.3 Simpson's Formula

This rule corresponds to a quadratic interpolant using the end-points and the mid-point of the interval I as nodes. To simplify things a bit let $I = [0, l]$ be the interval of integration and let $g(x) = c_0 + c_1 x + c_2 x^2$ be the interpolant. Since g interpolates f at the points $(0, f(0))$, $(\frac{l}{2}, f(\frac{l}{2}))$, and $(l, f(l))$ (i.e., its graph passes trough these points) their coordinates must satisfy the equation for g. This yields the following linear system for c_0, c_1, and c_2.

$$\begin{bmatrix} 0 & 0 & 1 \\ \frac{1}{4}l^2 & \frac{1}{2}l & 1 \\ l^2 & l & 1 \end{bmatrix} \begin{bmatrix} c_0 \\ c_1 \\ c_2 \end{bmatrix} = \begin{bmatrix} f(0) \\ f(\frac{l}{2}) \\ f(l) \end{bmatrix} \tag{1.62}$$

Solving this one readily finds

$$c_0 = 2(f(0) - 2f(\tfrac{l}{2}) + f(l))/l^2, \quad c_1 = -(3f(0) - 4f(\tfrac{l}{2}) + f(l))/l, \quad c_2 = f(0) \tag{1.63}$$

Now, integrating g from 0 to l one eventually ends up with

$$\int_0^l g(x)\,dx = \frac{f(0) + 4f(\frac{1}{2}l) + f(l)}{6}l \tag{1.64}$$

which is Simpson's formula.

On the interval $I = [x_0, x_1]$ Simpson's formula takes the form

$$J \approx \frac{f(x_0) + 4f(m) + f(x_1)}{6}h \tag{1.65}$$

with $m = \frac{1}{2}(x_0 + x_1)$ and $h = x_1 - x_0$.

Simpson's formula is exact for third order polynomials.

1.5 Computer Implementation

1.5.1 Assembly of the Mass Matrix

Having studied various quadrature rules, let us now go through the nitty gritty details of how to assemble the mass matrix M and load vector b. We begin by calculating the entries $M_{ij} = \int_I \varphi_i\varphi_j\,dx$ of the mass matrix. Because each hat φ_i is a linear polynomial the product of two hats is a quadratic polynomial. Thus, Simpson's formula can be used to integrate M_{ij} exactly. In doing so, since the hats φ_i and φ_j lack common support for $|i - j| > 1$ only M_{ii}, M_{ii+1}, and M_{i+1i} need to be calculated. All other matrix entries are zero by default. This is clearly seen from Fig. 1.9 showing two neighbouring hat functions and their support. This leads to the observation that the mass matrix M is tridiagonal.

Starting with the diagonal entries M_{ii} and using Simpson's formula we have

$$M_{ii} = \int_I \varphi_i^2\,dx \tag{1.66}$$

$$= \int_{x_{i-1}}^{x_i} \varphi_i^2\,dx + \int_{x_i}^{x_{i+1}} \varphi_i^2\,dx \tag{1.67}$$

$$= \frac{0 + 4\cdot(\frac{1}{2})^2 + 1}{6}h_i + \frac{1 + 4\cdot(\frac{1}{2})^2 + 0}{6}h_{i+1} \tag{1.68}$$

$$= \frac{h_i}{3} + \frac{h_{i+1}}{3}, \quad i = 1, 2, \dots, n-1 \tag{1.69}$$

where $x_i - x_{i-1} = h_i$ and $x_{i+1} - x_i = h_{i+1}$. The first and last diagonal entry are $M_{00} = h_1/3$ and $M_{nn} = h_n/3$, respectively, since the hat functions φ_0 and φ_n are only half.

Continuing with the subdiagonal entries M_{i+1i}, still using Simpson's formula, we have

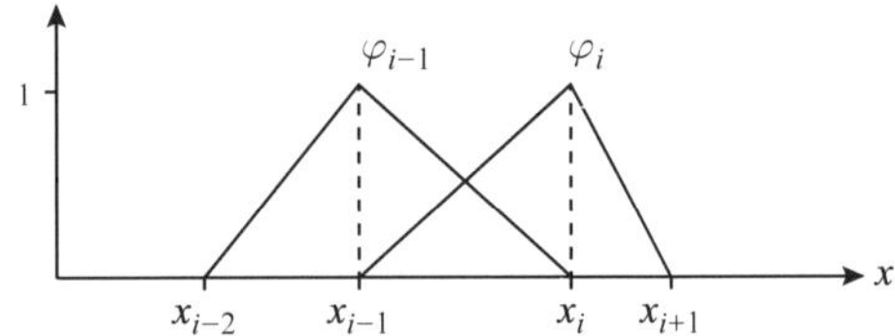

Fig. 1.9 Illustration of the hat functions φ_{i-1} and φ_i and their support

$$M_{i+1\,i} = \int_I \varphi_i \varphi_{i+1}\, dx \tag{1.70}$$

$$= \int_{x_i}^{x_{i+1}} \varphi_i \varphi_{i+1}\, dx \tag{1.71}$$

$$= \frac{1 \cdot 0 + 4 \cdot (\frac{1}{2})^2 + 0 \cdot 1}{6} h_{i+1} \tag{1.72}$$

$$= \frac{h_{i+1}}{6}, \quad i = 0, 1, \dots, n \tag{1.73}$$

A similar calculation shows that the superdiagonal entries are $M_{i\,i+1} = h_{i+1}/6$.

Hence, the mass matrix takes the form

$$M = \begin{bmatrix} \frac{h_1}{3} & \frac{h_1}{6} & & & & \\ \frac{h_1}{6} & \frac{h_1}{3} + \frac{h_2}{3} & \frac{h_2}{6} & & & \\ & \frac{h_2}{6} & \frac{h_2}{3} + \frac{h_3}{3} & \frac{h_3}{6} & & \\ & & \ddots & \ddots & \ddots & \\ & & & \frac{h_{n-1}}{6} & \frac{h_{n-1}}{3} + \frac{h_n}{3} & \frac{h_n}{6} \\ & & & & \frac{h_n}{6} & \frac{h_n}{3} \end{bmatrix} \tag{1.74}$$

From (1.74) it is evident that the global mass matrix M can be written as a sum of n simple matrices, viz.,

$$M = \begin{bmatrix} \frac{h_1}{3} & \frac{h_1}{6} & & \\ \frac{h_1}{6} & \frac{h_1}{3} & & \\ & & & \\ & & & \end{bmatrix} + \begin{bmatrix} & & & \\ & \frac{h_2}{3} & \frac{h_2}{6} & \\ & \frac{h_2}{6} & \frac{h_2}{3} & \\ & & & \end{bmatrix} + \dots + \begin{bmatrix} & & & \\ & & & \\ & & \frac{h_n}{3} & \frac{h_n}{6} \\ & & \frac{h_n}{6} & \frac{h_n}{3} \end{bmatrix} \tag{1.75}$$

$$= M^{I_1} + M^{I_2} + \dots + M^{I_n} \tag{1.76}$$

Each matrix M^{I_i}, $i = 1, 2 \dots, n$, is obtained by restricting integration to one subinterval, or element, I_i and is therefore called a global element mass matrix. In practice, however, these matrices are never formed, since it suffice to compute their 2×2 blocks of non-zero entries. From the sum (1.75) we see that on each element I this small block takes the form

$$M^I = \frac{1}{6}\begin{bmatrix} 2 & 1 \\ 1 & 2 \end{bmatrix} h \tag{1.77}$$

where h is the length of I. We refer to M^I as the local element mass matrix.

The summation of the element mass matrices into the global mass matrix is called assembling. The assembly process lies at the very heart of finite element programming because it allows the forming of the mass matrix through the use of a single loop over the elements. It also generalizes to higher dimensions.

The following algorithm summarizes how to assemble the mass matrix M:

Algorithm 2 Assembly of the mass matrix

Allocate memory for the $(n+1)\times(n+1)$ matrix M and initialize all matrix entries to zero.
for $i = 1, 2, \dots, n$ **do**
  Compute the 2×2 local element mass matrix M^I given by

$$M^I = \frac{1}{6}\begin{bmatrix} 2 & 1 \\ 1 & 2 \end{bmatrix} h \tag{1.78}$$

  where h is the length of element I_i.
  Add M^I_{11} to M_{ii}
  Add M^I_{12} to M_{ii+1}
  Add M^I_{21} to M_{i+1i}
  Add M^I_{22} to M_{i+1i+1}
end for

The following MATLAB routine assembles the mass matrix.

```
function M = MassAssembler1D(x)
n = length(x)-1; % number of subintervals
M = zeros(n+1,n+1); % allocate mass matrix
for i = 1:n % loop over subintervals
    h = x(i+1) - x(i); % interval length
    M(i,i) = M(i,i) + h/3; % add h/3 to M(i,i)
    M(i,i+1) = M(i,i+1) + h/6;
    M(i+1,i) = M(i+1,i) + h/6;
    M(i+1,i+1) = M(i+1,i+1) + h/3;
end
```

Input to this routine is a vector `x` holding the node coordinates. Output is the global mass matrix.

1.5.2 Assembly of the Load Vector

We next calculate the load vector b. Because the entries $b_i = \int_i f\varphi_i\, dx$ depend on the function f, we can not generally expect to calculate them exactly. However, we can approximate entry b_i using a quadrature rule. Using the Trapezoidal rule, for instance, we have

$$
\begin{aligned}
b_i &= \int_I f\varphi_i\, dx && (1.79)\\
&= \int_{x_{i-1}}^{x_{i+1}} f\varphi_i\, dx && (1.80)\\
&= \int_{x_{i-1}}^{x_i} f\varphi_i\, dx + \int_{x_i}^{x_{i+1}} f\varphi_i\, dx && (1.81)\\
&\approx (f(x_{i-1})\varphi_i(x_{i-1}) + f(x_i)\varphi_i(x_i))h_i/2 && (1.82)\\
&\quad + (f(x_i)\varphi_i(x_i) + f(x_{i+1})\varphi_i(x_{i+1}))h_{i+1}/2 && (1.83)\\
&= (0 + f(x_i))h_i/2 + (f(x_i) + 0)h_{i+1}/2 && (1.84)\\
&= f(x_i)(h_i + h_{i+1})/2 && (1.85)
\end{aligned}
$$

The approximate load vector then takes the form

$$
b = \begin{bmatrix} f(x_0)h_1/2 \\ f(x_1)(h_1 + h_2)/2 \\ f(x_2)(h_2 + h_3)/2 \\ \vdots \\ f(x_{n-1})(h_{n-1} + h_n)/2 \\ f(x_n)h_n/2 \end{bmatrix} \tag{1.86}
$$

Splitting b into a sum over the elements yields the n global element load vectors b^{I_i}

$$
b = \begin{bmatrix} f(x_0) \\ f(x_1) \\ \\ \\ \\ \end{bmatrix} h_1/2 + \begin{bmatrix} \\ f(x_1) \\ f(x_2) \\ \\ \\ \end{bmatrix} h_2/2 + \ldots + \begin{bmatrix} \\ \\ \\ f(x_{n-1}) \\ f(x_n) \end{bmatrix} h_n/2 \tag{1.87}
$$

$$
= b^{I_1} + b^{I_2} + \ldots + b^{I_n}. \tag{1.88}
$$

Each vector b^{I_i}, $i = 1, 2, \ldots, n$, is obtained by restricting integration to element I_i. The assembly of the load vector b is very similar to that of the mass matrix as the following algorithm shows:

Algorithm 3 Assembly of the load vector

1: Allocate memory for the $(n + 1) \times 1$ vector b and initialize all vector entries to zero.
2: **for** $i = 1, 2, \ldots, n$ **do**
3: Compute the 2×1 local element load vector b^I given by

$$b^I = \frac{1}{2} \begin{bmatrix} f(x_{i-1}) \\ f(x_i) \end{bmatrix} h \tag{1.89}$$

where h is the length of element I_i.
4: Add b_1^I to b_{i-1}
5: Add b_2^I to b_i
6: **end for**

A MATLAB routine for assembling the load vector is listed below.

```
function b = LoadAssembler1D(x,f)
n = length(x)-1;
b = zeros(n+1,1);
for i = 1:n
  h = x(i+1) - x(i);
  b(i) = b(i) + f(x(i))*h/2;
  b(i+1) = b(i+1) + f(x(i+1))*h/2;
end
```

Here, f is assumed to be a separate routine specifying the function f. This needs perhaps a little bit of explanation. MATLAB has a something called function handles, which provide a way of passing a routine as argument to another routine. For example, suppose we have written a routine called Foo1 to specify the function $f(x) = x \sin(x)$

```
function y = Foo1(x)
y=x.*sin(x)
```

To assemble the corresponding load vector, we type

```
b = LoadAssembler1D(x,@Foo1)
```

This passes the routine Foo1 as argument to LoadAssembler1D and allows it to be evaluated inside the assembler. The at sign @ creates the function handle. Indeed, function handles provide means for writing flexible and reusable code.

In this context we mention that if Foo1 is a small routine, then it can be inlined and called, viz.,

```
Foo1 = inline('x.*sin(x)','x')
b = LoadAssembler1D(x,Foo1)
```

Note that there is no at sign in the call to the load vector assembler.

Putting it all together we get the following main routine for computing L^2-projections.

```
function L2Projector1D()
n = 5; % number of subintervals
h = 1/n; % mesh size
x = 0:h:1; % mesh
M = MassAssembler1D(x); % assemble mass
b = LoadAssembler1D(x,@Foo1); % assemble load
Pf = M\b; % solve linear system
plot(x,Pf) % plot L^2 projection
```

1.6 Problems

Exercise 1.1. Let $I = [x_0, x_1]$. Verify by direct calculation that the basis functions

$$\lambda_0(x) = \frac{x_1 - x}{x_1 - x_0}, \qquad \lambda_1(x) = \frac{x - x_0}{x_1 - x_0}$$

for $P_1(I)$ satisfies $\lambda_0(x) + \lambda_1(x) = 1$ and $x_0\lambda_0(x) + x_1\lambda_1(x) = x$. Give a geometrical interpretation by drawing $\lambda_0(x)$, $\lambda_1(x)$, $\lambda_0(x) + \lambda_1(x)$, $x_0\lambda_0(x)$, $x_1\lambda_1(x)$ and $x_0\lambda_0(x) + x_1\lambda_1(x)$.

Exercise 1.2. Let $0 = x_0 < x_1 < x_2 < x_3 = 1$, where $x_1 = 1/6$ and $x_2 = 1/2$, be a partition of the interval $I = [0, 1]$ into three subintervals, and let V_h be the space of continuous piecewise linear functions on this partition.

(a) Determine analytical expressions for the hat function $\varphi_1(x)$ and draw it.
(b) Draw the function $v(x) = -\varphi_0(x) + \varphi_2(x) + 2\varphi_3(x)$ and its derivative $v'(x)$.
(c) Draw the piecewise constant mesh function $h(x) = h_i$ on each subinterval I_i.
(d) What is the dimension of V_h?

Exercise 1.3. Determine the linear interpolant $\pi f \in P_1(I)$ on the single interval $I = [0, 1]$ to the following functions f.

(a) $f(x) = x^2$.
(b) $f(x) = 3\sin(2\pi x)$.

Make plots of f and πf in the same figure.

Exercise 1.4. Let V_h be the space of all continuous piecewise linears on a uniform mesh with four nodes of $I = [0, 1]$. Draw the interpolant $\pi f \in V_h$ to the following functions f.

(a) $f(x) = x^2 + 1$.
(b) $f(x) = \cos(\pi x)$.

Can you think of a better partition of I assuming we are restricted to three subintervals?

Exercise 1.5. Calculate $\|f\|_\infty$ with $f = x(x-1/2)(x-1/3)$ on the interval $I = [0,1]$.

Exercise 1.6. Let $I = [0,1]$ and $f(x) = x^2$ for $x \in I$.

(a) Calculate $\int_I f\,dx$ analytically.
(b) Compute $\int_I f\,dx$ using the Mid-point rule.
(c) Compute $\int_I f\,dx$ using the Trapezoidal rule.
(d) Compute the quadrature errors in (b) and (c).

Exercise 1.7. Let $I = [0,1]$ and $f(x) = x^4$ for $x \in I$.

(a) Calculate $\int_I f\,dx$ analytically.
(b) Compute $\int_I f\,dx$ using Simpson's formula on the single interval I.
(c) Divide I into two equal subintervals and compute $\int_I f\,dx$ using Simpson's formula on each subinterval.
(d) Compute the quadrature errors in (b) and (c). By what factor has the error decreased?

Exercise 1.8. Let $I = [0,1]$ and let $f(x) = x^2$ for $x \in I$.

(a) Let V_h be the space $P_1(I)$ of linear functions on I. Compute the L^2-projection $P_h f \in V_h$ of f.
(b) Divide I into two subintervals of equal length and let V_h be the corresponding space V_h of continuous piecewise linear functions. Compute the L^2-projection $P_h f \in V_h$ of f.
(c) Plot your results and compare with the nodal interpolant πf.

Exercise 1.9. Show that $\int_\Omega (f - P_h f)v\,dx = 0$ for all $v \in V_h$, if and only if $\int_\Omega (f - P_h f)\varphi_i\,dx = 0$, for $i = 0, 1, \dots, n$, where $\{\varphi_i\}_{i=0}^n \subset V_h$ is the usual basis of hat functions.

Exercise 1.10. Let $(f, g) = \int_I fg\,dx$ and $\|f\|_{L^2(I)}^2 = (f, f)$ denote the L^2-scalar product and norm, respectively. Also, let $I = [0,\pi]$, $f = x$, $g = \cos(x)$, and $h = 2\cos(3x)$ for $x \in I$.

(a) Calculate (f, g).
(b) Calculate (g, h). Are g and h orthogonal?
(c) Calculate $\|f\|_{L^2(I)}$ and $\|g\|_{L^2(I)}$.

Exercise 1.11. Let V be a linear subspace of $\mathbb{R}^n$ with basis $\{v_1, \dots, v_m\}$ with $m < n$. Let $Px \in V$ be the orthogonal projection of $x \in \mathbb{R}^n$ onto the subspace V. Derive a linear system of equations that determines Px. Note that your results are analogous to the L^2-projection when the usual scalar product in $\mathbb{R}^n$ is replaced by the scalar product in $L^2(I)$. Compare this method of computing the projection Px to the method used for computing the projection of a three dimensional vector onto a two dimensional subspace. What happens if the basis $\{v_1, \dots, v_m\}$ is L^2-orthogonal?

Exercise 1.12. Show that $\{1, x, (3x^2-1)/2\}$ form a basis for the space of quadratic polynomials $P_2(I)$, on $I = [-1, 1]$. Then compute and draw the L^2-projections $P_h f \in P_2(I)$ on I for the following two functions f.

(a) $f(x) = 1 + 2x$.
(b) $f(x) = x^3$.

Exercise 1.13. Show that the hat function basis $\{\varphi_j\}_{j=0}^n$ of V_h is almost orthogonal. How can we see that it is almost orthogonal by looking at the non-zero elements of the mass matrix? What can we say about the mass matrix if we had a fully orthogonal basis?

Exercise 1.14. Modify `L2Projector1D` and compute the L^2-projection $P_h f$ of the following functions f.

(a) $f(x) = 1$.
(b) $f(x) = x^3(x-1)(1-2x)$.
(c) $f(x) = \arctan((x-0.5)/\epsilon)$, with $\epsilon = 0.1$ and 0.01.

Use a uniform mesh $\mathcal{I}$ of the interval $I = [0, 1]$ with $n = 5, 25$, and 100 subintervals.

Chapter 2
The Finite Element Method in 1D

Abstract In this chapter we shall introduce the finite element method as a general tool for the numerical solution of two-point boundary value problems. In doing so, the basic idea is to first rewrite the boundary value problem as a variational equation, and then seek a solution approximation to this equation from the space of continuous piecewise linears. We prove basic error estimates and show how to use these to formulate adaptive algorithms that can be used to automatically improve the accuracy of the computed solution. The derivation and areas of application of the studied boundary value problems are also discussed.

2.1 The Finite Element Method for a Model Problem

2.1.1 A Two-point Boundary Value Problem

Let us consider the following two-point boundary value problem: find u such that

$$-u'' = f, \quad x \in I = [0, L] \tag{2.1a}$$

$$u(0) = u(L) = 0 \tag{2.1b}$$

where f is a given function. Sometimes this problem is easy to solve analytically. For example, if $f = 1$, then we readily find $u = x(L-x)/2$ by integrating f twice and using the boundary conditions $u(0) = u(L) = 0$. However, for a general f it may be difficult or even impossible to find u with analytical techniques. Thus, we see that even a very simple differential equation like this one may be difficult to solve analytically. We take this as a good motivation for introducing the finite element method, which is a general numerical technique for solving differential equations.

M.G. Larson and F. Bengzon, *The Finite Element Method: Theory, Implementation, and Applications*, Texts in Computational Science and Engineering 10,
DOI 10.1007/978-3-642-33287-6_2,

2.1.2 Variational Formulation

The derivation of a finite element method always starts by rewriting the differential equation under consideration as a variational equation. This so-called variational formulation is in our case obtained by multiplying $f = -u''$ by a test function v, which is assumed to vanish at the end-points of the interval I, and integrating by parts.

$$\int_0^L f v \, dx = -\int_0^L u'' v \, dx \tag{2.2}$$

$$= \int_0^L u' v' \, dx - u'(L)v(L) + u'(0)v(0) \tag{2.3}$$

$$= \int_0^L u' v' \, dx \tag{2.4}$$

The last line follows from the assumption $v(0) = v(L) = 0$. For this calculation to make sense we must assert that the test function v is not too badly behaved so that the involved integrals do indeed exist. More specific, we require that both v and v' be square integrable on I. Of course, v must also vanish at $x = 0$ and $x = L$. Now, the largest collection of functions with these properties is given by the function space

$$V_0 = \{v : \|v\|_{L^2(I)} < \infty,\ \|v'\|_{L^2(I)} < \infty,\ v(0) = v(L) = 0\} \tag{2.5}$$

Obviously, this space contains many functions, and any of them can be used as test function v. In fact, there are infinitely many functions in V_0, and we therefore say that V_0 has infinite dimension.

Not just v, but also u, is a member of V_0. To see this, note that u is twice differentiable, which implies that u' is smooth, and satisfies the boundary conditions $u(0) = u(L) = 0$. This allows us to write down the following variational formulation of (2.2): find $u \in V_0$ such that

$$\int_I u' v' \, dx = \int_I f v dx, \quad \forall v \in V_0 \tag{2.6}$$

By analogy with the name test function for v, the solution u is sometimes called trial function.

2.1.3 Finite Element Approximation

We next try to approximate u by a continuous piecewise linear function. To this end, we introduce a mesh on the interval I consisting of n subintervals, and the corresponding space V_h of all continuous piecewise linears. Since we are dealing with functions vanishing at the end-points of I, we also introduce the following subspace $V_{h,0}$ of V_h that satisfies the boundary conditions

$$V_{h,0} = \{v \in V_h : v(0) = v(L) = 0\} \tag{2.7}$$

In other words, $V_{h,0}$ contains all piecewise linears which are zero at $x = 0$ and $x = L$. In terms of hat functions this means that a basis for $V_{h,0}$ is obtained by deleting the half hats φ_0 and φ_n from the usual set $\{\varphi_j\}_{j=0}^n$ of hat functions spanning V_h.

Replacing the large space V_0 with the much smaller subspace $V_{h,0} \subset V_0$ in the variational formulation (2.6), we obtain the following finite element method: find $u_h \in V_{h,0}$ such that

$$\int_I u_h' v' \, dx = \int_I f v \, dx, \quad \forall v \in V_{h,0} \tag{2.8}$$

We mention that this type of finite element method, with similar trial and test space, is sometimes called a Galerkin method, named after a famous Russian mathematician and engineer.

2.1.4 Derivation of a Linear System of Equations

In order to actually compute the finite element approximation u_h we first note that (2.8) is equivalent to

$$\int_I u_h' \varphi_i' \, dx = \int_I f \varphi_i \, dx, \quad i = 1, 2, \dots, n-1 \tag{2.9}$$

where, as said before, φ_i, $i = 1, 2, \dots, n-1$ are the hat functions spanning $V_{h,0}$. This is a consequence of the fact that if (2.9) is satisfied for all hat functions $\{\varphi_j\}_{j=1}^{n-1}$, then it is also satisfied for a linear combination of hats.

Then, since u_h belongs to $V_{h,0}$ we can write it as the linear combination

$$u_h = \sum_{j=1}^{n-1} \xi_j \varphi_j \tag{2.10}$$

where ξ_j, $j = 1, 2 \dots, n-1$, are $n-1$ unknown coefficients to be determined.

Inserting the ansatz (2.10) into (2.9) we get

$$\int_I f \varphi_i \, dx = \int_I \left(\sum_{j=1}^{n-1} \xi_j \varphi_j' \right) \varphi_i' \, dx \tag{2.11}$$

$$= \sum_{j=1}^{n-1} \xi_j \int_I \varphi_j' \varphi_i' \, dx, \quad i = 1, 2, \dots, n-1 \tag{2.12}$$

Further, introducing the notation

$$A_{ij} = \int_I \varphi_j' \varphi_i' \, dx, \quad i, j = 1, 2, \ldots, n-1 \tag{2.13}$$

$$b_i = \int_I f \varphi_i \, dx, \quad i = 1, 2, \ldots, n-1 \tag{2.14}$$

we have

$$b_i = \sum_{j=1}^{n-1} A_{ij} \xi_j, \quad i = 1, 2, \ldots, n-1 \tag{2.15}$$

which is an $(n-1) \times (n-1)$ linear system for the $n-1$ unknown coefficients ξ_j, $j = 1, 2, \ldots, n-1$. In matrix form, we write this

$$A\xi = b \tag{2.16}$$

where the entries of the $(n-1) \times (n-1)$ matrix A and the $(n-1) \times 1$ vector b are defined by (2.13) and (2.14), respectively.

We, thus, conclude that the coefficients ξ_j, $j = 1, 2, \ldots, n-1$ in the ansatz (2.10) satisfy a square linear system, which must be solved in order to obtain the finite element solution u_h.

We refer to A as the stiffness matrix and to b as the load vector.

2.1.5 Basic Algorithm to Compute the Finite Element Solution

The following algorithm summarizes the basic steps for computing the finite element solution u_h:

Algorithm 4 Basic finite element algorithm

1: Create a mesh with n elements on the interval I and define the corresponding space of continuous piecewise linear functions $V_{h,0}$.

2: Compute the $(n-1) \times (n-1)$ matrix A and the $(n-1) \times 1$ vector b, with entries

$$A_{ij} = \int_I \varphi_j' \varphi_i' \, dx, \qquad b_i = \int_I f \varphi_i \, dx \tag{2.17}$$

3: Solve the linear system

$$A\xi = b \tag{2.18}$$

4: Set

$$u_h = \sum_{j=1}^{n-1} \xi_j \varphi_j \tag{2.19}$$

2.1.6 A Priori Error Estimate

Because u_h generally only approximates u, estimates of the error $e = u - u_h$ are necessary to judge the quality and usability of u_h. To this end, we make the following key observation.

Theorem 2.1 (Galerkin orthogonality). *The finite element approximation u_h, defined by* (2.9)*, satisfies the orthogonality*

$$\int_I (u - u_h)' v' \, dx = 0, \quad \forall v \in V_{h,0} \tag{2.20}$$

Proof. From the variational formulation we have

$$\int_I u' v' \, dx = \int_I f v \, dx, \quad \forall v \in V_0 \tag{2.21}$$

and from the finite element method we further have

$$\int_I u_h' v' \, dx = \int_I f v \, dx, \quad \forall v \in V_{h,0} \tag{2.22}$$

Subtracting these and using the fact that $V_{h,0} \subset V_0$ immediately proves the claim. □

The next theorem is a best approximation result.

Theorem 2.2. *The finite element solution u_h, defined by* (2.9)*, satisfies the best approximation result*

$$\|(u - u_h)'\|_{L^2(I)} \leq \|(u - v)'\|_{L^2(i)}, \quad \forall v \in V_{h,0} \tag{2.23}$$

Proof. Writing $u - u_h = u - v + v - u_h$ for any $v \in V_{h,0}$, we have

$$\|(u - u_h)'\|_{L^2(i)}^2 = \int_I (u - u_h)'(u - v + v - u_h)' \, dx \tag{2.24}$$

$$= \int_I (u - u_h)'(u - v)' \, dx + \int_I (u - u_h)'(v - u_h)' \, dx \tag{2.25}$$

$$= \int_I (u - u_h)'(u - v)' \, dx \tag{2.26}$$

$$\leq \|(u - u_h)'\|_{L^2(I)} \|(u - v)'\|_{L^2(I)} \tag{2.27}$$

where we used the Galerkin orthogonality to conclude that

$$\int_I (u - u_h)'(v - u_h)'\, dx = 0 \tag{2.28}$$

since $v - u_h \in V_h$. Dividing by $\|(v - u_h)'\|_{L^2(I)}$ concludes the proof. □

There are two types of error estimates, namely, a priori and a posteriori error estimates. The difference between the two types is that a priori error estimates express the error in terms of the exact, unknown, solution u, while a posteriori error estimates express the error in terms of the computable finite element approximation u_h. The mesh size h_i is usually present in both types of estimates. This is important to be able to show convergence of the numeric methods.

We have the following basic a priori error estimate.

Theorem 2.3. *The finite element solution u_h, defined by* (2.9)*, satisfies the estimate*

$$\|(u - u_h)'\|^2_{L^2(I)} \leq C \sum_{i=1}^{n} h_i^2 \|u''\|^2_{L^2(I_i)} \tag{2.29}$$

where C is a constant.

Proof. Starting from the best approximation result, choosing $v = \pi u$ the interpolant of u, and using the interpolation error estimate of Proposition 1.1, the a priori estimate immediately follows. □

Recalling the definition $h = \max_{1 \leq i \leq n} h_i$ we conclude that

$$\|(u - u_h)'\|_{L^2(I)} \leq Ch\|u''\|_{L^2(I)} \tag{2.30}$$

Thus, the derivative of the error tends to zero as the maximum mesh size h tends to zero.

2.2 Mathematical Modeling

A fundamental tool for deriving the equations of applied mathematics and physics is the idea that some quantities can be tracked within a physical system. This idea is used to first create some balance laws for the system, and then to express these with equations. Common examples include conservation of mass, energy, and balance of momentum (i.e., force). To familiar ourselves with this way of thinking we shall now derive two differential equations, one governing heat transfer in a rod, and one governing the elastic deformation of a bar. As we shall see, the modeling of both these physical systems leads to the two-point boundary value problem (2.1). This might seem a little surprising, but it is generally so that many different physical phenomena are described by the same type of equations. As a consequence, the numerical methods and mathematical theory can often be developed for a few model problems, but still be applicable to a wide range of real-world problems.

2.2.1 Derivation of the Stationary Heat Equation

Consider a thin metal rod of length L occupying the interval $I = [0, L]$. The rod is heated by a heat source (e.g., an electric current) of intensity f [J/(sm)], which has been acting for a long time so that the transfer of heat within the rod is at steady state. We wish to find the temperature T [K] of the rod. To this end, we first use the first law of thermodynamics, which expresses conservation of energy, and says that the amount of heat produced by the heat source equals the flow of heat out of the rod. In the language of mathematics, this is equivalent to

$$A(L)q(L) - A(0)q(0) = \int_I f\,dx \tag{2.31}$$

where A [m^2] is the cross section area of the rod, and q [J/(sm^2)] is the heat flux in the direction of increasing x. Dividing (2.31) by L and letting $L \to 0$ we obtain the differential equation

$$(Aq)' = f \tag{2.32}$$

Then, since heat flows from hot to cold regions, it is reasonable to assume that the heat flux is proportional to the negative temperature gradient. This empirical observation is expressed by Fourier's law

$$q = -kT' \tag{2.33}$$

where k [J/(Kms)] is the thermal conductivity of the rod, and T the sought temperature.

Combining (2.32) and (2.33) we finally obtain

$$-(AkT')' = f \tag{2.34}$$

which is the stationary so-called Heat equation.

We note that this is a problem with variable coefficients, since A, k, and f might vary.

2.2.2 Boundary Conditions for the Heat Equation

Generally, there are many functions T, which satisfies the Heat equation (2.34) for a given right hand side. For example, if $A = k = 1$ and $f = 0$, then any linear function T satisfies $T'' = 0$. Thus, to obtain a unique solution T it is necessary to impose some auxiliary constraints on the equation. As we know, these are the boundary conditions, which specify T at the end-points 0 and L of the rod.

There are three types of boundary that occur frequently, namely:

- Dirichlet,
- Neumann, and
- Robin

boundary conditions. We shall describe these next.

2.2.2.1 Dirichlet Boundary Conditions

Dirichlet, or strong, boundary conditions prescribe the value of the solution at the boundary. For example, $T(L) = 0$. From a physical point of view this corresponds to cooling the right end-point $x = L$ of the rod so that it is always kept at constant zero temperature.

2.2.2.2 Neumann Boundary Conditions

Neumann, or natural, boundary conditions prescribe the value of the solution derivative at the boundary. Because $T' = -q/k$, this corresponds to prescribing the heat flux q at the boundary. Indeed, $T'(0) = 0$ means that the left end-point of the rod is thermally isolated.

2.2.2.3 Robin Conditions

Robin boundary conditions is a mixture of Dirichlet and Neumann ditto. For the rod they typically take the form

$$AkT' = \kappa(T - T_\infty) + q_\infty \tag{2.35}$$

where $\kappa > 0$, T_∞, and q_∞ are given boundary data.

In real-world applications this is perhaps the most realistic boundary condition.

A nice and useful thing with Robin boundary conditions is that they can be used to approximate boundary conditions of either Dirichlet or Neumann type. Indeed, choosing $\kappa = 0$ immediately yields the Neumann boundary condition $AkT' = q_\infty$. Choosing, on the other hand, κ very large yields the approximate Dirichlet boundary condition $T \approx T_\infty$.

2.2.3 *Derivation of a Differential Equation for the Deformation of a Bar*

A bar is a mechanical structure that is only subjected to axial loads.

Consider a bar occupying the interval $I = [0, L]$, and subjected to a line load f [N/m]. We wish to find the vertical displacement u [m] of the bar. To this end, we first assume equilibrium of forces in the bar, which implies

$$A(L)\sigma(L) - A(0)\sigma(0) + \int_I f\,dx = 0 \tag{2.36}$$

where A [m^2] is the cross section area of the bar, and σ [N/m^2] is the stress in the bar. By definition, $A(x)\sigma(x) = F(x)$ [N] is the vertical force $F(x)$ acting on the bar at any given point $0 \leq x \leq L$. Dividing (2.36) by L and letting $L \to 0$ we obtain the differential equation

$$-(A\sigma)' = f \tag{2.37}$$

Then, assuming that the bar is made of a linear elastic material, a relation between the stress and the deformation is given by Hooke's law

$$\sigma = E\varepsilon \tag{2.38}$$

where E is the elastic modulus (i.e., stiffness), and $\varepsilon = u'$ is the strain, with u the sought displacement.

Combining (2.37) and (2.38) we finally obtain

$$-(AEu')' = f \tag{2.39}$$

We note that this is, also, a problem with variable coefficients, since A, E, and f might vary.

2.2.4 Boundary Conditions for the Bar

2.2.4.1 Dirichlet Boundary Conditions

Dirichlet boundary conditions take the form $u = g_D$. They are used to model a given displacement g_D of the bar. For example, $u(0) = 0$ means that the bar is rigidly clamped to the surrounding at its left end-point $x = 0$.

2.2.4.2 Neumann Boundary Conditions

Neumann boundary conditions take the form $AEu' = g_N$, and model a situation when a given force of strength g_N acts on the boundary.

2.2.4.3 Robin Boundary Conditions

Robin boundary conditions is a mixture of Dirichlet and Neumann boundary ditto. They typically take the form $AEu' = \kappa(u - g_D)$. This models a situation where the bar is connected to a spring with spring constant $\kappa > 0$ such that $\kappa(u - g_D)$ is the force from the spring acting on the bar.

2.3 A Model Problem with Variable Coefficients

Considering two real-world applications just presented it is clear that we must be able to treat differential equations with variable coefficients and different types of boundary conditions. Therefore, let us consider the model problem: find u such that

$$-(au')' = f, \quad x \in I = [0, L] \tag{2.40a}$$

$$au'(0) = \kappa_0(u(0) - g_0) \tag{2.40b}$$

$$-au'(L) = \kappa_L(u(L) - g_L) \tag{2.40c}$$

where $a > 0$ and f are given functions, and $\kappa_0 \geq 0$, $\kappa_L \geq 0$, g_0, and g_L are given parameters.

We remark that positiveness assumptions on a, κ_0, and κ_L are necessary to assert existence and uniqueness of the solution u.

2.3.1 Variational Formulation

Multiplying $f = -(au')'$ by a test function v and integrating by parts, we have

$$\int_0^L f v \, dx = \int_0^L -(au')' v \, dx \tag{2.41}$$

$$= \int_0^L au'v' \, dx + a(L)u'(L)v(L) + a(0)u'(0)v(0) \tag{2.42}$$

$$= \int_0^L au'v' \, dx + \kappa_L(u(L) - g_L)v(L) + \kappa_0(u(0) - g_0)v(0) \tag{2.43}$$

where we used the boundary conditions to rewrite the boundary terms. As we do not make any assumptions about the specific values of v and u at $x = 0$ and $x = L$, the appropriate test and trial space is given by

$$V = \{v : \|v\|_{L^2(I)} < \infty, \ \|v'\|_{L^2(I)} < \infty\} \tag{2.44}$$

Collecting terms involving the unknown solution u on the left hand side, and terms involving given data on the right hand side, we obtain the following variational formulation of (2.40): find $u \in V$ such that

$$\int_I au'v'\,dx + \kappa_L u(L)v(L) + \kappa_0 u(0)v(0)$$
$$= \int_I fv\,dx + \kappa_L g_L v(L) + \kappa_0 g_0 v(0), \quad \forall v \in V \tag{2.45}$$

2.3.2 Finite Element Approximation

Replacing the continuous space V with the discrete space of continuous piecewise linears V_h in the variational formulation (2.45) we obtain the following finite element method: find $u_h \in V_h$ such that

$$\int_I au_h'v'\,dx + \kappa_L u_h(L)v(L) + \kappa_0 u_h(0)v(0)$$
$$= \int_I fv\,dx + \kappa_L g_L v(L) + \kappa_0 g_0 v(0), \quad \forall v \in V_h \tag{2.46}$$

We next show how to implement this finite element method.

2.4 Computer Implementation

2.4.1 Assembly of the Stiffness Matrix and Load Vector

A basis for V_h is given by the set of $n+1$ hat functions $\{\varphi_i\}_{i=0}^n$. Note that all hats are present, including the half hats φ_0 and φ_n at the interval end-points $x = 0$ and $x = L$.

Inserting the ansatz

$$u_h = \sum_{j=0}^{n} \xi_j \varphi_j \tag{2.47}$$

into the finite element method (2.46), and choosing $v = \varphi_i$, $i = 0, 1, \dots, n$, we end up with the linear system

$$(A + R)\xi = b + r \tag{2.48}$$

where the entries of the $(n+1) \times (n+1)$ matrices A and R, and the $n+1$ vectors b and r are given by

$$A_{ij} = \int_I a\varphi_j'\varphi_i'\,dx \tag{2.49}$$

$$R_{ij} = \kappa_L\varphi_j(L)\varphi_i(L) + \kappa_0\varphi_j(0)\varphi_i(0) \tag{2.50}$$

$$b_i = \int_I f\varphi_i\,dx \tag{2.51}$$

$$r_i = \kappa_L g_L\varphi_i(L) + \kappa_0 g_0\varphi_i(0) \tag{2.52}$$

To assemble A and b we recall that the explicit expression for a hat function φ_i is given by

$$\varphi_i = \begin{cases} (x - x_{i-1})/h_i, & \text{if } x \in I_i \\ (x_{i+1} - x)/h_{i+1}, & \text{if } x \in I_{i+1} \\ 0, & \text{otherwise} \end{cases} \tag{2.53}$$

Hence, the derivative φ_i' is either one of the constants $1/h_i$, $-1/h_{i+1}$, or 0 depending on the subinterval.

Using (2.53) it is straight forward to calculate the entries of A. For $|i - j| > 1$, we have $A_{ij} = 0$, since φ_i and φ_j lack common support. Thus, only A_{ii}, $A_{i\,i+1}$, and $A_{i+1\,i}$ need to be calculated. Let us use mid-point quadrature to do so. To simplify the notation, let a_i denote the value of the function a at the mid-point of element I_i. Then, when $i = j$ we have the diagonal entries

$$A_{ii} = \int_I a\varphi_i'^2\,dx \tag{2.54}$$

$$= \int_{x_{i-1}}^{x_i} a\varphi_i'^2\,dx + \int_{x_i}^{x_{i+1}} a\varphi_i'^2\,dx \tag{2.55}$$

$$\approx a_i\frac{1}{h_i^2}h_i + a_{i+1}\frac{(-1)^2}{h_{i+1}^2}h_{i+1} \tag{2.56}$$

$$= \frac{a_i}{h_i} + \frac{a_{i+1}}{h_{i+1}}, \quad i = 1, 2, \ldots, n-1 \tag{2.57}$$

The integrals of the first and last diagonal entries are a_1/h_1 and a_n/h_n since φ_0 and φ_n are only half.

Further, when $j = i + 1$ we have the subdiagonal entries

$$A_{i\,i+1} = \int_I a\varphi_{i+1}'\varphi_i'\,dx \tag{2.58}$$

$$= \int_{x_i}^{x_{i+1}} a\varphi_{i+1}'\varphi_i'\,dx \tag{2.59}$$

$$\approx a_{i+1}\frac{(-1)}{h_{i+1}}\cdot\frac{1}{h_{i+1}}h_{i+1} \tag{2.60}$$

$$= -\frac{a_{i+1}}{h_{i+1}}, \quad i = 0, 1, \dots, n \tag{2.61}$$

The superdiagonal entries are obviously the same as the subdiagonal entries.

The entries $R_{ij} = \kappa_0\varphi_j(0)\varphi_i(0) + \kappa_L\varphi_j(L)\varphi_i(L)$ are all zero, except when $i = j = 0$ or $i = j = n$, in which case we have $R_{00} = \kappa_0$ and $R_{nn} = \kappa_L$.

Hence, the stiffness matrix $A + R$ takes the form

$$A + R = \begin{bmatrix} \frac{a_1}{h_1} & -\frac{a_1}{h_1} & & & & \\ -\frac{a_1}{h_1} & \frac{a_1}{h_1}+\frac{a_2}{h_2} & -\frac{a_2}{h_2} & & & \\ & -\frac{a_2}{h_2} & \frac{a_2}{h_2}+\frac{a_3}{h_3} & -\frac{a_3}{h_3} & & \\ & & \ddots & \ddots & \ddots & \\ & & & -\frac{a_{n-1}}{h_{n-1}} & \frac{a_{n-1}}{h_{n-1}}+\frac{a_n}{h_n} & -\frac{a_n}{h_n} \\ & & & & -\frac{a_n}{h_n} & \frac{a_n}{h_n} \end{bmatrix} + \begin{bmatrix} \kappa_0 & & \\ & & \\ & & \kappa_L \end{bmatrix} \tag{2.62}$$

The computation of the load vector $b + r$ is done exactly as shown for the L^2-projection, apart from the addition of the terms $r_1 = \kappa_0 g_0 \varphi_i(0)$ and $r_n = \kappa_L g_L \varphi_i(L)$ to the first and last vector entry. Hence, we have

$$b + r = \begin{bmatrix} f(x_0)h_1/2 \\ f(x_1)(h_1+h_2)/2 \\ f(x_2)(h_2+h_3)/2 \\ \vdots \\ f(x_{n-1})(h_{n-1}+h_n)/2 \\ f(x_n)h_n/2 \end{bmatrix} + \begin{bmatrix} \kappa_0 g_0 \\ \\ \\ \vdots \\ \\ \kappa_L g_L \end{bmatrix} \tag{2.63}$$

The global stiffness matrix $A + R$ can be split into a sum of global element stiffness matrices

$$A + R = \frac{a_1}{h_1}\begin{bmatrix} 1 & -1 & & \\ -1 & 1 & & \\ & & & \\ & & & \end{bmatrix} + \frac{a_2}{h_2}\begin{bmatrix} & & & \\ & 1 & -1 & \\ & -1 & 1 & \\ & & & \end{bmatrix} + \dots + \frac{a_n}{h_n}\begin{bmatrix} & & & \\ & & & \\ & & 1 & -1 \\ & & -1 & 1 \end{bmatrix} \tag{2.64}$$

$$+\begin{bmatrix} \kappa_0 & & \\ & \ddots & \\ & & \kappa_L \end{bmatrix}$$

$$= A^{I_1} + A^{I_2} + \ldots + A^{I_n} + R \tag{2.65}$$

Each global element stiffness matrix A^{I_i}, $i = 1, 2, \ldots, n$ is found by restricting integration to a single element I_i. The following algorithm summarizes the assembly process of $A + R$:

Algorithm 5 Assembly of the stiffness matrix

1: Allocate memory for the $(n + 1) \times (n + 1)$ matrix A and initialize all matrix entries to zero.
2: **for** $i = 1, 2, \ldots, n$ **do**
3: Compute the 2×2 local element stiffness matrix A^I given by

$$A^I = \frac{a_i}{h} \begin{bmatrix} 1 & -1 \\ -1 & 1 \end{bmatrix} \tag{2.66}$$

where h is the length of element $I_i = [x_{i-1}, x_i]$, and $a_i = a((x_{i-1} + x_i)/2)$.
4: Add A^I_{11} to A_{ii}.
5: Add A^I_{12} to A_{ii+1}.
6: Add A^I_{21} to A_{i+1i}.
7: Add A^I_{22} to A_{i+1i+1}.
8: **end for**
9: Add κ_0 to a_{00}.
10: Add κ_L to a_{n+1n+1}.

A MATLAB routine for assembling this stiffness matrix is listed below.

```
function A = StiffnessAssembler1D(x,a,kappa)
n = length(x)-1;
A = zeros(n+1,n+1);
for i = 1:n
  h = x(i+1) - x(i);
  xmid = (x(i+1) + x(i))/2; % interval mid-point
  amid = a(xmid); % value of a(x) at mid-point
  A(i,i) = A(i,i) + amid/h; % add amid/h to A(i,i)
  A(i,i+1) = A(i,i+1) - amid/h;
  A(i+1,i) = A(i+1,i) - amid/h;
  A(i+1,i+1) = A(i+1,i+1) + amid/h;
end
A(1,1) = A(1,1) + kappa(1);
A(n+1,n+1) = A(n+1,n+1) + kappa(2);
```

Input to this routine is a vector **x** holding node coordinates, a function handle **a** to a routine specifying the function a, and a vector **kappa** for the boundary condition parameters κ_0 and κ_L. Output is the assembled stiffness matrix $A + R$.

The load, or source, vector $b + r$ is computed in a similar manner. In fact, b can be assembled by reusing the routine LoadAssembler1D. Thus, we only have to write code to assemble r. This is almost trivial, since only the first and last entry of r are non-zero.

```
function b = SourceAssembler1D(x,f,kappa,g)
b = LoadAssembler1D(x,f);
b(1) = b(1) + kappa(1)*g(1);
b(end) = b(end) + kappa(2)*g(2);
```

The inputs **x**, **f**, and **kappa** are as before. The vector **g**, holds the boundary parameters g_0 and g_L. Output is the assembled source vector $b + r$.

2.4.2 A Finite Element Solver for a General Two-Point Boundary Value Problem

With the above pieces of code it is easy to write a finite element solver for our general two-point boundary value problem. For fun sake let us use it to compute the temperature T in a rod of length $L = 6$ m, cross section $A = 0.1\,\text{m}^2$, thermal conductivity $k = 5 - 0.6x$ [J/(Ksm)], internal heat source $f = 0.03(x-6)^4$ [J/sm], held at constant temperature $T = -1$ [K] at $x = 2$, and thermally insulated at $x = 8$. Thus, we want to solve

$$-(0.5 + 0.7x)T'' = 0.3x^2, \quad 2 < x < 8, \quad T(2) = -1, \quad T'(8) = 0 \tag{2.67}$$

To approximate the Dirichlet condition $T(2) = 7$ we use the Robin condition (2.40b) with parameters $\kappa_0 = 10^6$ and $g_0 = -1$. Similarly, to impose the Neumann condition $T'(8) = 0$ we let $\kappa_L = 0$ in (2.40c). The value of g_L does not matter.

The main solver routine takes the following form.

```
function PoissonSolver1D()
h = 0.1; % mesh size
x = 2:h:8; % mesh
kappa = [1.e+6 0];
g = [-1 0];
A = StiffnessAssembler1D(x, @Conductivity, kappa);
b = SourceAssembler1D(x, @Source, kappa, g);
u = A\b;
plot(x,u)
```

Here, the heat conductivity and source are specified by the following two routines.

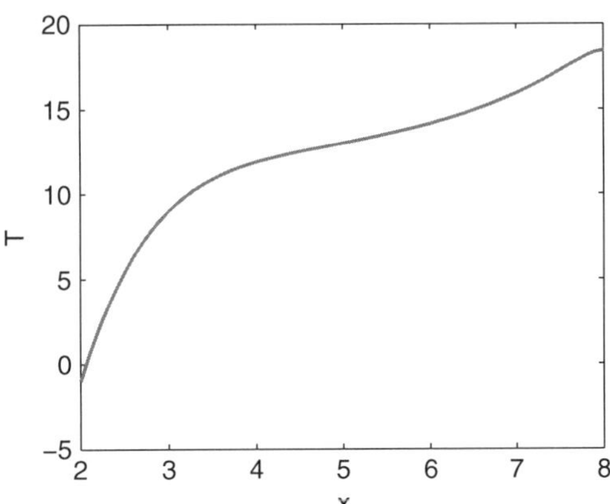

Fig. 2.1 Computed temperature on a uniform mesh with 25 elements

```
function y = Conductivity(x)
y = 0.1*(5 - 0.6*x); % heat conductivity times area

function y = Source(x)
y = 0.03*(x-6)^4; % heat source
```

Running this code we get the temperature distribution shown in Fig. 2.1.

2.5 Adaptive Finite Element Methods

Smart, so-called adaptive, finite element methods uses information extracted from earlier computations to locally refine or modify the mesh in order to obtain a better solution approximation u_h. The necessary information is obtained using a posteriori error estimates. The aim is to get u_h to be optimal in the sense that maximal accuracy is achieved at minimal computational cost.

2.5.1 *A Posteriori Error Estimate*

Let us return to the simple model problem (2.1) (i.e., $-u'' = f$, $x \in I$, $u(0) = u(L) = 0$). We have the following a posteriori error estimate for its finite element solution u_h.

Proposition 2.1. *The finite element solution* u_h*, defined by* (2.9)*, satisfies the estimate*

$$\|(u - u_h)'\|^2_{L^2(I)} \leq C \sum_{i=1}^{n} \eta_i^2(u_h) \tag{2.68}$$

where the so-called element residual $\eta_i(u_h)$ is defined by

$$\eta_i(u_h) = h_i \| f + u_h'' \|_{L^2(I_i)} \tag{2.69}$$

We observe that, since u_h is linear on element I_i, $u_h''|_{I_i} = 0$, the element residual can be simplified to just

$$\eta_i(u_h) = h_i \| f \|_{L^2(I_i)} \tag{2.70}$$

Proof. Let $e = u - u_h$ be the error. We then have

$$\|e'\|^2_{L^2(I)} = \int_I e'^2 \, dx \tag{2.71}$$

$$= \int_I e'(e - \pi e)' \, dx \tag{2.72}$$

$$= \sum_{i=1}^n \int_{x_{i-1}}^{x_i} e'(e - \pi e)' \, dx \tag{2.73}$$

$$= \sum_{i=1}^n \int_{x_{i-1}}^{x_i} (-e'')(e - \pi e) \, dx + \left[e'(e - \pi e) \right]_{x_{i-1}}^{x_i} \tag{2.74}$$

$$= \sum_{i=1}^n \int_{x_{i-1}}^{x_i} (-e'')(e - \pi e) \, dx \tag{2.75}$$

where we have used the Galerkin orthogonality (2.20) to subtract the interpolant $\pi e \in V_{h,0}$ to e, integration by parts on each element I_i, and that e and πe coincide at the nodes to get rid of all boundary terms.

Now, examining $-e''$ on I_i we find

$$-e'' = -(u - u_h)'' = -u'' + u_h'' = f + u_h'' \tag{2.76}$$

Using this, the Cauchy-Schwarz inequality, and a standard interpolation error estimate we finally have

$$\|e'\|^2_{L^2(I)} = \sum_{i=1}^n \int_{x_{i-1}}^{x_i} (f + u_h'')(e - \pi e) \, dx \tag{2.77}$$

$$\leq \sum_{i=1}^n \|f + u_h''\|_{L^2(I_i)} \|e - \pi e\|_{L^2(I_i)} \tag{2.78}$$

$$\leq \sum_{i=1}^n \|f + u_h''\|_{L^2(I_i)} C h_i \|e'\|_{L^2(I_i)} \tag{2.79}$$

$$= C \sum_{i=1}^{n} h_i \|f + u_h''\|_{L^2(I_i)} \|e'\|_{L^2(I_i)} \tag{2.80}$$

$$\leq C \left(\sum_{i=1}^{n} h_i^2 \|f + u_h''\|_{L^2(I_i)}^2 \right)^{1/2} \left(\sum_{i=1}^{n} \|e'\|_{L^2(I_i)}^2 \right)^{1/2} \tag{2.81}$$

$$= C \left(\sum_{i=1}^{n} h_i^2 \|f + u_h''\|_{L^2(I_i)}^2 \right)^{1/2} \|e'\|_{L^2(I)} \tag{2.82}$$

Dividing both sides by $\|e'\|_{L^2(I)}$ concludes the proof. □

Perhaps needless to say, there are two variants of the Cauchy-Schwarz inequality, namely, the continuous one $\int_I uv\,dx \leq \|u\|_{L^2(I)}\|v\|_{L^2(I)}$ for any functions $u, v \in L^2(I)$, and the discrete one $a \cdot b = a_1b_1 + \dots + a_nb_n \leq (a_1^2 + \dots + a_n^2)^{1/2} (b_1^2 + \dots + b_n^2)^{1/2} = \|a\|\,\|b\|$ for any $n \times 1$ vectors a and b. The former is used in estimating (2.77) and the latter in estimating (2.80), for example.

2.5.2 *Adaptive Mesh Refinement*

The result of Proposition 2.1 is perhaps not so surprising, since we expect the error to be small if either the mesh is fine, or if our differential equation is well satisfied by u_h. Indeed, if the discrete solution u_h was the exact solution u, then $f + u_h'' = 0$. Hence, the element residual $\eta_i(u_h)$ is in a sense proportional to the error e on element I_i. To increase the accuracy of u_h it is therefore tempting to selectively split the elements with the largest residuals into smaller ones, since this decreases h_i and, thus, also $\eta_i(u_h)$. Moreover, in doing so, it is natural to strive for a uniform distribution of the error among the elements. This line of reasoning leads us to adaptive finite element methods, which automatically controls the error using a posteriori estimates in combination with local mesh refinement. The following algorithm summarizes a prototype adaptive finite element method:

Algorithm 6 Prototype adaptive finite element method

```
Given a (coarse) mesh with n elements.
while n is not too large do
    Compute the finite element approximation u_h.
    Evaluate the element residuals η_i(u_h), i = 1, 2, ..., n.
    Select and refine the the most error prone elements.
end while
```

The adaptive algorithm above consists of four main components:

1. Computation of the element residuals η_i.
2. Selection of elements to be refined.

3. A refinement procedure.
4. A stopping criterion.

Let us discuss the computer implementation of these four steps.

In practice, we calculate the element residuals η_i using quadrature. Let us store them in a vector, say, eta.

```
eta = zeros(n,1); % allocate element residuals
for i = 1:n % loop over elements
  h = x(i+1) - x(i); % element length
  a = f(x(i)); % temporary variables
  b = f(x(i+1));
  t = (a^2+b^2)*h/2; % integrate f^2. Trapezoidal rule
  eta(i) = h*sqrt(t); % element residual
end
```

As usual **x** is a vector of node coordinates and **n** is the number of elements.

There are different possibilities for selecting the elements to be refined given the element residuals η_i. The popular so-called fixed-rate strategy is to refine element I_i if

$$\eta_i > \alpha \max_{i=1,2,\dots,n} \eta_i, \tag{2.83}$$

where $0 \leq \alpha \leq 1$ is a parameter to be chosen. The choice $\alpha = 0$ gives a uniform refinement, while $\alpha = 1$ gives no refinement at all.

The refinement procedure consists of inserting a new node at the mid-point of each element chosen for refinement. In other words, if we are refining element $I_i = [x_i, x_{i+1}]$, then we replace it by $[x_i, (x_i + x_{i+1})/2] \cup [(x_i + x_{i+1})/2, x_{i+1}]$. This is easily implemented by looping over the elements and inserting the mid-point coordinate of any element chosen for refinement at the end of the vector **x**, and then sort this vector.

```
alpha = 0.9 % refinement parameter
for i = 1:length(eta)
  if eta(i) > alpha*max(eta) % if large residual
    x = [x (x(i+1)+x(i))/2]; % insert new node point
  end
end
x = sort(x); % sort node points accendingly
```

The stopping criterion determines when the adaptive algorithm should stop. It can, for instance, take the form of a maximum bound on the number of nodes or elements, the memory usage, the time of the computation, the total size of the residual, or a combination of these.

Adaptive mesh refinement is particularly useful for problems with solutions containing high localized gradients, such as shocks or kinks, for instance. One such problem is $-u'' = \delta$, $0 < x < 1$, $u(0) = u(1) = 0$, where δ is the narrow pulse $\delta = \exp(-c|x - 0.5|^2)$, with $c = 100$. The solution to this problem looks like a single triangle wave with its peak at $x = 0.5$. In Fig. 2.2 we show the computed

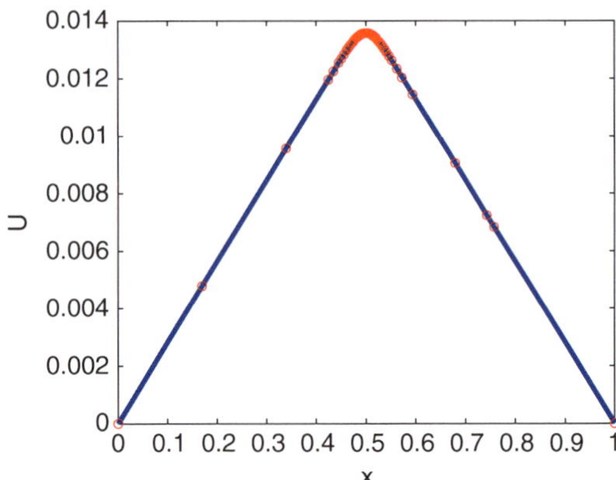

Fig. 2.2 Adaptively computed solution u_h. Each *red ring* symbolizes a nodal value

solution u_h to this problem after 25 mesh refinement loops, starting from a coarse mesh with five nodes distributed randomly over the computational domain. Clearly, the adaptive algorithm has identified and resolved the difficult region with high gradients near the peak of the triangle wave. This allows for high accuracy while at the same time saving computational resources.

2.6 Further Reading

There are many introductory texts on finite element methods for two-point boundary value problems, and it is virtually impossible to give a just overview of all these. For a very nice and accessible text explaining both theory and implementation we recommend Gockenbach [36]. Also, [35] by the same author is recommended, since it has broad scope and shows the important link between finite element and spectral methods (i.e., Fourier analysis). For a more theoretically advanced text, including a deeper discussion on the function spaces underlying the variational formulation, we refer to Axelsson and Barker [3].

2.7 Problems

Exercise 2.1. Solve the model problem (2.1) analytically with

(a) $f(x) = 1$.
(b) $f(x) = x - u$.

Exercise 2.2. Let $0 = x_0 < x_1 < x_2 < x_3 = 1$, where $x_1 = 1/6$ and $x_2 = 1/2$ be a partition of the interval $I = [0, 1]$ into three subintervals. Furthermore, let $V_{h,0}$

be the space of continuous piecewise linear functions on this partition that vanish at the end-points $x = 0$ and $x = 1$.

(a) Compute the stiffness matrix A defined by (2.13).
(b) Compute the load vector b, with $f = 1$, defined by (2.14).
(c) Solve the linear system $A\xi = b$ and compute the finite element solution u_h. Plot u_h.

Exercise 2.3. Consider the problem

$$-u'' = 7, \quad x \in I = [0, 1]$$
$$u(0) = 2, \; u(1) = 3$$

(a) What is a suitable finite element space V_h?
(b) Formulate a finite element method for this problem.
(c) Derive the discrete system of equations using a uniform mesh with 4 nodes.

Exercise 2.4. Consider the problem

$$-((1 + x)u')' = 0, \quad x \in I = [0, 1]$$
$$u(0) = 0, \; u'(1) = 1$$

Divide the interval I into three subintervals of equal length $h = 1/3$ and let V_h be the corresponding space of continuous piecewise linear functions vanishing at $x = 0$.

(a) Determine the analytical solution u.
(b) Use V_h to formulate a finite element method.
(c) Verify that the stiffness matrix A and load vector b are given by

$$A = \frac{1}{2}\begin{bmatrix} 16 & -9 & 0 \\ -9 & 20 & -11 \\ 0 & -11 & 11 \end{bmatrix}, \qquad b = \begin{bmatrix} 0 \\ 0 \\ 2 \end{bmatrix}$$

(d) Verify that A is symmetric and positive definite.

Exercise 2.5. Compute the stiffness matrix to the problem

$$-u'' = f, \quad x \in I = [0, 1]$$
$$u'(0) = u'(1) = 0$$

on a uniform mesh of I with two elements. Why is this matrix singular?

Exercise 2.6. Consider the problem

$$-u'' + u = f, \quad x \in I = [0, 1]$$
$$u(0) = u(1) = 0$$

(a) Choose a suitable finite element space V_h.
(b) Formulate a finite element method.
(c) Derive the discrete system of equations.

Exercise 2.7. Let u be defined on $I = [0, 1]$ and such that $u(0) = 0$. Prove the so-called Poincaré inequality

$$\|u\|_{L^2(I)} \leq C\|u'\|_{L^2(I)}$$

Exercise 2.8. Derive an a posteriori error estimate for the problem

$$-u'' + u = f, \quad x \in I = [0, L]$$
$$u(0) = u(L) = 0$$

Exercise 2.9. Consider the problem

$$-\epsilon u'' + xu' + u = f, \quad x \in I = [0, L]$$
$$u(0) = u'(L) = 0$$

where $\epsilon > 0$ is a constant. Prove that the solution satisfies

$$\|\epsilon u''\|_{L^2(I)} \leq \|f\|_{L^2(I)}$$

Exercise 2.10. Consider the model problem

$$-u'' = f, \quad x \in I = [0, L]$$
$$u(0) = u(L) = 0$$

Its variational formulation reads: find $u \in V_0$ such that

$$\int_I u'v'\,dx = \int_I fv\,dx, \quad \forall v \in V_0$$

Show that the solution $u \in V_0$ to the variational formulation minimizes the functional

$$F(w) = \frac{1}{2}\int_I w'^2\,dx - \int_I fw\,dx$$

over the space V_0. *Hint:* Write $w = u+v$ and show that $F(w) = F(u) + \ldots \geq F(u)$.

Chapter 3
Piecewise Polynomial Approximation in 2D

Abstract In this chapter we extend the concept of piecewise polynomial approximation to two dimensions. As before, the basic idea is to construct spaces of piecewise polynomial functions that are easy to represent in a computer and to show that they can be used to approximate more general functions. A difficulty with the construction of piecewise polynomials in higher dimension is that first the underlying domain must be partitioned into elements, such as triangles, which may be a nontrivial task if the domain has complex shape. We present a methodology for building representations of piecewise polynomials on triangulations that is efficient and suitable for computer implementation and study the approximation properties of these spaces.

3.1 Meshes

3.1.1 Triangulations

Let $\Omega \subset \mathbb{R}^2$ be a bounded two-dimensional domain with smooth or polygonal boundary $\partial\Omega$. A triangulation, or mesh, $\mathcal{K}$ of Ω is a set $\{K\}$ of triangles K, such that $\Omega = \cup_{K\in\mathcal{K}} K$, and such that the intersection of two triangles is either an edge, a corner, or empty. No triangle corner is allowed to be hanging, that is, lie on an edge of another triangle. The corners of the triangles are called the nodes. Figure 3.1 shows a triangle mesh of the Greek letter π.

The set of triangle edges E is denoted by $\mathcal{E} = \{E\}$. We distinguish between edges lying within the domain Ω and edges lying on the boundary $\partial\Omega$. The former belongs to the set of interior edges $\mathcal{E}_I$, and the latter to the set of boundary edges $\mathcal{E}_B$, respectively.

To measure the size of a triangle K we introduce the local mesh size h_K, defined as the length of the longest edge in K. See Fig. 3.5. Moreover, to measure the quality

M.G. Larson and F. Bengzon, *The Finite Element Method: Theory, Implementation, and Applications*, Texts in Computational Science and Engineering 10,
DOI 10.1007/978-3-642-33287-6_3, © Springer-Verlag Berlin Heidelberg 2013

Fig. 3.1 A mesh of π

of K, let d_K be the diameter of the inscribed circle and introduce the so-called chunkiness parameter c_K, defined by

$$c_K = h_K/d_K \tag{3.1}$$

We say that a triangulation $\mathcal{K}$ is shape regular if there is a constant $c_0 > 0$ such that

$$c_K \geq c_0, \quad \forall K \in \mathcal{K} \tag{3.2}$$

This condition means that the shape of the triangles can not be too extreme in the sense that the angles of any triangle can neither be very wide nor very narrow. We always assume that the meshes we work with are shape regular. As we shall see, this has implications for the approximation properties of the piecewise polynomial spaces to be defined on these meshes.

A global mesh size is given by $h = \max_{K \in \mathcal{K}} h_K$.

Also, occasionally we will refer to a quasi-uniform mesh, which means that the mesh size h_K is roughly the same for all elements. In particular, there exist a constant $\rho > 0$ such that $\rho < h_K/h_{K'} < \rho^{-1}$ for any two elements K and K'.

3.1.2 Data Structures for a Triangulation

The standard way of representing a triangle mesh with n_p nodes and n_t elements in a computer is to store it as two matrices, P and T, called the point matrix, and the connectivity matrix, respectively. The point matrix P is of size $2 \times n_p$ and column j contains the coordinates $x_1^{(j)}$ and $x_2^{(j)}$ of node N_j. The connectivity matrix T is of size $3 \times n_t$ and column j contains the numbers of the three nodes in triangle K_j. In the following we shall adopt the common convention of ordering these three nodes in a counter clockwise sense. It does not, however, matter on which of the nodes the ordering starts.

Fig. 3.2 A *triangle* mesh of the L-shaped domain

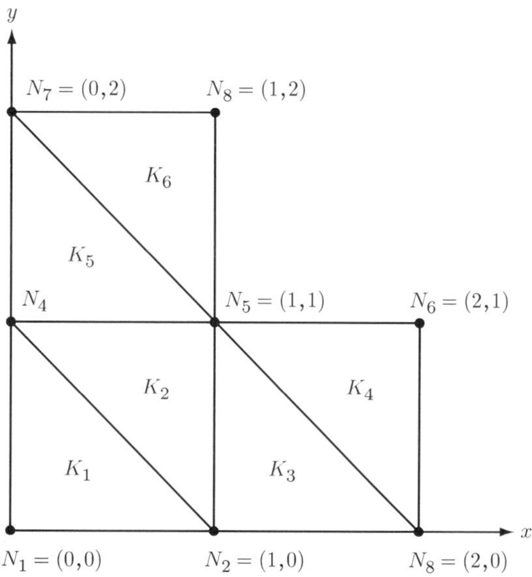

Figure 3.2 shows a small triangulation of a domain shaped like the letter L. The mesh has eight nodes and six triangles. The point matrix and connectivity matrix for this mesh are given by

$$P = \begin{bmatrix} 0.0 \; 1.0 \; 2.0 \; 0.0 \; 1.0 \; 2.0 \; 0.0 \; 1.0 \\ 0.0 \; 0.0 \; 0.0 \; 1.0 \; 1.0 \; 1.0 \; 2.0 \; 2.0 \end{bmatrix}, \qquad T = \begin{bmatrix} 1 \; 2 \; 5 \; 3 \; 4 \; 5 \\ 2 \; 5 \; 2 \; 6 \; 5 \; 8 \\ 4 \; 4 \; 8 \; 5 \; 7 \; 7 \end{bmatrix} \tag{3.3}$$

Thus, for example, the coordinates $(x_1^{(3)}, x_2^{(3)}) = (2,0)$ of node N_3 are given by the matrix entries P_{13} and P_{23}, respectively. In the connectivity matrix T column 2 contains the numbers 2, 5, and 4 of the three nodes N_2, N_5, and N_4 making up triangle K_2. Note that the nodes are ordered in a counter clockwise sense.

We remark that the representation of a mesh via a point and connectivity matrix is very common and generalizes to almost any element type and space dimension. In particular, tetrahedral meshes used to partition domains in three dimensions can be stored in this way. In this case, P has 3 rows for the three node coordinates, and T has 4 rows containing the four nodes of a tetrahedron.

3.1.3 Mesh Generation

Over the past decades advanced computer algorithms for the automatic construction of meshes have been developed. However, depending on the complexity of the domain it is still more or less difficult to generate a mesh. In particular, difficulties may arise for three-dimensional geometries, since in practice these often have complicated shape. By contrast, in two dimensions there are efficient algorithms for creating a mesh on quite general domains. One of these is the Delaunay algorithm, which given a set of points can determine a triangulation with the given points as triangle nodes. Theoretically, Delaunay triangulations are optimal in the sense that the angles of all triangles are maximal. In practice, however, this is not always so due to the necessity for the triangulation to respect the domain boundaries, which may lead to triangles with poor quality.

MATLAB has a non-standard set of routines called the PDE-Toolbox which includes a Delaunay mesh generator for creating high quality triangulations of two dimensional geometries. Let us illustrate its use by creating a mesh of the L-shaped domain.

In MATLAB the geometry of the L-shaped domain is defined by a geometry matrix `g`, given by

```
g=[2  0  2  0  0  1  0
   2  2  2  0  1  1  0
   2  2  1  1  1  1  0
   2  1  1  1  2  1  0
   2  1  0  2  2  1  0
   2  0  0  2  0  1  0
];
```

Each column of `g` describes one of the six line segments making up the boundary of the L-shaped domain. In each such column rows two and three contain the starting and ending x_1-coordinate, and rows four and five the corresponding x_2-coordinate. Rows six and seven indicate if the geometry is on the left or right side of the line segment when traversing it in the direction induced by the start- and end-points. The fact that we are defining a line segment is indicated by the number 2 in the first column.

To generate a mesh of the domain `g` we type

```
[p,e,t] = initmesh(g,'hmax',0.1)
```

The call to the `initmesh` routine invokes the mesh generator, which triangulates the domain in `g`. The final two arguments `'hmax',0.1` specifies that the maximum edge length of the triangles to be generated may not exceed one tenth. Output is the point matrix `p`, the connectivity matrix `t`, and the so-called edge matrix `e` containing the node numbers of the triangle edges making up the boundary of the mesh. We will return to discuss the `e` matrix later on.

We remark that `t` has four rows with the last row containing subdomain numbers for each triangle in case the domain `g` has subdomains.

In the PDE-Toolbox there are a few built-in geometries, including:

- `cicrcleg`, the unit radius circle centered at origo.
- `squareg`, the square $[-1, 1]^2$.

For future use we extend this list of geometries with a rectangle, defined by

```
function r = Rectg(xmin,ymin,xmax,ymax)
r=[2 xmin xmax ymin ymin 1 0;
   2 xmax xmax ymin ymax 1 0;
   2 xmax xmin ymax ymax 1 0;
   2 xmin xmin ymax ymin 1 0]';
```

Note the transpose on `g`.

To view the generated mesh one can type

```
pdemesh(p,e,t)
```

More general geometries can be drawn in the PDE-Toolbox GUI. It is initialized by typing

```
pdetool
```

at the MATLAB prompt.

In this context we remind about the extensive help available in MATLAB. Help about any command can be obtained by typing

```
help command
```

where `command` is the name of the command (e.g., `jigglemesh`).

3.2 Piecewise Polynomial Spaces

The reason for meshing a domain is that it allows for a simple construction of piecewise polynomial function spaces, which is otherwise a very difficult task. We shall now discuss how this is done in the special case of linear polynomials on triangles. The concepts to be introduced generalizes to higher dimensions and other types of elements.

3.2.1 The Space of Linear Polynomials

Let K be a triangle and let $P_1(K)$ be the space of linear functions on K, defined by

$$P_1(K) = \{v : v = c_0 + c_1x_1 + c_2x_2,\ (x_1, x_2) \in K,\ c_0, c_1, c_2 \in \mathbb{R}\} \tag{3.4}$$

In other words $P_1(K)$ contains all functions of the form $v = c_0 + c_1x_1 + c_2x_2$ on K.

We observe that any v in $P_1(K)$ is uniquely determined by its nodal values $\alpha_i = v(N_i)$, $i = 1, 2, 3$. This follows by assuming α_i to be given and evaluating v at the three nodes $N_i = (x_1^{(i)}, x_2^{(i)})$. In doing so, we end up with the linear system

$$\begin{bmatrix} 1 & x_1^{(1)} & x_2^{(1)} \\ 1 & x_1^{(2)} & x_2^{(2)} \\ 1 & x_1^{(3)} & x_2^{(3)} \end{bmatrix} \begin{bmatrix} c_0 \\ c_1 \\ c_2 \end{bmatrix} = \begin{bmatrix} \alpha_1 \\ \alpha_2 \\ \alpha_3 \end{bmatrix} \tag{3.5}$$

Here, computing the determinant of the system matrix we find that its absolute value equals $2|K|$, where $|K|$ is the area of K, which means that (3.5) has a unique solution as long as K is not degenerate in any way.

The natural basis $\{1, x_1, x_2\}$ for $P_1(K)$ is not suitable for us, since we wish to use the nodal values as degrees of freedom. Therefore, we introduce a nodal basis $\{\lambda_1, \lambda_2, \lambda_3\}$, defined by

$$\lambda_j(N_i) = \begin{cases} 1, & i = j \\ 0, & i \neq j \end{cases}, \quad i, j = 1, 2, 3 \tag{3.6}$$

Using the new basis we can express any function v in $P_1(K)$ as

$$v = \alpha_1\lambda_1 + \alpha_2\lambda_2 + \alpha_3\lambda_3 \tag{3.7}$$

where $\alpha_i = v(N_i)$.

On the reference triangle $\bar{K}$ with nodes at origo, $(1, 0)$, and $(0, 1)$, the nodal basis functions for $P_1(\bar{K})$ are given by

$$\lambda_1 = 1 - x_1 - x_2, \quad \lambda_2 = x_1, \quad \lambda_3 = x_2 \tag{3.8}$$

3.2.2 *The Space of Continuous Piecewise Linear Polynomials*

The construction of piecewise linear functions on a mesh $\mathcal{K} = \{K\}$ is straight forward. On each triangle K any such function v is simply required to belong to $P_1(K)$. Requiring also continuity of v between neighbouring triangles, we obtain the space of all continuous piecewise linear polynomials V_h, defined by

$$V_h = \{v : v \in C^0(\Omega),\ v|_K \in P_1(K),\ \forall K \in \mathcal{K}\} \tag{3.9}$$

Here, $C^0(\Omega)$ denotes the space of all continuous functions on Ω.

An example of a continuous piecewise linear function is shown in Fig. 3.3.

To construct a basis for V_h we first show that a function v in V_h is uniquely determined by its nodal values

$$\{v(N_j)\}_{j=1}^{n_p} \tag{3.10}$$

and, conversely, that for each set of nodal values there is a unique function v in V_h with these nodal values. To prove this claim we first note that the nodal values determines a function in $P_1(K)$ uniquely for each $K \in \mathcal{K}$, and, thus, a function in

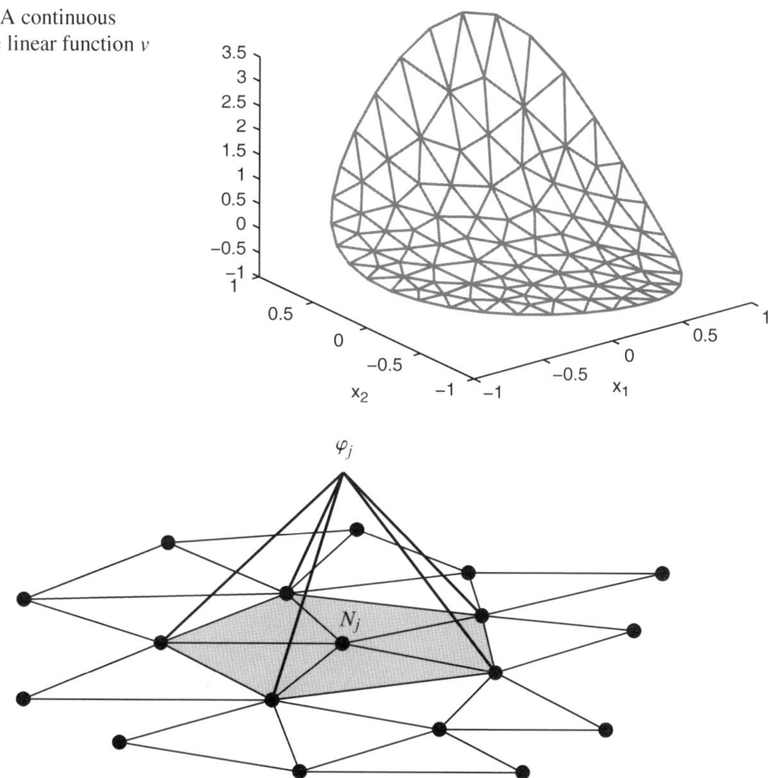

Fig. 3.3 A continuous piecewise linear function v

Fig. 3.4 A two-dimensional hat function φ_j on a general *triangle* mesh

V_h is uniquely determined by its values in the nodes. Then, consider two triangles K_1 and K_2 sharing edge $E = K_1 \cap K_2$. Let v_1 and v_2 be the two unique linear polynomials in $P_1(K_1)$ and $P_1(K_2)$, respectively, determined by the nodal values on K_1 and K_2. Because v_1 and v_2 are linear polynomials on K_1 and K_2, they are also linear polynomials on E, and because they coincide at the end-points of E we conclude that $v_1 = v_2$ on E. Therefore, for any set of nodal values there is a continuous piecewise linear polynomial with these nodal values.

Motivated by this result we let the nodal values be our degrees of freedom and define a corresponding basis $\{\varphi_j\}_{j=1}^{n_p} \subset V_h$ such that

$$\varphi_j(N_i) = \begin{cases} 1, & i = j \\ 0, & i \neq j \end{cases}, \quad i, j = 1, 2, \ldots, n_p \tag{3.11}$$

Figure 3.4 illustrates a typical basis function φ_j.

From the figure it is clear that each basis function φ_j is continuous, piecewise linear, and with support only on the small set of triangles sharing node N_j. Similar to

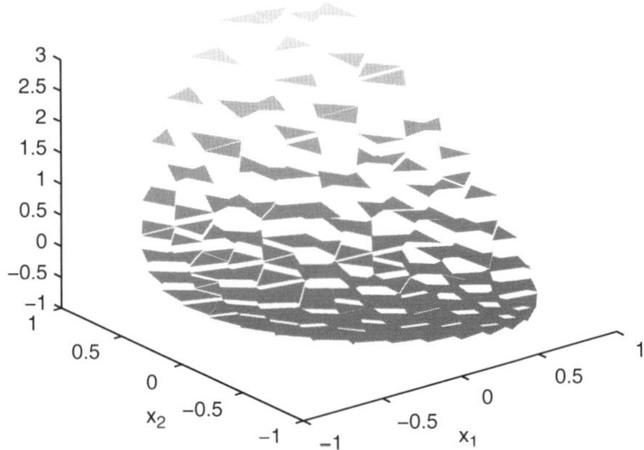

Fig. 3.5 A discontinuous piecewise constant function w

the one-dimensional case, these two dimensional basis functions are also called hat functions.

We remark that hat functions can also be defined analogously in three dimensions on tetrahedra.

Now, using the hat function basis we note that any function v in V_h can be written

$$v = \sum_{i=1}^{n_p} \alpha_i \varphi_i \tag{3.12}$$

where $\alpha_i = v(N_i)$, $i = 1, 2, \dots, n_p$, are the nodal values of v.

3.2.3 The Space of Piecewise Constants

Differentiating a continuous piecewise linear function with respect to either of the space coordinates x_1 or x_2 gives a discontinuous piecewise constant function. This is clearly seen in Fig. 3.3, which shows the x_1-derivative of the function in Fig. 3.5.

The collection of functions that are constant on each element forms the space of piecewise constants W_h, defined by

$$W_h = \{w : w|_K \in P_0(K),\ \forall K \in \mathcal{K}\} \tag{3.13}$$

where $P_0(K)$ is the space of constant functions on element K.

To measure the size of $w \in W_h$ on the edges it is customary to use the so-called average and jump operators $\langle \cdot \rangle$ and $[\cdot]$, respectively. To define these consider the two adjacent elements K^+ and K^- that share the edge E in Fig. 3.6. The average of w on E is defined by

$$\langle w \rangle = \frac{w^+ + w^-}{2}, \quad x \in E \tag{3.14}$$

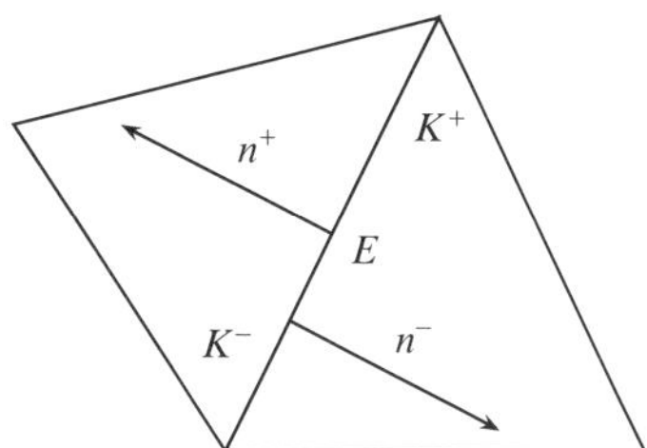

Fig. 3.6 Two adjacent elements K^+ and K^- sharing edge E, with normal $n_E = n^+ = -n^-$

where $w^\pm = w|_{K^\pm}$. By analogy, $n^\pm$ is the outward unit normals on the element boundaries $\partial K^\pm$. Note that $n^+ = -n^-$ on E, since n^+ and n^- point in opposite directions. To E we assign a fixed normal $n_E = n^+ = -n^-$, so that $w^\pm = \lim_{\epsilon \to 0} w(x \mp \epsilon n_E)$ for any point x on E. The jump of w on E is defined by

$$[w] = w^+ - w^-, \quad x \in E \tag{3.15}$$

If w happens to be a vector with components w_i in W_h, then

$$[n \cdot w] = n^+ \cdot w^+ + n^- \cdot w^- = n_E \cdot (w^+ - w^-) = n_E \cdot [w], \quad x \in E \tag{3.16}$$

When E lies on the boundary of the domain Ω, its normal n_E is chosen to coincide with the outward unit normal on $\partial\Omega$, and we define $\langle w \rangle = w^+$ and $[u] = w^+$.

The average and jump easily generalize to matrices W with components W_{ij} in W_h.

3.3 Interpolation

3.3.1 Linear Interpolation

Now, let us return to the problem of approximating functions. Given a continuous function f on a triangle K with nodes N_i, $i = 1, 2, 3$, the linear interpolant $\pi f \in P_1(K)$ to f is defined by

$$\pi f = \sum_{i=1}^{3} f(N_i)\varphi_i \tag{3.17}$$

The interpolant $\pi f \in P_1(K)$ is a plane, which coincides with f at the three node points. Thus, by definition we have N_i $\pi f(N_i) = f(N_i)$. See Fig. 3.7.

To estimate the interpolation error $f - \pi f$ we need to introduce some measure of the size of the first and second order derivatives of f. To this end, let D and D^2 be the differential operators

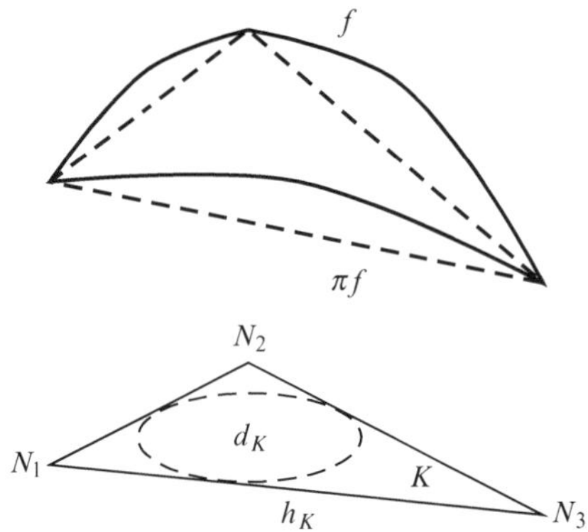

Fig. 3.7 The linear interpolant πf of a function f on a *triangle* K with nodes N_1, N_2, and N_3. Also shown is the longest edge length h_K, and the diameter d_K of the inscribed *circle*

$$Df = \left(\left| \frac{\partial f}{\partial x_1} \right|^2 + \left| \frac{\partial f}{\partial x_2} \right|^2 \right)^{1/2}, \quad D^2 f = \left(\left| \frac{\partial^2 f}{\partial x_1^2} \right|^2 + 2 \left| \frac{\partial^2 f}{\partial x_1 \partial x_2} \right|^2 + \left| \frac{\partial^2 f}{\partial x_2^2} \right|^2 \right)^{1/2} \tag{3.18}$$

Because D and D^2 include all first and second partial derivatives, we say that Df and $D^2 f$ are the total first and second derivatives of f, respectively.

In this context we recall that the $L^2(\Omega)$-norm of a function $f = f(x_1, x_2)$ of two variables x_1 and x_2 on a domain Ω is given by

$$\|f\|_{L^2(\Omega)} = \left(\int_\Omega f^2 \, dx \right)^{1/2} \tag{3.19}$$

Using these notations we have the following estimate of the interpolation error.

Proposition 3.1. *The interpolant πf satisfies the estimates*

$$\|f - \pi f\|_{L^2(K)} \leq C h_K^2 \|D^2 f\|_{L^2(K)} \tag{3.20}$$

$$\|D(f - \pi f)\|_{L^2(K)} \leq C h_K \|D^2 f\|_{L^2(K)} \tag{3.21}$$

We refer to Brenner and Scott [60], or Elman and co-authors [26] for proofs of this proposition.

In Proposition 3.1, it is possible to show that the occurring constants C are proportional to the inverse of $\sin(\theta_K)$, where θ_K is the smallest angle in triangle K. From this it follows that C blows up if θ_K becomes very small, which renders the interpolation error estimates practically useless. This explains why it is critical that K has neither too narrow nor too wide angles. Indeed, we speak about the angles as a measure of the triangle quality. Recall that the chunkiness parameter c_K of (3.1) is used as a measure of the quality of K.

3.3.2 Continuous Piecewise Linear Interpolation

The concept of continuous piecewise linear interpolation easily extends from one to two and even three dimensions. Indeed, given a continuous function f on the domain Ω, we define its continuous piecewise linear interpolant $\pi f \in V_h$ on a mesh $\mathcal{K}$ of Ω by

$$\pi f = \sum_{i=1}^{n_p} f(N_i)\varphi_i \tag{3.22}$$

Again, πf approximates f by taking on the values of f at the nodes N_i.

In MATLAB it is easy to draw πf given f. For example, to plot the interpolant to $f = x_1 x_2$ on the square domain $\Omega = [-1, 1]^2$ it takes only four lines of code.

```
[p,e,t] = initmesh('squareg','hmax',0.1); % mesh
x = p(1,:); y = p(2,:); % node coordinates
pif = x.*y; % nodal values of interpolant
pdesurf(p,t,pif) % plot interpolant
```

Looking at the code let us make a remark about out programming style. The conversion of mathematical symbols to computer code is not always obvious and self explanatory. In this book we have tried to keep a close correlation between the notation introduced in the formulas, and the names of the variables used in the code. However, attempting to write as short, yet clear, code as as possible has unavoidable lead to a few inconsistencies in this respect. For example, we have throughout used the variables **x** and **y** to denote the space coordinates x_1 and x_2. We hope that the code comments and the context shall make it clear what is meant.

The size of the interpolation error $f - \pi f$ can be estimated with the help of the following proposition.

Proposition 3.2. *The interpolant πf satisfies the estimates*

$$\|f - \pi f\|^2_{L^2(\Omega)} \leq C \sum_{K\in\mathcal{K}} h_K^4 \|D^2 f\|^2_{L^2(K)} \tag{3.23}$$

$$\|D(f - \pi f)\|^2_{L^2(\Omega)} \leq C \sum_{K\in\mathcal{K}} h_K^2 \|D^2 f\|^2_{L^2(K)} \tag{3.24}$$

Proof. Using the Triangle inequality and Proposition 3.1, we have

$$\|f - \pi f\|^2_{L^2(\Omega)} = \sum_{K\in\mathcal{K}} \|f - \pi f\|^2_{L^2(K)} \tag{3.25}$$

$$\leq \sum_{K\in\mathcal{K}} C h_K^4 \|D^2 f\|^2_{L^2(K)} \tag{3.26}$$

which proves the first estimate. The second follows similarly.

3.4 L^2-Projection

3.4.1 Definition

Given a function $f \in L^2(\Omega)$ the L^2-projection $P_h f \in V_h$ of f is defined by

$$\int_\Omega (f - P_h f) v \, dx = 0, \quad \forall v \in V_h \tag{3.27}$$

3.4.2 Derivation of a Linear System of Equations

In order to actually compute the L^2-projection $P_h f$, we first note that the definition (3.27) is equivalent to

$$\int_\Omega (f - P_h f)\varphi_i \, dx = 0, \quad i = 1, 2, \dots, n_p \tag{3.28}$$

where φ_i are the hat basis functions spanning V_h.

Then, since $P_h f$ belongs to V_h it can be written as the linear combination

$$P_h f = \sum_{j=1}^{n_p} \xi_j \varphi_j \tag{3.29}$$

where ξ_j, $j = 1, 2, \dots, n_p$, are n_p unknown coefficients to be determined.

Inserting the ansatz (3.29) into (3.28) we get

$$\int_\Omega f \varphi_i \, dx = \int_\Omega \left(\sum_{j=1}^{n_p} \xi_j \varphi_j \right) \varphi_i \, dx \tag{3.30}$$

$$= \sum_{j=1}^{n_p} \xi_j \int_\Omega \varphi_j \varphi_i \, dx \tag{3.31}$$

Further, using the notation

$$M_{ij} = \int_\Omega \varphi_j \varphi_i \, dx, \quad i, j = 1, 2, \dots, n_p \tag{3.32}$$

$$b_i = \int_\Omega f \varphi_i \, dx, \quad i = 1, 2 \dots, n_p \tag{3.33}$$

we have

$$b_i = \sum_{j=1}^{n_p} M_{ij}\xi_j, \quad i = 1, 2 \ldots, n_p \tag{3.34}$$

which is an $n_p \times n_p$ linear system for the unknowns ξ_j. In matrix form, we write this

$$M\xi = b \tag{3.35}$$

where the entries of the $n_p \times n_p$ mass matrix M and the $n_p \times 1$ load vector b are defined by (3.32) and (3.33), respectively. Solving the linear system (3.35) we obtain the unknowns ξ_j, and, thus, $P_h f$.

3.4.3 Basic Algorithm to Compute the L^2-Projection

The following algorithm summarizes the basic steps in computing the L^2-projection $P_h f$:

Algorithm 7 Basic Algorithm to Compute the L^2-Projection.

1: Create a mesh $\mathcal{K}$ of Ω and define the corresponding space of continuous piecewise linear functions V_h with hat function basis $\{\varphi_i\}_{i=1}^{n_p}$.
2: Assemble the $n_p \times n_p$ mass matrix M and the $n_p \times 1$ load vector b, with entries

$$M_{ij} = \int_\Omega \varphi_j \varphi_i \, dx, \quad b_i = \int_\Omega f \varphi_i \, dx \tag{3.36}$$

3: Solve the linear system

$$M\xi = b \tag{3.37}$$

4: Set

$$P_h f = \sum_{j=1}^{n_p} \xi_j \varphi_j \tag{3.38}$$

3.4.4 Existence and Uniqueness of the L^2-Projection

Theorem 3.1. *The L^2-projection $P_h f$, defined by* (3.27), *exists and is unique.*

Proof. We first show the uniqueness claim. The argument is by contradiction. Assume that there are two L^2-projections $P_h f$ and $\widetilde{P_h f}$ satisfying (3.27). Then, we have

$$\int_\Omega P_h f v \, dx = \int_\Omega f v \, dx, \quad \forall v \in V_h \tag{3.39}$$

$$\int_\Omega \widetilde{P_h f} v \, dx = \int_\Omega f v \, dx, \quad \forall v \in V_h \tag{3.40}$$

Subtracting these equations we get

$$\int_\Omega (P_h f - \widetilde{P_h f}) v \, dx = 0, \quad \forall v \in V_h \tag{3.41}$$

Now, choosing $v = P_h f - \widetilde{P_h f} \in V_h$ we further get

$$\int_\Omega |P_h f - \widetilde{P_h f}|^2 \, dx = 0 \tag{3.42}$$

From this identity we conclude that $P_h f - \widetilde{P_h f}$ must be zero.

To prove existence we recall that $P_h f$ is determined by a square linear system. The existence of a solution to a linear system follows from the uniqueness of the solution. □

3.4.5 A Priori Error Estimate

The next theorem is a best approximation result.

Theorem 3.2. *The L^2-projection $P_h f$, defined by* (3.27)*, satisfies the best approximation result*

$$\|f - P_h f\|_{L^2(\Omega)} \le \|f - v\|_{L^2(\Omega)}, \quad \forall v \in V_h \tag{3.43}$$

Proof. Using the definition of the L^2-norm and writing $f - P_h f = f - v + v - P_h f$ for any $v \in V_h$, we have

$$\|f - P_h f\|^2_{L^2(\Omega)} = \int_\Omega (f - P_h f)(f - v + v - P_h f) \, dx \tag{3.44}$$

$$= \int_\Omega (f - P_h f)(f - v) \, dx + \int_\Omega (f - P_h f)(v - P_h f) \, dx \tag{3.45}$$

$$= \int_\Omega (f - P_h f)(f - v) \, dx \tag{3.46}$$

$$\le \|f - P_h f\|_{L^2(\Omega)} \|f - v\|_{L^2(\Omega)} \tag{3.47}$$

where we used the definition of the L^2-projection to conclude that

$$\int_{\Omega} (f - P_h f)(v - P_h f)\, dx = 0 \tag{3.48}$$

since $v - P_h f \in V_h$. Dividing by $\|f - P_h f\|_{L^2(\Omega)}$ concludes the proof. □

Theorem 3.3. *The L^2-projection $P_h f$ satisfies the estimate*

$$\|f - P_h f\|^2_{L^2(\Omega)} \leq C \sum_{K \in \mathcal{K}} h_K^4 \|D^2 f\|^2_{L^2(K)} \tag{3.49}$$

Proof. Starting from the best approximation result, choosing $v = \pi f$ the interpolant of f, and using the interpolation error estimate of Proposition 3.1, we have

$$\|f - P_h f\|^2_{L^2(\Omega)} \leq \|f - \pi f\|^2_{L^2(\Omega)} \tag{3.50}$$

$$\leq \sum_{K \in \mathcal{K}} \|f - \pi f\|^2_{L^2(K)} \tag{3.51}$$

$$\leq \sum_{K \in \mathcal{K}} C h_K^4 \|D^2 f\|^2_{L^2(K)} \tag{3.52}$$

which proves the estimate. □

Hence, we conclude that

$$\|f - P_h f\|_{L^2(\Omega)} \leq C h^2 \|D^2 f\|_{L^2(\Omega)} \tag{3.53}$$

In other words, the L^2-error tends to zero as the mesh size h tends to zero.

3.5 Quadrature and Numerical Integration

Quadrature in two and three dimensions works in principle the same as in one. The integral under consideration is approximated with a sum of weights times the values of the integrand at a set of carefully selected quadrature points. Indeed, a general quadrature rule on a triangle K takes the form

$$\int_K f\, dx \approx \sum_j w_j f(q_j) \tag{3.54}$$

where $\{q_j\}$ is the set of quadrature points in K, and $\{w_j\}$ the corresponding quadrature weights. Below we list a few quadrature formulas for integrating a continuous function f over a general triangle K with nodes N_1, N_2, and N_3.

The simplest quadrature formula is the center of gravity rule

$$\int_K f\,dx \approx f\left(\frac{N_1 + N_2 + N_3}{3}\right)|K| \tag{3.55}$$

where $|K|$ denotes the area of K. The center of gravity formula is a two-dimensional variant of the Mid-point rule.

There is also a two-dimensional analog to the Trapezoidal rule, namely, the so-called corner quadrature formula

$$\int_K f\,dx \approx \sum_{i=1}^{3} f(N_i)\frac{|K|}{3} \tag{3.56}$$

A better quadrature formula is the two-dimensional Mid-point rule

$$\int_K f\,dx \approx \sum_{1 \le i < j \le 3}^{3} f\left(\frac{N_i + N_j}{2}\right)\frac{|K|}{3} \tag{3.57}$$

where $(N_i + N_j)/2$ denotes the mid-point of the edge between node number i and j.

We remark that there are numerous other quadrature rules with different kinds of precision and computational cost. Indeed, each element type and space dimension generally requires its set of quadrature points and weights. However, for many computations primitive rules, such as the center of gravity rule, for instance, suffice. We refer the interested reader to the book by Evans [28] for a more thorough description of this subject.

3.6 Computer Implementation

3.6.1 Assembly of the Mass Matrix

We next show how to compute the mass matrix M in two dimensions. This is quite a bit more complicated than in one dimension and we therefore do this by example. To this end, consider the small mesh of the rectangle Ω shown in Fig. 3.8.

On this mesh we wish to compute the mass matrix M, given by

$$M = \int_\Omega \begin{bmatrix} \varphi_1\varphi_1 & \varphi_2\varphi_1 & \varphi_3\varphi_1 & \varphi_4\varphi_1 & \varphi_5\varphi_1 \\ \varphi_1\varphi_2 & \varphi_2\varphi_2 & \varphi_3\varphi_2 & \varphi_4\varphi_2 & \varphi_5\varphi_2 \\ \varphi_1\varphi_3 & \varphi_2\varphi_3 & \varphi_3\varphi_3 & \varphi_4\varphi_3 & \varphi_5\varphi_3 \\ \varphi_1\varphi_4 & \varphi_2\varphi_4 & \varphi_3\varphi_4 & \varphi_4\varphi_4 & \varphi_5\varphi_4 \\ \varphi_1\varphi_5 & \varphi_2\varphi_5 & \varphi_3\varphi_5 & \varphi_4\varphi_5 & \varphi_5\varphi_5 \end{bmatrix} dx \tag{3.58}$$

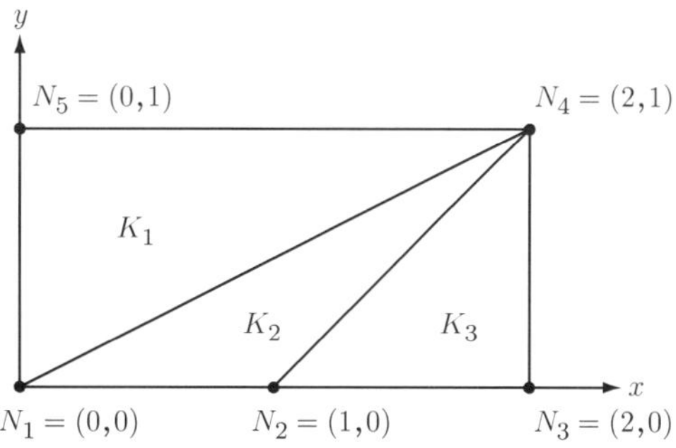

Fig. 3.8 A small mesh of the *rectangle* $\Omega = [0, 2] \times [0, 1]$

To do so, we first break the integral over the whole domain Ω into a sum of integrals over the triangles K_i, $i = 1, 2, 3$. We then have

$$M = \sum_{i=1}^{3} \int_{K_i} \begin{bmatrix} \varphi_1\varphi_1 & \varphi_2\varphi_1 & \varphi_3\varphi_1 & \varphi_4\varphi_1 & \varphi_5\varphi_1 \\ \varphi_1\varphi_2 & \varphi_2\varphi_2 & \varphi_3\varphi_2 & \varphi_4\varphi_2 & \varphi_5\varphi_2 \\ \varphi_1\varphi_3 & \varphi_2\varphi_3 & \varphi_3\varphi_3 & \varphi_4\varphi_3 & \varphi_5\varphi_3 \\ \varphi_1\varphi_4 & \varphi_2\varphi_4 & \varphi_3\varphi_4 & \varphi_4\varphi_4 & \varphi_5\varphi_4 \\ \varphi_1\varphi_5 & \varphi_2\varphi_5 & \varphi_3\varphi_5 & \varphi_4\varphi_5 & \varphi_5\varphi_5 \end{bmatrix} dx = \sum_{i=1}^{3} M^{K_i} \tag{3.59}$$

As we know there are only three non-zero hat functions on each triangle. For example, the only non-zero hats on K_1 are φ_1, φ_4, and φ_5. Integrating the product of these we see that K_1, or any triangle for that matter, gives rise to a total of $3 \cdot 3 = 9$ integral contributions to M. Moreover, for a given triangle, the index on the non-zero hat functions are the same as the node numbers for that triangle. Thus, inspecting which hats are non-zero on which triangle, we can therefore beforehand say which rows and columns are non-zero in each global element matrix M^{K_i}. For example, the only non-zero entries of M^{K_1} are $M_{11}^{K_1}$, $M_{14}^{K_1}$, $M_{15}^{K_1}$, $M_{41}^{K_1}$, $M_{44}^{K_1}$, $M_{45}^{K_1}$, $M_{51}^{K_1}$, $M_{54}^{K_1}$, and $M_{55}^{K_1}$. Proceeding similarly, we find

$$M = \int_{K_1} \begin{bmatrix} \varphi_1\varphi_1 & 0 & 0 & \varphi_4\varphi_1 & \varphi_5\varphi_1 \\ 0 & 0 & 0 & 0 & 0 \\ 0 & 0 & 0 & 0 & 0 \\ \varphi_1\varphi_4 & 0 & 0 & \varphi_4\varphi_4 & \varphi_5\varphi_4 \\ \varphi_1\varphi_5 & 0 & 0 & \varphi_4\varphi_5 & \varphi_5\varphi_5 \end{bmatrix} dx \tag{3.60}$$

$$+ \int_{K_1} \begin{bmatrix} \varphi_1\varphi_1 & \varphi_2\varphi_1 & 0 & \varphi_4\varphi_1 & 0 \\ \varphi_1\varphi_2 & \varphi_2\varphi_2 & 0 & \varphi_4\varphi_2 & 0 \\ 0 & 0 & 0 & 0 & 0 \\ \varphi_1\varphi_4 & \varphi_2\varphi_4 & 0 & \varphi_4\varphi_4 & 0 \\ 0 & 0 & 0 & 0 & 0 \end{bmatrix} dx \tag{3.61}$$

$$+ \int_{K_1} \begin{bmatrix} 0 & 0 & 0 & 0 & 0 \\ 0 & \varphi_2\varphi_2 & \varphi_3\varphi_2 & \varphi_4\varphi_2 & 0 \\ 0 & \varphi_2\varphi_3 & \varphi_3\varphi_3 & \varphi_4\varphi_3 & 0 \\ 0 & \varphi_2\varphi_4 & \varphi_3\varphi_4 & \varphi_4\varphi_4 & 0 \\ 0 & 0 & 0 & 0 & 0 \end{bmatrix} dx \tag{3.62}$$

$$= M^{K_1} + M^{K_2} + M^{K_3} \tag{3.63}$$

In practice the global element matrices M^{K_i} are never formed, but only their small 3×3 local element matrices necessary for storing the non-zero entries.

Having reduced the computation of the mass matrix M to a series of operations on the triangles, we next consider a single triangle K with its three nodes N_1, N_2, and N_3, and corresponding hat functions φ_1, φ_2, and φ_3. These nodes will almost certainly have a different node numbering, say N_r, N_s, and N_t, in the mesh as a whole, but let us label them 1, 2, and 3 for now.

The computation of the element masses could of course be done using quadrature. However, there is a much easier way. Using induction it is possible to show the integration formula

$$\int_K \varphi_1^m \varphi_2^n \varphi_3^p \, dx = \frac{2m!n!p!}{(m+n+p+2)!} |K| \tag{3.64}$$

where $|K|$ is the area of K and m, n, and p are positive integers. From this we immediately have

$$M_{ij}^K = \int_K \varphi_i \varphi_j \, dx = \frac{1}{12}(1 + \delta_{ij})|K| \quad i, j = 1, 2, 3 \tag{3.65}$$

where δ_{ij} is the Kronecker symbol, that is, 1 if $i = j$, and 0 if $i \neq j$. Writing out the entries M_{ij}^K explicitly, we, thus, have the local element mass matrix

$$M^K = \frac{1}{12} \begin{bmatrix} 2 & 1 & 1 \\ 1 & 2 & 1 \\ 1 & 1 & 2 \end{bmatrix} |K| \tag{3.66}$$

The mapping $\{1, 2, 3\} \mapsto \{r, s, t\}$ between the global node numbers r, s, and t and the local node numbers 1, 2, and 3 is called the local-to-global mapping. It is used when adding the entries of the local element mass matrix M^K to their appropriate positions in the global mass matrix M. This is done by cycling the index i and j over 1, 2, and 3, while adding M_{ij}^K to $M_{\text{loc2glb}_i, \text{loc2glb}_j}$, where $\text{loc2glb} = [r, s, t]$. This gives a simple, yet flexible, way of organizing the assembly. We summarize this assembly in the following algorithm:

Algorithm 8 Assembly of the Mass Matrix.

Let n_p be the number of nodes and n_t the number of elements in a mesh described by its point matrix P and connectivity matrix T.
Allocate memory for the $n_p \times n_p$ matrix M and initialize all matrix entries to zero.
for $K = 1, 2, \dots, n_t$ **do**
    Compute the 3×3 local element mass matrix M^K given by

$$M^K = \frac{1}{12}\begin{bmatrix} 2 & 1 & 1 \\ 1 & 2 & 1 \\ 1 & 1 & 2 \end{bmatrix} |K| \tag{3.67}$$

    Set up the local to global mapping, loc2glb $= [r, s, t]$.
    for $i = 1, 2, 3$ **do**
        for $j = 1, 2, 3$ **do**

$$M_{\text{loc2glb}_i \, \text{loc2glb}_j} = M_{\text{loc2glb}_i \, \text{loc2glb}_j} + M_{ij}^K \tag{3.68}$$

        end for
    end for
end for

The conversion of this algorithm into MATLAB code is straight forward.

```
function M = MassAssembler2D(p,t)
np = size(p,2); % number of nodes
nt = size(t,2); % number of elements
M = sparse(np,np); % allocate mass matrix
for K = 1:nt % loop over elements
    loc2glb = t(1:3,K); % local-to-global map
    x = p(1,loc2glb); % node x-coordinates
    y = p(2,loc2glb); %             y
    area = polyarea(x,y); % triangle area
    MK = [2 1 1;
          1 2 1;
          1 1 2]/12*area; % element mass matrix
    M(loc2glb,loc2glb) = M(loc2glb,loc2glb) ...
        + MK; % add element masses to M
end
```

Input to this routine is the point matrix `p` and connectivity matrix `t` given by `initmesh`. Output is the assembled global mass matrix M. Note that the allocation of the mass matrix is done using the `sparse` command, which tells MATLAB to store only non-zero matrix entries. This is important in order to save memory, since the number of nodes and consequently the matrix size might be large.

Running this routine on our mesh of the rectangle, which has the point and connectivity matrix

```
p=[0 1 2 2 0;
   0 0 0 1 1]
t=[1 1 2;
   4 2 3;
   5 4 4];
```

we get the 5×5 global mass matrix

```
M =

    0.2500    0.0417         0    0.1250    0.0833
    0.0417    0.1667    0.0417    0.0833         0
         0    0.0417    0.0833    0.0417         0
    0.1250    0.0833    0.0417    0.3333    0.0833
    0.0833         0         0    0.0833    0.1667
```

We remark that the assembly procedure is similar for other types of elements and higher space dimension. In particular, on a tetrahedral mesh the local element mass matrix M^K is of size 4×4, and can be computed using the integration formula

$$\int_K \varphi_1^m \varphi_2^n \varphi_3^p \varphi_4^q \, dx = \frac{6m!n!p!q!}{(m+n+p+q+6)!} |K| \tag{3.69}$$

where $|K|$ is the volume of tetrahedron K. This gives us

$$\int_K \varphi_i \varphi_j \, dx = \frac{1}{20}(1+\delta_{ij})|K| \quad i,j = 1,2,3,4 \tag{3.70}$$

Hence, we have

$$M^K = \frac{1}{20} \begin{bmatrix} 2 & 1 & 1 & 1 \\ 1 & 2 & 1 & 1 \\ 1 & 1 & 2 & 1 \\ 1 & 1 & 1 & 2 \end{bmatrix} |K| \tag{3.71}$$

3.6.2 Assembly of the Load Vector

The load vector b is assembled using the same technique as the mass matrix M, that is, by summing element load vectors b^K over the mesh. On each element K we get a local 3×1 element load vector b^K with entries

$$b_i^K = \int_K f \varphi_i \, dx, \quad i = 1,2,3 \tag{3.72}$$

Using node quadrature, for instance, to compute these integrals we end up with

$$b_i^K \approx \frac{1}{3} f(N_i)|K|, \quad i = 1, 2, 3 \tag{3.73}$$

We summarize the computation of the load vector in the following algorithm:

Algorithm 9 Assembly of the Load Vector.

1: Let n_p be the number of nodes and n_t the number of elements in a mesh described by its point matrix P and connectivity matrix T.
2: Allocate memory for the $n_p \times 1$ vector b and initialize all vector entries to zero.
3: **for** $K = 1, 2, \dots, n_t$ **do**
4: Compute the 3×1 local element load vector b^K given by

$$b^K = \frac{1}{3} \begin{bmatrix} f(N_1) \\ f(N_2) \\ f(N_3) \end{bmatrix} |K| \tag{3.74}$$

5: Set up the local-to-global mapping, loc2glb $= [r, s, t]$.
6: **for** $i = 1, 2, 3$ **do**
7:

$$b_{\text{loc2glb}_i} = b_{\text{loc2glb}_i} + b_i^K \tag{3.75}$$

8: **end for**
9: **end for**

Translated into MATLAB code the algorithm takes the following form.

```
function b = LoadAssembler2D(p,t,f)
np = size(p,2);
nt = size(t,2);
b = zeros(np,1);
for K = 1:nt
  loc2glb = t(1:3,K);
  x = p(1,loc2glb);
  y = p(2,loc2glb);
  area = polyarea(x,y);
  bK = [f(x(1),y(1));
        f(x(2),y(2));
        f(x(3),y(3))]/3*area; % element load vector
  b(loc2glb) = b(loc2glb) ...
    + bK; % add element loads to b
end
```

Here, we assume that `f` is a function handle to a routine specifying f, for example,

```
function f = Foo2(x, y)
f = x.*y;
```

A main routine for computing the L^2-projection πf of $f = x_1 x_2$ on the unit square $\Omega = [0, 1]^2$ is given below.

```
function L2Projector2D()
g = Rectg(0,0,1,1); % unit square
[p,e,t] = initmesh(g,'hmax',0.1); % create mesh
M = MassAssembler2D(p,t); % assemble mass matrix
b = LoadAssembler2D(p,t,@Foo2); % assemble load vector
Pf = M\b; % solve linear system
pdesurf(p,t,Pf) % plot projection
```

3.6.3 Properties of the Mass Matrix

Theorem 3.4. *The mass matrix M is symmetric and positive definite.*

Proof. M is obviously symmetric, since $M_{ij} = M_{ji}$.

To prove that M is positive definite we must show that

$$\xi^T M \xi > 0 \tag{3.76}$$

for all non-zero $n_p \times 1$ vectors ξ.

Now, a straight forward calculation reveals that

$$\xi^T M \xi = \sum_{ij=1}^{n_p} \xi_i M_{ij} \xi_j \tag{3.77}$$

$$= \sum_{ij=1}^{n_p} \xi_i \left(\int_\Omega \varphi_j \varphi_i \, dx \right) \xi_j \tag{3.78}$$

$$= \int_\Omega \left(\sum_{i=1}^{n_p} \xi_i \varphi_i \right) \left(\sum_{j=1}^{n_p} \xi_j \varphi_j \right) dx \tag{3.79}$$

$$= \left\| \sum_{i=1}^{n_p} \xi_i \varphi_i \right\|_{L^2(\Omega)}^2 \tag{3.80}$$

The last norm is equal to zero if and only if the sum $s = \sum_{i=1}^{n_p} \xi_i \varphi_i = 0$. However, this sum can be viewed as a function in V_h, and if s is zero then all coefficients ξ_i must also be zero. □

The condition number $\kappa(M)$ of a matrix M is a measure of how sensitive the solution ξ to the linear system $M\xi = b$ is to perturbations of the right hand side b, and is defined by

$$\kappa(M) = \|M\| \, \|M^{-1}\| \tag{3.81}$$

where we define the matrix norm $\|\cdot\|$ by

$$\|M\| = \max_{\xi \neq 0} \frac{\xi^T M \xi}{\xi^T \xi} \tag{3.82}$$

Now, if M is symmetric and positive definite, then κ can be expressed in terms of the largest and smallest eigenvalue of M. Indeed, if λ is an eigenvalue of M such that $M\xi = \lambda\xi$ for some eigenvector ξ, then $\lambda = \xi^T M \xi / \xi^T \xi$, which is positive by virtue of Theorem 3.4. Taking the maximum over all eigenvalues and corresponding eigenvectors it follows that $\|M\| = \lambda_{\max}(M)$. Further, since the eigenvalues of M^{-1} is the inverse of those of M, it also follows that $\|M^{-1}\| = 1/\lambda_{\min}(M)$. Thus, the condition number is the quotient

$$\kappa(M) = \frac{\lambda_{\max}(M)}{\lambda_{\min}(M)} \tag{3.83}$$

Theorem 3.5. *The condition number of the mass matrix M satisfies the estimate*

$$\kappa(M) \leq C \tag{3.84}$$

where C is a constant.

Proof. Let us assume that the mesh $\mathcal{K}$ is quasi-uniform and write $\xi^T M \xi$ as the sum $\sum_{K \in \mathcal{K}} \xi^T|_K M^K \xi|_K$. On each element K, the element mass matrix M^K is given by (3.66), and, thus, proportional to the area $|K|$ of K. Now, $|K|$ is in turn proportional to the local mesh size h_K^2, so defining the global mesh size $h = \max_K h_K$, we have $C_1 h^2 I \leq M^K \leq C_2 h^2 I$ with I the identity element matrix. These bounds on M^K imply

$$C_3 h^2 \leq \frac{\xi^T M \xi}{\xi^T \xi} \leq C_4 h^2 \tag{3.85}$$

which shows that the extremal eigenvalues of M are bounded by $C_3 h^2 \leq \lambda_{\min}(M)$ and $\lambda_{\max}(M) \leq C_4 h^2$. Hence, $\kappa(M) = C_4/C_3$. □

3.7 Further Reading

Partitioning a domain into geometric simplices is generally a difficult task, and we refer the interested reader to the book by George and Frey [33] for a comprehensive survey of the different algorithms and data structures used for mesh generation.

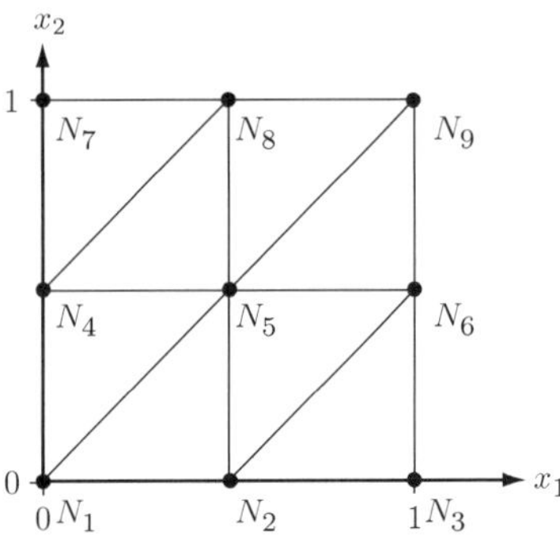

Fig. 3.9 Triangulation of the unit square

Also, the papers by Shewchuk [64, 65] thoroughly explains the ideas behind Delaunay triangulation and tetrahedralization.

3.8 Problems

Exercise 3.1. Read the help for the PDE-Toolbox commands `initmesh`, `pdemesh`, and `pdesurf`.

Exercise 3.2. Write down the geometry matrix g for the unit square $\Omega = [0, 1]^2$.

Exercise 3.3. Express the area of an arbitrary triangle in terms of its corner coordinates $(x_1^{(1)}, x_2^{(1)})$, $(x_1^{(2)}, x_2^{(2)})$, and $(x_1^{(3)}, x_2^{(3)})$.

Exercise 3.4. Derive explicit expressions for the hat functions on the triangle with corners at $(-1, -1)$, $(1, 0)$, and $(-1, 1)$.

Exercise 3.5. Determine the basis functions for piecewise linear functions on an arbitrary triangle with corner coordinates (x_1, y_1), (x_2, y_2) and (x_3, y_3).

Exercise 3.6. Determine a linear coordinate transform which maps an arbitrary triangle onto the reference triangle $\bar{K}$ with corners at origo, $(1, 0)$, and $(0, 1)$.

Exercise 3.7. Given the triangulation of Fig. 3.9.

(a) Write down the point matrix P and the connectivity matrix T.
(b) Determine the mesh function $h(x)$, (i.e., the piecewise constant function giving the mesh size on each element).

Exercise 3.8. Draw the hat functions φ_1 and φ_5 associated with nodes N_1 and N_5 in Fig. 3.9.

Exercise 3.9. Consider again the mesh of the unit square Ω shown in Fig. 3.9.

(a) Determine the sparsity pattern of the mass matrix on this mesh.
(b) Compute the integrals $\int_\Omega \phi_1\phi_2\,dx$, $\int_\Omega \phi_7\phi_4\,dx$, $\int_\Omega \phi_7\phi_8\,dx$, and $\int_\Omega x_1\phi_1\,dx$.

Exercise 3.10. Let $f = x_1 x_2$ and let $\Omega = [0, 1]^2$ be the unitsquare.

(a) Calculate $\int_\Omega f\, dx$ analytically.
(b) Compute $\int_\Omega f\, dx$ by using the center of gravity rule on each triangle of the mesh in Fig. 3.9.
(c) Compute $\int_\Omega f\, dx$ by using the corner quadrature rule on each triangle of the mesh in Fig. 3.9.

Exercise 3.11. Compute the L^2-projection $P_h f \in V_h$ to $f = x_1^2$ on the mesh shown in Fig. 3.9. Use the corner quadrature rule to evaluate the integrals of the mass matrix and the load vector.

Chapter 4
The Finite Element Method in 2D

Abstract In this chapter we develop finite element methods for numerical solution of partial differential equations in two dimensions. The approach taken is the same as before, that is, we first rewrite the equation in variational form, and then seek an approximate solution in the space of continuous piecewise linear functions. Although the numerical methods presented are general, we focus on linear second order elliptic equations with the Poisson equation as our main model problem. We prove basic error estimates, discuss the implementation of the involved algorithms, and study some examples of application.

4.1 Green's Formula

At the outset let us recall a few mathematical preliminaries that will be of frequent use.

Let Ω be a domain in $\mathbb{R}^2$, with boundary $\partial\Omega$ and exterior unit normal n. We recall the following form of the divergence theorem.

$$\int_\Omega \frac{\partial f}{\partial x_i}\, dx = \int_{\partial\Omega} f n_i\, ds, \quad i = 1, 2 \tag{4.1}$$

where n_i is component i of n.

Setting $f = fg$ we get the partial integration formula

$$\int_\Omega \frac{\partial f}{\partial x_i} g dx = -\int_\Omega f \frac{\partial g}{\partial x_i}\, dx + \int_{\partial\Omega} f g n_i\, ds, \quad i = 1, 2 \tag{4.2}$$

Applying (4.2) with $f = w_i$, the components of a vector field w on Ω, and $g = v$, and taking the sum over $i = 1, 2$ we obtain

M.G. Larson and F. Bengzon, *The Finite Element Method: Theory, Implementation, and Applications*, Texts in Computational Science and Engineering 10,
DOI 10.1007/978-3-642-33287-6_4, © Springer-Verlag Berlin Heidelberg 2013

$$\int_\Omega (\nabla \cdot w) v \, dx = -\int_\Omega w \cdot \nabla v \, dx + \int_{\partial\Omega} (w \cdot n) v ds \tag{4.3}$$

Finally, choosing $w = -\nabla u$ in (4.3) we obtain the so-called Green's formula

$$\int_\Omega -\Delta u v \, dx = \int_\Omega \nabla u \cdot \nabla v \, dx - \int_{\partial\Omega} n \cdot \nabla u v ds \tag{4.4}$$

We remark that Green's formula also holds in higher space dimensions.

4.2 The Finite Element Method for Poisson's Equation

4.2.1 Poisson's Equation

Let us consider Poisson's equation: find u such that

$$-\Delta u = f, \quad \text{in } \Omega \tag{4.5a}$$

$$u = 0, \quad \text{on } \partial\Omega \tag{4.5b}$$

where $\Delta = \partial^2/\partial x_1^2 + \partial^2/\partial x_2^2$ is the Laplace operator, and f is a given function in, say, $L^2(\Omega)$.

4.2.2 Variational Formulation

To derive a variational formulation of Poisson's equation (4.5) we multiply $f = -\Delta u$ by a function v, which is assumed to vanish on the boundary, and integrate using Green's formula.

$$\int_\Omega f v dx = -\int_\Omega \Delta u v \, dx \tag{4.6}$$

$$= \int_\Omega \nabla u \cdot \nabla v \, dx - \int_{\partial\Omega} n \cdot \nabla u v ds \tag{4.7}$$

$$= \int_\Omega \nabla u \cdot \nabla v \, dx \tag{4.8}$$

The last line follows due to the assumption $v = 0$ on $\partial\Omega$. Introducing the spaces

$$V = \{v : \|v\|_{L^2(\Omega)} + \|\nabla v\|_{L^2(\Omega)} < \infty\} \tag{4.9}$$

$$V_0 = \{v \in V : v|_{\partial\Omega} = 0\} \tag{4.10}$$

we have the following variational formulation of (4.5): find $u \in V_0$ such that

$$\int_\Omega \nabla u \cdot \nabla v \, dx = \int_\Omega f v \, dx, \quad \forall v \in V_0 \tag{4.11}$$

With this choice of test and trial space V_0 the integrals $\int_\Omega \nabla u \cdot \nabla v \, dx$ and $\int_\Omega f v \, dx$ make sense. To see this, note that due to the Cauchy-Schwarz inequality, we have $\int_\Omega f v \, dx \leq \|f\|_{L^2(\Omega)} \|v\|_{L^2(\Omega)}$, which is less than infinity by the assumptions on v and f. A similar line of reasoning applies to $\int_\Omega \nabla u \cdot \nabla v \, dx$.

In this context we would like to a point out a subtlety that we have not yet touched upon. Even though the solution to Poisson's equation (4.5) is also a solution to the variational formulation (4.11), the converse is generally not true. To see this, it suffice to note that the solution to the variational formulation does not need to be two times differentiable. For this reason the variational formulation is sometimes called the weak form, as opposed to the original, or strong, form. Proving that a weak solution is also a strong solution generally depends on the shape of the domain Ω and the regularity (i.e., smoothness) of the coefficients (i.e., f).

4.2.3 Finite Element Approximation

Let $\mathcal{K}$ be a triangulation of Ω, and let V_h be the space of continuous piecewise linears on $\mathcal{K}$. To satisfy the boundary conditions, let also $V_{h,0} \subset V_h$ be the subspace

$$V_{h,0} = \{v \in V_h : v|_{\partial\Omega} = 0\} \tag{4.12}$$

Replacing V_0 with $V_{h,0}$ in the variational formulation (4.11) we obtain the following finite element method: find $u_h \in V_{h,0}$ such that

$$\int_\Omega \nabla u_h \cdot \nabla v \, dx = \int_\Omega f v \, dx, \quad \forall v \in V_{h,0} \tag{4.13}$$

4.2.4 Derivation of a Linear System of Equations

In order to actually compute the finite element approximation u_h, let $\{\varphi_i\}_{i=1}^{n_i}$ be the basis for $V_{h,0}$ consisting of the hat functions associated with the n_i interior nodes within the mesh. We do not allow any hat functions on the mesh boundary, since the functions in $V_{h,0}$ must vanish there. Now, using this basis, we first note that the finite element method (4.13) is equivalent to

$$\int_\Omega \nabla u_h \cdot \nabla \varphi_i \, dx = \int_\Omega f \varphi_i \, dx, \quad i = 1, 2, \ldots, n_i \tag{4.14}$$

Then, since u_h belongs to $V_{h,0}$ it can be written as the linear combination

$$u_h = \sum_{j=1}^{n_i} \xi_j \varphi_j \tag{4.15}$$

with n_i unknowns ξ_j, $j = 1, 2, \dots, n_i$, to be determined.

Inserting the ansatz (4.15) into (4.14) we get

$$\int_\Omega f \varphi_i \, dx = \int_\Omega \nabla u_h \cdot \nabla \varphi_i \, dx \tag{4.16}$$

$$= \int_\Omega \nabla \left(\sum_{j=1}^{n_i} \xi_j \varphi_j \right) \cdot \nabla \varphi_i \, dx \tag{4.17}$$

$$= \sum_{j=1}^{n_i} \xi_j \int_\Omega \nabla \varphi_j \cdot \nabla \varphi_i \, dx, \quad i = 1, 2, \dots, n_i \tag{4.18}$$

Further, using the notation

$$A_{ij} = \int_\Omega \nabla \varphi_j \cdot \nabla \varphi_i \, dx, \quad i, j = 1, 2, \dots, n_i \tag{4.19}$$

$$b_i = \int_\Omega f \varphi_i \, dx, \quad i = 1, 2, \dots, n_i \tag{4.20}$$

we have

$$b_i = \sum_{j=1}^{n_i} A_{ij} \xi_j, \quad i = 1, 2, \dots, n_i \tag{4.21}$$

which is an $n_i \times n_i$ linear system for the unknowns ξ_j. In matrix form, we write this

$$A\xi = b \tag{4.22}$$

where the entries of the $n_i \times n_i$ stiffness matrix A, and the $n_i \times 1$ load vector b are defined by (4.19) and (4.20), respectively. Solving the linear system (4.22) we obtain the unknowns ξ_j, and, thus, u_h.

4.2.5 Basic Algorithm to Compute the Finite Element Solution

The following algorithm summarizes the basic steps in computing the finite element solution u_h:

Algorithm 10 Basic Finite Element Algorithm.

1: Create a triangulation $\mathcal{K}$ of Ω and define the corresponding space of continuous piecewise linear functions $V_{h,0}$ hat function basis $\{\varphi_i\}_{i=1}^{n_i}$.

2: Assemble the $n_i \times n_i$ stiffness matrix A and the $n_i \times 1$ load vector b, with entries

$$A_{ij} = \int_\Omega \nabla\varphi_j \cdot \nabla\varphi_i \, dx, \quad b_i = \int_\Omega f\varphi_i \, dx \tag{4.23}$$

3: Solve the linear system

$$A\xi = b \tag{4.24}$$

4: Set

$$u_h = \sum_{j=1}^{n_i} \xi_j \varphi_j \tag{4.25}$$

4.3 Some Useful Inequalities

In this section we cover some standard inequalities that are very useful and of frequent use.

Theorem 4.1 (Poincaré Inequality). *Let $\Omega \subset \mathbb{R}^2$ be a bounded domain. Then, there is constant $C = C(\Omega)$, such that for any $v \in V_0$,*

$$\|v\|_{L^2(\Omega)} \leq C\,\|\nabla v\|_{L^2(\Omega)} \tag{4.26}$$

Proof. Let ϕ be a function satisfying $-\Delta\phi = 1$ in Ω with $\sup_{x\in\Omega} |\nabla\phi(x)| < C$. Such a function exist if the boundary of Ω is sufficiently smooth.

Multiplying $1 = -\Delta\phi$ with v^2, and integrating by parts using Green's formula, we have

$$\int_\Omega v^2 \, dx = -\int_\Omega v^2 \Delta\phi \, dx \tag{4.27}$$

$$= -\int_{\partial\Omega} v^2 n \cdot \nabla\phi \, ds + \int_\Omega 2v\nabla v \cdot \nabla\phi \, dx \tag{4.28}$$

$$= \int_\Omega 2v\nabla v \cdot \nabla\phi \, dx \tag{4.29}$$

since $v = 0$ on $\partial\Omega$. Using the Cauchy-Schwarz inequality we further have

$$\|v\|^2_{L^2(\Omega)} = \int_\Omega v^2 \, dx = \int_\Omega 2v\nabla v \cdot \nabla\phi \, dx \leq 2\max|\nabla\phi|\,\|v\|_{L^2(\Omega)}\|\nabla v\|_{L^2(\Omega)} \tag{4.30}$$

Finally, we set $C = 2\max|\nabla\phi|$, and divide by $\|v\|_{L^2(\Omega)}$. □

The above result also holds if v is zero on part of the boundary or if the average $|\Omega|^{-1}\int_\Omega v\,dx$ of v is zero.

The restriction $v|_{\partial\Omega}$ of v to the boundary $\partial\Omega$ of Ω is called the trace of v. The following so-called Trace inequality estimates the L^2-norm of the trace in terms of the L^2-norm of v and ∇v on Ω.

Theorem 4.2 (Trace Inequality). *Let $\Omega \subset \mathbb{R}^2$ be a bounded domain with smooth or convex polygonal boundary $\partial\Omega$. Then, there is constant $C = C(\Omega)$, such that for any $v \in V$,*

$$\|v\|_{L^2(\partial\Omega)} \leq C\left(\|v\|^2_{L^2(\Omega)} + \|\nabla v\|^2_{L^2(\Omega)}\right)^{1/2} \tag{4.31}$$

Proof. Let ϕ be a function satisfying $-\Delta\phi = -|\partial\Omega|/|\Omega|$ in Ω with $n\cdot\nabla\phi = 1$ on $\partial\Omega$.

Multiplying $-|\partial\Omega|/|\Omega| = -\Delta\phi$ with v^2, and integrating by parts using Green's formula, we have

$$-\frac{|\partial\Omega|}{|\Omega|}\int_\Omega v^2\,dx = -\int_\Omega v^2\Delta\phi\,dx \tag{4.32}$$

$$= -\int_{\partial\Omega} v^2 n\cdot\nabla\phi\,ds + \int_\Omega 2v\nabla v\cdot\nabla\phi\,dx \tag{4.33}$$

$$= -\int_{\partial\Omega} v^2\,ds + \int_\Omega 2v\nabla v\cdot\nabla\phi\,dx \tag{4.34}$$

Using the Cauchy-Schwarz inequality, we get

$$\|v\|^2_{L^2(\partial\Omega)} = \int_{\partial\Omega} v^2\,ds \tag{4.35}$$

$$= \frac{|\partial\Omega|}{|\Omega|}\int_\Omega v^2\,dx + 2\int_\Omega v\nabla v\cdot\nabla\phi\,dx \tag{4.36}$$

$$\leq C_1\|v\|^2_{L^2(\Omega)} + 2C_2\|v\|_{L^2(\Omega)}\|\nabla v\|_{L^2(\Omega)} \tag{4.37}$$

$$\leq \max(C_1, C_2)\left(\|v\|^2_{L^2(\Omega)} + \|\nabla v\|^2_{L^2(\Omega)}\right) \tag{4.38}$$

where we have set $C_1 = |\partial\Omega|/|\Omega|$, and $C_2 = \sup_{x\in\Omega}|\nabla\phi|$. Finally, we have used the arithmetic-geometric inequality $2ab \leq a^2 + b^2$, with $a = \|v\|_{L^2(\Omega)}$ and $b = \|\nabla v\|_{L^2(\Omega)}$. □

For the Trace inequality to hold it is necessary that the boundary of the domain is not too rough. In fact, the shape of the domain often plays a crucial role in establishing results concerning the regularity of functions. Irregularly shaped domains tend to give rise to irregular functions, and vice versa. For example, if the domain is convex, then the total derivative D^2 can be estimated in terms of only the Laplacian Δ. This result is called elliptic regularity.

Theorem 4.3 (Elliptic Regularity). *Let $\Omega \subset \mathbb{R}^2$ be a bounded convex domain with polygonal boundary or a general domain with smooth boundary. Then, there is a constant $C = C(\Omega)$, such that for any sufficiently smooth function v with $v = 0$ or $n \cdot \nabla v = 0$ on $\partial\Omega$,*

$$\|D^2 v\|_{L^2(\Omega)} \leq C\|\Delta v\|_{L^2(\Omega)} \tag{4.39}$$

If Ω is convex, then $0 < C \leq 1$. Otherwise $C > 1$.

We refer to Eriksson and co-authors [27] for a proof of this theorem.

In a polynomial space such as V_h with finite dimension, all norms are equivalent. That is, for any two norms $\|\cdot\|_\alpha$ and $\|\cdot\|_\beta$ there are constants C_1 and C_2, such that $C_1\|\cdot\|_\alpha \leq \|\cdot\|_\beta \leq C_2\|\cdot\|_\alpha$. As a consequence, we can invent more or less any inequality we want on V_h. However, the constant depends in general on the mesh size parameter h, that is, we have $\|v\|_\alpha \leq C(h)\|v\|_\beta$. Therefore, the dependence of C on h must be established. This can be done using scaling arguments where we basically make sure that we have the same units on both sides of the estimates. Since $\partial/\partial x_i$, $1, 2$, has the unit of inverse length it corresponds to a factor h^{-1}. For example, for a linear polynomial $v \in P_1(K)$ on a triangle K, we have

$$\|\nabla v\|_{L^2(K)} \leq C h_K^{-1}\|v\|_{L^2(K)} \tag{4.40}$$

Summing over all triangles $K \in \mathcal{K}$, and assuming a quasi-uniform mesh $\mathcal{K}$, we obtain the following so-called inverse estimate.

Theorem 4.4 (Inverse Estimate). *On a quasi-uniform mesh any $v \in V_h$ satisfies the inverse estimate*

$$\|\nabla v\|_{L^2(\Omega)} \leq C h^{-1}\|v\|_{L^2(\Omega)} \tag{4.41}$$

4.4 Basic Analysis of the Finite Element Method

4.4.1 Existence and Uniqueness of the Finite Element Solution

Theorem 4.5. *The finite element solution u_h, defined by* (4.13)*, exists and is unique.*

Proof. We first show the uniqueness claim. The argument is by contradiction. Assume that there are two finite element solutions u_h and $\tilde{u}_h$ satisfying (4.13). Then, we have

$$\int_{\Omega} \nabla u_h \cdot \nabla v \, dx = \int_{\Omega} f v \, dx, \quad \forall v \in V_{h,0} \tag{4.42}$$

$$\int_{\Omega} \nabla \tilde{u}_h \cdot \nabla v \, dx = \int_{\Omega} f v \, dx, \quad \forall v \in V_{h,0} \tag{4.43}$$

Subtracting these equations we get

$$\int_{\Omega} \nabla (u_h - \tilde{u}_h) \cdot \nabla v \, dx = 0, \quad \forall v \in V_{h,0} \tag{4.44}$$

Now, choosing $v = u_h - \tilde{u}_h \in V_{h,0}$ we further get

$$\int_{\Omega} |\nabla (u_h - \tilde{u}_h)|^2 \, dx = 0 \tag{4.45}$$

From this identity we conclude that $u_h - \tilde{u}_h$ must be a constant function. However, using the boundary conditions we see that this constant must be zero, since $u_h = \tilde{u}_h = 0$ on $\partial\Omega$.

To prove existence we recall that u_h is determined by a square linear system. The existence of a solution to a linear system follows from the uniqueness of the solution. □

4.4.2 A Priori Error Estimates

Theorem 4.6 (Galerkin Orthogonality). *The finite element approximation u_h, defined by* (4.13)*, satisfies the orthogonality*

$$\int_{\Omega} \nabla (u - u_h) \cdot \nabla v \, dx = 0, \quad \forall v \in V_{h,0} \tag{4.46}$$

Proof. From the variational formulation we have

$$\int_{\Omega} \nabla u \cdot \nabla v \, dx = \int_{\Omega} f v \, dx, \quad \forall v \in V_0 \tag{4.47}$$

and from the finite element method we further have

$$\int_{\Omega} \nabla u_h \cdot \nabla v \, dx = \int_{\Omega} f v \, dx, \quad \forall v \in V_{h,0} \tag{4.48}$$

Subtracting these and using the fact that $V_{h,0} \subset V_0$ immediately proves the claim. □

To estimate the error we now introduce the following norm, called the energy norm on V_0,

$$|||v|||^2 = \int_\Omega \nabla v \cdot \nabla v dx \tag{4.49}$$

Note that $|||v||| = \|\nabla v\|_{L^2(\Omega)}$.

The next theorem is a best approximation result.

Theorem 4.7. *The finite element solution u_h, defined by* (4.13)*, satisfies the best approximation result*

$$|||u - u_h||| \leq |||u - v|||, \quad \forall v \in V_{h,0} \tag{4.50}$$

Proof. Writing $u - u_h = u - v + v - u_h$ for any $v \in V_{h,0}$, we have

$$|||u - u_h|||^2 = \int_\Omega \nabla(u - u_h) \cdot \nabla(u - u_h)\, dx \tag{4.51}$$

$$= \int_\Omega \nabla(u - u_h) \cdot \nabla(u - v)\, dx + \int_\Omega \nabla(u - u_h) \cdot \nabla(v - u_h)\, dx \tag{4.52}$$

$$= \int_\Omega \nabla(u - u_h) \cdot \nabla(u - v)\, dx \tag{4.53}$$

$$\leq |||u - u_h|||\,|||u - v||| \tag{4.54}$$

where we used the Galerkin orthogonality (4.46) to conclude that

$$\int_\Omega \nabla(u - u_h) \cdot \nabla(v - u_h)\, dx = 0 \tag{4.55}$$

since $v - u_h \in V_{h,0}$. Dividing by $|||u - u_h|||$ concludes the proof. □

Theorem 4.8. *The finite element solution u_h, defined by* (4.13)*, satisfies the estimate*

$$|||u - u_h|||^2 \leq C \sum_{K \in \mathcal{K}} h_K^2 \|D^2 u\|_{L^2(K)}^2 \tag{4.56}$$

Proof. Starting from the best approximation result, choosing $v = \pi u$, and using the interpolation error estimate of Proposition 3.1, we have

$$|||u - u_h|||^2 \leq |||u - \pi u|||^2 \tag{4.57}$$

$$= \sum_{K \in \mathcal{K}} \|D(u - \pi u)\|_{L^2(K)}^2 \tag{4.58}$$

$$\leq \sum_{K \in \mathcal{K}} C h_K^2 \|D^2 u\|_{L^2(K)}^2 \tag{4.59}$$

which proves the estimate. □

Here, we tacitly assume that u is two times differentiable, so that D^2u makes sense.

Hence, we conclude that

$$|||u - u_h||| \leq Ch\|D^2u\|_{L^2(\Omega)} \tag{4.60}$$

In other words, the gradient of the error tends to zero as the mesh size h tends to zero.

We remark that the a priori error estimate above also holds in three dimension on tetrahedral meshes.

The energy norm $|||\cdot|||$ allows a simple derivation of the above a priori error estimate since the energy norm is closely related to the variational formulation. However, we may be interested in estimating the error also in other norms, for instance the L^2 norm. Combining (4.60) and (4.26) we immediately obtain

$$\|u - u_h\|_{L^2(\Omega)} \leq C|||u - u_h||| \leq Ch\|D^2u\|_{L^2(\Omega)} \tag{4.61}$$

which is a non-optimal L^2 estimate since it only indicates first order convergence instead of the expected optimal second order convergence.

It is however possible to improve on this result using the so called Nitsche's trick [50].

Theorem 4.9. *The finite element solution u_h, defined by* (4.13)*, satisfies the estimate*

$$\|u - u_h\|_{L^2(\Omega)} \leq Ch^2\|D^2u\|_{L^2(\Omega)} \tag{4.62}$$

Proof. Let $e = u - u_h$ be the error, and let ϕ be the solution of the so-called dual, or adjoint, problem

$$-\Delta\phi = e, \quad \text{in } \Omega \tag{4.63a}$$

$$\phi = 0, \quad \text{on } \partial\Omega \tag{4.63b}$$

Multiplying $e = -\Delta\phi$ by e, and integrating using Green's formula, we have

$$\int_\Omega e^2\,dx = -\int_\Omega e\Delta\phi\,dx \tag{4.64}$$

$$= \int_\Omega \nabla e \cdot \nabla\phi \cdot dx - \int_{\partial\Omega} en \cdot \nabla\phi\,ds \tag{4.65}$$

$$= \int_\Omega \nabla e \cdot \nabla\phi\,dx \tag{4.66}$$

$$= \int_\Omega \nabla e \cdot \nabla(\phi - \pi\phi)\,dx \tag{4.67}$$

where we have used Galerkin orthogonality (4.46) in the last line to subtract an interpolant $\pi\phi \in V_{h,0}$ to ϕ. Further, using the Cauchy-Schwarz inequality, we have

$$\|e\|^2_{L^2(\Omega)} \leq \|\nabla e\|_{L^2(\Omega)} \|\nabla(\phi - \pi\phi)\|_{L^2(\Omega)} \tag{4.68}$$

Here, $\|\nabla e\|_{L^2(\Omega)}$ can be estimated using Theorem 4.8. To estimate also $\|\nabla(\phi - \pi\phi)\|_{L^2(\Omega)}$ we first use a standard interpolation estimate, and then the elliptic regularity $\|D^2\phi\|_{L^2(\Omega)} \leq C\|\Delta\phi\|_{L^2(\Omega)}$, which yields

$$\|\nabla(\phi - \pi\phi)\|_{L^2(\Omega)} \leq Ch\|D^2\phi\|_{L^2(\Omega)} \leq Ch\|\Delta\phi\|_{L^2(\Omega)} = Ch\|e\|_{L^2(\Omega)} \tag{4.69}$$

since $-\Delta\phi = e$. Thus, we have

$$\|e\|^2_{L^2(\Omega)} \leq Ch\|D^2u\|_{L^2(\Omega)} Ch\|e\|_{L^2(\Omega)} \tag{4.70}$$

Dividing by $\|e\|_{L^2(\Omega)}$ concludes the proof. □

4.4.3 Properties of the Stiffness Matrix

Theorem 4.10. *The stiffness matrix A is symmetric and positive definite.*

Proof. A is obviously symmetric, since $A_{ij} = A_{ji}$.

To prove that A is positive definite we must show that

$$\xi^T A\xi > 0 \tag{4.71}$$

for all non-zero $n_i \times 1$ vectors ξ.

Now, a straight forward calculation reveals that

$$\xi^T A\xi = \sum_{i,j=1}^{n_i} \xi_i A_{ij} \xi_j \tag{4.72}$$

$$= \sum_{i,j=1}^{n_i} \xi_i \xi_j \int_\Omega \nabla\varphi_j \cdot \nabla\varphi_i \, dx \tag{4.73}$$

$$= \int_\Omega \nabla\left(\sum_{i=1}^{n_i} \xi_i\varphi_i\right) \cdot \nabla\left(\sum_{j=1}^{n_i} \xi_j\varphi_j\right) dx \tag{4.74}$$

$$= \left\|\nabla\left(\sum_{i=1}^{n_i} \xi_i\varphi_i\right)\right\|^2_{L^2(\Omega)} \tag{4.75}$$

The last norm is larger than zero as long as the sum $s = \sum_{i=1}^{n_i} \xi_i\varphi_i$, which may be viewed as a function in $V_{h,0}$, does not represent a constant function. However, the only constant function in $V_{h,0}$ is the zero function $s = 0$ with all coefficients ξ_i equal to zero. □

In the above proof we used the fact that the only constant function s in $V_{h,0}$ is the zero function. This is due to the Dirichlet boundary conditions $s|_{\partial\Omega} = 0$.

However, for Neumann conditions the situation is different since the constant function resides in V_h and thus A has a one dimensional null space. In this case A is only positive semi-definite.

Finally, we consider the condition number of the stiffness matrix. Recall that the condition number is defined by

$$\kappa(A) = \|A\| \, \|A^{-1}\| \tag{4.76}$$

and that $\kappa(A)$ is the quotient of the largest and smallest eigenvalues

$$\kappa(A) = \frac{\lambda_{\max}(A)}{\lambda_{\min}(A)} \tag{4.77}$$

since A is symmetric and positive definite.

Theorem 4.11. *The condition number of the stiffness matrix A satisfies the estimate*

$$\kappa(A) \leq C h^{-2} \tag{4.78}$$

Proof. From the Poincaré inequality (4.26), we have

$$\|s\|^2_{L^2(\Omega)} \leq C |||s|||^2 \tag{4.79}$$

or, in matrix notation,

$$\xi^T M \xi \leq C \xi^T A \xi \tag{4.80}$$

where M is the $n_i \times n_i$ mass matrix and ξ is the vector of nodal values of s. Also, from the inverse estimate (4.41), we have

$$|||s|||^2 \leq C h^{-2} \|s\|^2_{L^2(\Omega)} \tag{4.81}$$

or, in matrix notation,

$$\xi^T A \xi \leq C h^{-2} \xi^T M \xi \tag{4.82}$$

Thus,

$$\frac{\xi^T M \xi}{\xi^T \xi} \leq \frac{\xi^T A \xi}{\xi^T \xi} \leq C h^{-2} \frac{\xi^T M \xi}{\xi^T \xi} \tag{4.83}$$

Now, we recall from Theorem 3.5 that there are constants C_1 and C_2 such that

$$C_1 h^2 \leq \frac{\xi^T M \xi}{\xi^T \xi} \leq C_2 h^2 \tag{4.84}$$

Combining this knowledge with (4.83) we conclude that $C_1h^2 \leq \lambda_{\min}(A)$ and $\lambda_{\max}(A) \leq CC_2$. Hence, $\kappa(A) = C_3/h^2$. □

4.5 A Model Problem with Variable Coefficients

We next consider the following model problem, involving variable coefficients and more general boundary conditions: find u such that

$$-\nabla \cdot (a\nabla u) = f, \qquad \text{in } \Omega \tag{4.85a}$$

$$-n \cdot (a\nabla u) = \kappa(u - g_D) - g_N, \qquad \text{on } \partial\Omega \tag{4.85b}$$

where $a > 0$, f, $\kappa > 0$, g_D, and g_N are given functions.

We shall seek a solution to this problem in the space $V = \{v : \|v\|_{L^2(\Omega)} + \|\nabla v\|_{L^2(\Omega)} < \infty\}$.

Multiplying $f = -\nabla \cdot (a\nabla u)$ by a test function $v \in V$, and integrating using Green's formula, we have

$$\int_\Omega fv\,dx = \int_\Omega -\nabla \cdot (a\nabla u)v\,dx \tag{4.86}$$

$$= \int_\Omega a\nabla u \cdot \nabla v\,dx - \int_{\partial\Omega} n \cdot (a\nabla u)v ds \tag{4.87}$$

$$= \int_\Omega a\nabla u \cdot \nabla v\,dx + \int_{\partial\Omega} (\kappa(u - g_D) - g_N)v ds \tag{4.88}$$

where we used the boundary condition to replace $-n \cdot a\nabla u$ by $\kappa(u - g_D) - g_N$.

Collecting terms, we obtain the variational formulation: find $u \in V$ such that

$$\int_\Omega a\nabla u \cdot \nabla v\,dx + \int_{\partial\Omega} \kappa uv ds = \int_\Omega fv\,dx + \int_{\partial\Omega} (\kappa g_D + g_N)v ds, \quad \forall v \in V \tag{4.89}$$

Replacing V with V_h, we obtain the finite element method: find $u_h \in V_h \subset V$ such that

$$\int_\Omega a\nabla u_h \cdot \nabla v\,dx + \int_{\partial\Omega} \kappa u_h v ds = \int_\Omega fv\,dx + \int_{\partial\Omega} (\kappa g_D + g_N)v ds, \quad \forall v \in V_h \tag{4.90}$$

4.6 Computer Implementation

We are now going to describe how to implement the finite element method (4.90). The linear system resulting form this equation takes the form

$$(A + R)\xi = b + r \tag{4.91}$$

where the entries of the involved matrices and vectors are given by

$$A_{ij} = \int_{\Omega} a\nabla\varphi_i \cdot \nabla\varphi_j \, dx \tag{4.92}$$

$$R_{ij} = \int_{\partial\Omega} \kappa\varphi_i\varphi_j \, ds \tag{4.93}$$

$$b_i = \int_{\Omega} f\varphi_i \, dx \tag{4.94}$$

$$r_i = \int_{\partial\Omega} (\kappa g_D + g_N)\varphi_i \, ds \tag{4.95}$$

for $i, j = 1, 2, \dots, n_p$ with n_p the number of nodes.

4.6.1 Assembly of the Stiffness Matrix

The assembly of the stiffness matrix A is performed in the same manner as shown previously for the mass matrix M. Of course, the matrix entries of A are different than those of M. The local element stiffness matrix is given by

$$A_{ij}^K = \int_K a\nabla\varphi_i \cdot \nabla\varphi_j \, dx, \quad i, j = 1, 2, 3 \tag{4.96}$$

We shall now compute these nine integrals.

Consider a triangle K with nodes $N_i = (x_1^{(i)}, x_2^{(i)})$, $i = 1, 2, 3$. To each node N_i there is a hat function φ_i associated, which takes the value 1 at node N_i and 0 at the other two nodes. Each hat function is a linear function on K so it has the form

$$\varphi_i = a_i + b_i x_1 + c_i x_2 \tag{4.97}$$

where the coefficients a_i, b_i, and c_i, are determined by

$$\varphi_i(N_j) = \begin{cases} 1, & i = j \\ 0, & i \neq j \end{cases} \tag{4.98}$$

The explicit expressions for the coefficients a_i, b_i, and c_i are given by

$$a_i = \frac{x_1^{(j)} x_2^{(k)} - x_1^{(k)} x_2^{(j)}}{2|K|}, \quad b_i = \frac{x_2^{(j)} - x_2^{(k)}}{2|K|}, \quad c_i = \frac{x_1^{(k)} - x_1^{(j)}}{2|K|} \tag{4.99}$$

with cyclic permutation of the indices $\{i, j, k\}$ over $\{1, 2, 3\}$. Note that the gradient of φ_i is just the constant vector $\nabla\varphi_i = [b_i, c_i]^T$. As these gradients will occur frequently, let us write a separate routine for computing them.

```
function [area,b,c] = HatGradients(x,y)
area=polyarea(x,y);
b=[y(2)-y(3); y(3)-y(1); y(1)-y(2)]/2/area;
c=[x(3)-x(2); x(1)-x(3); x(2)-x(1)]/2/area;
```

Input `x` and `y` are two vectors holding the node coordinates of the triangle. Output are the vectors `b` and `c` holding the coefficients b_i and c_i of the gradients. As the area is computed as a by product we also return it in the variable `area`.

Once we have $\nabla\varphi_i$ it is easy to compute the local element stiffness matrix A^K. Using the center of gravity quadrature formula we have

$$A_{ij}^K = \int_K a\nabla\varphi_i \cdot \nabla\varphi_j \, dx \tag{4.100}$$

$$= (b_i b_j + c_i c_j) \int_K a \, dx \tag{4.101}$$

$$\approx \bar{a}\,(b_i b_j + c_i c_j)|K|, \quad i, j = 1, 2, 3 \tag{4.102}$$

where $\bar{a} = a(\frac{1}{3}(N_1 + N_2 + N_3))$ is the center of gravity value of A on K.

We summarize the assembly of the global stiffness matrix in the following algorithm:

Algorithm 11 Assembly of the Stiffness Matrix.

Let n be the number of nodes and m the number of elements in a mesh, and let the mesh be described by its point matrix P and connectivity matrix T.
Allocate memory for the $n \times n$ matrix A and initialize all matrix entries to zero.
for $K = 1, 2, \dots, m$ **do**
  Compute the gradients $\nabla\varphi_i = [b_i, c_i]$, $i = 1, 2, 3$ of the three hat functions φ_i on K.
  Compute the 3×3 local element mass matrix A^K given by

$$A^K = \bar{a}\begin{bmatrix} b_1^2 + c_1^2 & b_1b_2 + c_1c_2 & b_1b_3 + c_1c_3 \\ b_2b_1 + c_2c_1 & b_2^2 + c_2^2 & b_2b_3 + c_2c_3 \\ b_3b_1 + c_3c_1 & b_3b_2 + c_3c_2 & b_3^2 + c_3^2 \end{bmatrix} |K| \tag{4.103}$$

  Set up the local-to-global mapping, loc2glb $= [r, s, t]$.
  for $i = 1, 2, 3$ **do**
    for $j = 1, 2, 3$ **do**

$$A_{\text{loc2glb}_i\,\text{loc2glb}_j} = A_{\text{loc2glb}_i\,\text{loc2glb}_j} + A_{ij}^K \tag{4.104}$$

    end for
  end for
end for

It is straight forward to translate this algorithm into MATLAB code.

```
function A = StiffnessAssembler2D(p,t,a)
np = size(p,2);
nt = size(t,2);
A = sparse(np,np); % allocate stiffness matrix
for K = 1:nt
  loc2glb = t(1:3,K); % local-to-global map
  x = p(1,loc2glb); % node x-coordinates
  y = p(2,loc2glb); % node y-
  [area,b,c] = HatGradients(x,y);
  xc = mean(x); yc = mean(y); % element centroid
  abar = a(xc,yc); % value of a(x,y) at centroid
  AK = abar*(b*b'...
    +c*c')*area; % element stiffness matrix
  A(loc2glb,loc2glb) = A(loc2glb,loc2glb) ...
        + AK; % add element stiffnesses to A
   end
```

A few comments about this routine are perhaps in order. For each element we compute the area and the hat function gradient vectors using the `HatGradients` routine. The local element stiffness matrix A^K is then the sum of the outer product of the vectors b_i and c_i times the element area $|K|$ and $\bar{a}$ (i.e., `AK = abar*(b*b'+c*c')*area`. The function a is assumed to be defined by a separate subroutine. Finally, A^K is added to the appropriate places in A using the command `A(loc2glb,loc2glb) = A(loc2glb,loc2glb) + AK`. Input is the point and connectivity matrix describing the mesh, and a function handle `a` to a subroutine specifying a. Output is the assembled global stiffness matrix A.

The load vector b is exactly the same as for the L^2-projection and assembled as shown previously.

We remark that the stiffness and mass matrices and the load vector can also be assembled with the built-in routine `assema`. In the simplest case the syntax for doing so is

```
[A,M,b] = assema(p,t,1,1,1);
```

We also remark that on a tetrahedron the hat functions takes the form $\varphi_i = a_i + b_i x_1 + c_i x_2 + d_i x_3$, $i = 1, \dots, 4$, where the coefficients, a_i, b_i, etc., can be computed, as above, from the requirement $\varphi_i(N_j) = \delta_{ij}$.

4.6.2 Assembling the Boundary Conditions

We must also assemble the boundary matrix R and the boundary vector r containing line integrals originating from the Robin boundary condition. To this end, let N_i and N_j be two nodes on the boundary $\partial\Omega$, and let be the edge E between them.

Assuming, for simplicity, that κ, g_D, and g_N are constant on E, we have the local edge contributions

$$R_{ij}^E = \int_E \kappa \varphi_i \varphi_j \, ds = \frac{1}{6}\kappa(1 + \delta_{ij})|E|, \quad i, j = 1, 2 \tag{4.105}$$

$$r_i^E = \int_E (\kappa g_D + g_N)\varphi_i \, ds = \frac{1}{2}(\kappa g_D + g_N)|E|, \quad i = 1, 2 \tag{4.106}$$

where $|E|$ is the length of E.

We can think of R as a one-dimensional mass matrix on a mesh with nodes located along $\partial\Omega$ instead of along the x-axis.

MATLAB stores starting and ending nodes for the edges on the mesh boundary in the first two rows of the edge matrix `e`, which is output from `initmesh`. Consequently, to assemble R we loop over these edges and for each edge we add the entries of the local boundary matrix R^E to the appropriate entries in the global boundary matrix R.

```
function R = RobinMassMatrix2D(p,e,kappa)
np = size(p,2); % number of nodes
ne = size(e,2); % number of boundary edges
R = sparse(np,np); % allocate boundary matrix
for E = 1:ne
  loc2glb = e(1:2,E); % boundary nodes
  x = p(1,loc2glb); % node x-coordinates
  y = p(2,loc2glb); % node y-
  len = sqrt((x(1)-x(2))^2+(y(1)-y(2))^2); % edge length
  xc = mean(x); yc = mean(y); % edge mid-point
  k = kappa(xc,yc); % value of kappa at mid-point
  RE = k/6*[2 1; 1 2]*len; % edge boundary matrix
  R(loc2glb,loc2glb) = R(loc2glb,loc2glb) + RE;
end
```

Input is the point and edge matrix describing the mesh, and a function handle to a subroutine specifying the function κ. Output is the assembled global boundary matrix R.

The boundary vector r is assembled similarly.

```
function r = RobinLoadVector2D(p,e,kappa,gD,gN)
np = size(p,2);
ne = size(e,2);
r = zeros(np,1);
for E = 1:ne
  loc2glb = e(1:2,E);
  x = p(1,loc2glb);
  y = p(2,loc2glb);
  len = sqrt((x(1)-x(2))^2+(y(1)-y(2))^2);
```

```
  xc = mean(x); yc = mean(y);
  tmp = kappa(xc,yc)*gD(xc,yc)+gN(xc,yc);
  rE = tmp*[1; 1]*len/2;
  r(loc2glb) = r(loc2glb) + rE;
end
```

For convenience, let us write a routine for computing both R and r at the same time.

```
function [R,r] = RobinAssembler2D(p,e,kappa,gD,gN)
R = RobinMassMatrix2D(p,e,kappa);
r = RobinLoadVector2D(p,e,kappa,gD,gN);
```

4.6.3 A Finite Element Solver for Poisson's Equation

Next we present a physical application that can be simulated with the code written so far.

4.6.3.1 Potential Flow Over a Wing

When designing aircrafts it is very important to know the aerodynamic properties of the wings to assess various performance features, such as the lift force, for instance. In the simplest case, the flow of air over a wing can be simulated by solving a Poisson type equation. In doing so, by assuming that the wing is much longer than its width, the problem can be reduced to two dimensions. Figure 4.1 shows the computational domain, which consists of a cross section of a wing surrounded by a rectangular channel.

A potential equation for the airflow around the wing follows from the somewhat unphysical assumption that the velocity vector field $u = [u_1, u_2]$ of the air is irrational, that is, $\nabla \times u = 0$. In this case, there exists a scalar function Φ, such that $u = -\nabla\Phi$. This is called the flow potential and is governed by the Laplace equation

$$-\Delta\Phi = 0, \quad \text{in } \Omega \tag{4.107}$$

We impose the boundary conditions

$$n \cdot \nabla\Phi = 1, \quad \text{on } \Gamma_{\text{in}} \tag{4.108}$$

$$\Phi = 0, \quad \text{on } \Gamma_{\text{out}} \tag{4.109}$$

$$n \cdot \nabla\Phi = 0, \quad \text{elsewhere} \tag{4.110}$$

which corresponds to a constant flow of air from left to right.

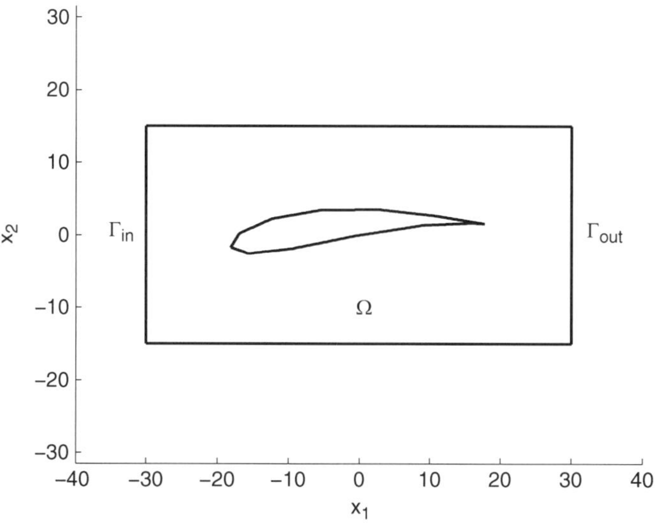

Fig. 4.1 Cross-section of a channel surrounding a wing

A slight complication with the boundary conditions is that the occurring Dirichlet and Neumann conditions must be approximated with the Robin condition we have written code for. To this end, we set $\kappa = 10^6$ on Γ_{out}, which penalizes any deviation of the solution from zero on this boundary segment. On Γ_{in} we set $\kappa = 0$ and $g_N = 1$. The following subroutines specify κ, g_D and g_N.

```
function z = Kappa1(x,y)
z=0;
if (x>29.99), z=1.e+6; end

function z = gD1(x,y)
z=0;

function z = gN1(x,y)
z=0;
if (x<-29.99), z=1; end
```

The velocity potential can now be computed with just a couple of lines of code.

```
function WingFlowSolver2D()
g = Airfoilg();
[p,e,t] = initmesh(g,'hmax',0.5);
A = StiffnessAssembler2D(p,t,inline('1','x','y'));
[R,r] = RobinAssembler2D(p,e,@Kappa1,@gD1,@gN1);
phi = (A+R)\r;
pdecont(p,t,phi)
```

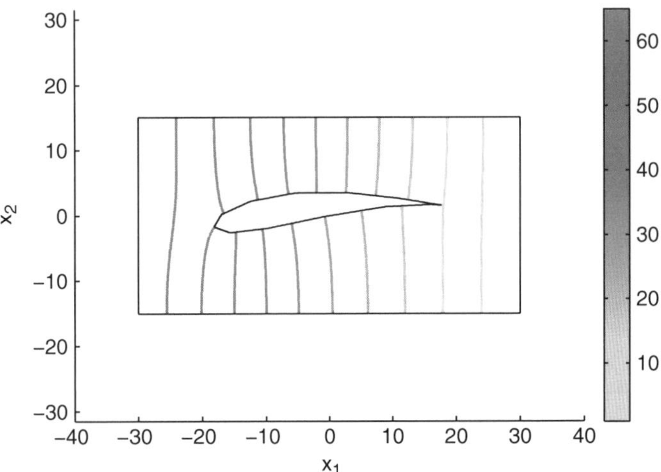

Fig. 4.2 Isocontours of the computed finite element velocity potential Φ_h

Here, **Airfoilg** is a subroutine specifying the geometry matrix. It is listed in the Appendix.

Figure 4.2 shows the computed finite element approximation Φ_h to the velocity potential.

The velocity field u is defined by $u = -\nabla\Phi$. Its computed counterpart u_h can be obtained and visualized by using the built-in routines `pdegrad` and `pdesurf`. To do so, we type

```
[phix,phiy] = pdegrad(p,t,phi); % derivatives of 'phi'
u = -phix';
v = -phiy';
pdeplot(p,e,t,'flowdata',[u v])
```

which gives the velocity glyphs of Fig. 4.3.

Finally, a pressure, the so-called Bernoulli pressure, around the wing can be defined by $p = -|\nabla\Phi|^2$. Indeed, typing `-sqrt(u.*u+v.*v)` we obtain the computed pressure p_h shown in Fig. 4.4. We remark that p stems from the Bernoulli equation, which says that the quantity $\frac{1}{2}|u|^2 + p$ is constant along the streamlines in a flowing fluid.

In the next three sections we shall study three Poisson type problems that demand special attention, namely, the so-called pure Dirichlet problem, the pure Neumann problem, and the eigenvalue problem.

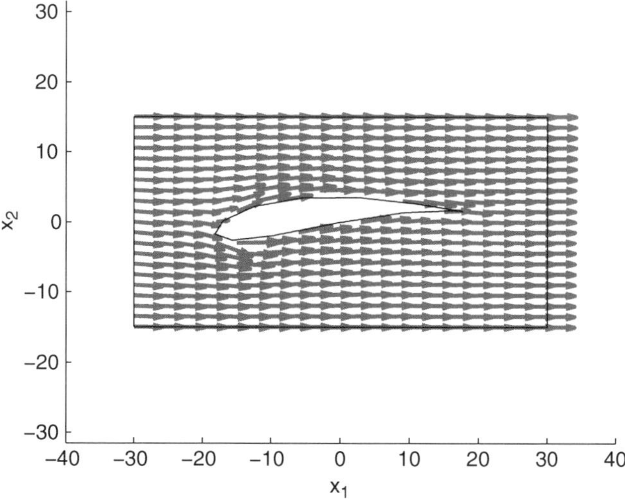

Fig. 4.3 Glyphs of the computed velocity u_h around the wing

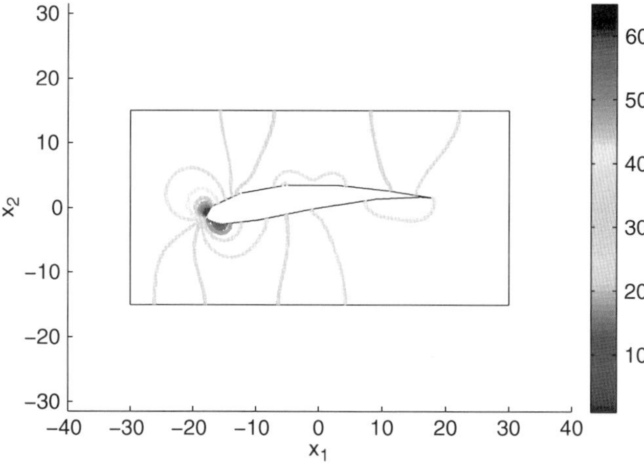

Fig. 4.4 Isocontours of the computed Bernoulli pressure p_h around the wing

4.7 The Dirichlet Problem

We consider the following model problem with inhomogeneous boundary conditions: find u such that

$$-\Delta u = f, \quad \text{in } \Omega \tag{4.111a}$$

$$u = g_D, \quad \text{on } \partial\Omega \tag{4.111b}$$

where f and g_D are given functions.

This problem has different trial and test space due to the boundary condition $u = g_D$. The trial space is given by

$$V_{g_D} = \{v : \|v\|_{L^2(\Omega)} + \|\nabla v\|_{L^2(\Omega)} < \infty, \ v|_{\partial\Omega} = g_D\} \tag{4.112}$$

whereas the test space is given by V_0.

Multiplying $f = -\Delta u$ by a test function $v \in V_0$, and integrating using Green's formula, we have

$$\int_\Omega f v \, dx = -\int_\Omega \Delta u v \, dx \tag{4.113}$$

$$= \int_\Omega \nabla u \cdot \nabla v \, dx - \int_{\partial\Omega} n \cdot \nabla u v ds \tag{4.114}$$

$$= \int_\Omega \nabla u \cdot \nabla v \, dx \tag{4.115}$$

since $v = 0$ on $\partial\Omega$. Thus, the variational formulation reads: find $u \in V_{g_D}$ such that

$$\int_\Omega \nabla u \cdot \nabla v \, dx = \int_\Omega f v \, dx, \quad \forall v \in V_0 \tag{4.116}$$

Now, let us assume that g_D is the restriction of a continuous piecewise linear function to the boundary. In other words, there is a function $u_{h,g_D} \in V_h$ such that $u_{h,g_D} = g_D$ on $\partial\Omega$. If this is not so, we have to first approximate g by a function in V_h, for instance using linear interpolation on the boundary.

Introducing the affine subspace

$$V_{h,g_D} = \{v \in V_h : v|_{\partial\Omega} = g_D\} \tag{4.117}$$

the finite element method reads: find $u_h \in V_{h,g_D}$ such that

$$\int_\Omega \nabla u_h \cdot \nabla v \, dx = \int_\Omega f v \, dx, \quad \forall v \in V_{h,0} \tag{4.118}$$

To derive an equation for u_h we write it in the form

$$u_h = u_{h,0} + u_{h,g_D} \tag{4.119}$$

where u_{h,g_D} is a given function in V_{h,g_D}, and $u_{h,0}$ a sought function in $V_{h,0}$. This construction of u_h will satisfy the boundary conditions, since $u_{h,g_D} = g_D$ and $u_{h,0} = 0$ on the boundary. Because u_{h,g_D} is known it only remains to determine $u_{h,0}$. In doing so, we obtain the equation: find $u_{h,0} \in V_{h,0}$ such that

$$\int_\Omega \nabla u_{h,0} \cdot \nabla v \, dx = \int_\Omega f v \, dx - \int_\Omega \nabla u_{h,g_D} \cdot \nabla v \, dx, \quad \forall v \in V_{h,0} \tag{4.120}$$

This is a problem of the same kind as the original, but with a modified right hand side. Using Galerkin orthogonality is possible to show that $u_{h,0}$ is independent of the particular choice of u_{h,g_D}. Therefore, u_{h,g_D} is often chosen to be zero at all interior nodes.

To solve Eq. (4.120), let n_p be the number of nodes, and assume that the first n_i of these are interior, while the remaining $n_g = n_p - n_i$ are exterior, and lie on the mesh boundary. Further, let A and b be the usual $n_p \times n_p$ stiffness matrix and $n_p \times 1$ load vector, output from `assema`. We have the $n_p \times n_p$ linear system

$$\begin{bmatrix} A_{00} & A_{0g} \\ 0 & I \end{bmatrix} \begin{bmatrix} \xi_0 \\ \xi_g \end{bmatrix} = \begin{bmatrix} b_0 \\ b_g \end{bmatrix} \tag{4.121}$$

where A_{00} is the upper left $n_i \times n_i$ block of A, A_{0g} is the $n_i \times n_g$ upper right block of A, I is the $n_g \times n_g$ identity matrix, b_0 is the first $n_i \times 1$ block of b, b_g is the $n_g \times 1$ vector with nodal values of g_D, ξ_0 the $n_i \times 1$ vector with nodal values of $u_{h,0}$, and ξ_g the $n_g \times 1$ vector with nodal values of u_{h,g_D}. Rearranging the first n_i equations we obtain the $n_i \times n_i$ linear system

$$A_{00}\xi_0 = b_0 - A_{0g}\xi_g = b_0 - A_{0g}g \tag{4.122}$$

from which ξ_0 can be obtained.

The translation of this into MATLAB code is straight forward. Suppose we have a vector `fixed` holding the numbers of all boundary nodes, and another vector `g` holding the corresponding nodal values. Then, we can set up and solve (4.122) with the following piece of code.

```
[A,unused,b] = assema(p,t,...); % assemble A and b
np = size(p,2); % total number of nodes
fixed = unique([e(1,:) e(2,:)]); % boundary nodes
free = setdiff([1:np],fixed); % interior nodes
b = b(free)-A(free,fixed)*g; % modify A
A = A(free,free); % modify b
xi = zeros(np,1); % allocate solution
xi(fixed) = g; % insert fixed node values
xi(free) = A\b; % solve for free node values
```

4.8 The Neumann Problem

Next we consider the following model problem: find u such that

$$-\Delta u = f, \quad \text{in } \Omega \tag{4.123a}$$

$$n \cdot \nabla u = g_N, \quad \text{on } \partial\Omega \tag{4.123b}$$

where f and g_N are given functions.

Let us try to seek a solution to this problem in the space $V = \{v : \|v\|_{L^2(\Omega)} + \|\nabla v\|_{L^2(\Omega)} < \infty\}$.

Multiplying $f = -\Delta u$ by a test function $v \in V$, and integrating using Green's formula, as usual, we have

$$\int_\Omega f v \, dx = -\int_\Omega \Delta u v \, dx \tag{4.124}$$

$$= \int_\Omega \nabla u \cdot \nabla v \, dx - \int_{\partial\Omega} n \cdot \nabla u v ds \tag{4.125}$$

$$= \int_\Omega \nabla u \cdot \nabla v \, dx - \int_{\partial\Omega} g_N v ds \tag{4.126}$$

Thus, the variational formulation reads: find $u \in V$ such that

$$\int_\Omega \nabla u \cdot \nabla v \, dx = \int_\Omega f v \, dx + \int_{\partial\Omega} g_N v ds, \quad \forall v \in V \tag{4.127}$$

Here, by choosing the test function v as any constant, the gradient ∇v can be made to vanish. As a consequence, (4.127) is solvable if and only if f and g_N satisfy the so-called conservation property

$$\int_\Omega f dx + \int_{\partial\Omega} g_N \, ds = 0 \tag{4.128}$$

The solution u is, however, only determined up to a constant. To fix this constant and obtain a unique solution a common trick is to impose the additional constraint

$$\int_\Omega u \, dx = 0 \tag{4.129}$$

In doing so, the appropriate place to look for the solution u to (4.127) is the space

$$\bar{V} = \left\{ v \in V : \int_\Omega v \, dx = 0 \right\} \tag{4.130}$$

which contains only functions with a zero mean value. This is a called a quotient space.

Now, the finite element method takes the form: find $u_h \in \bar{V}_h \subset \bar{V}$ such that

$$\int_\Omega \nabla u_h \cdot \nabla v \, dx = \int_\Omega f v \, dx + \int_{\partial\Omega} g_N v ds, \quad \forall v \in \bar{V}_h \tag{4.131}$$

where $\bar{V}_h$ is the space of all continuous piecewise linears with a zero mean.

The quotient space $\bar{V}_h$ is mostly of theoretical interest as the assembly of the stiffness matrix and load vector is done on the unconstrained space V_h in practice. Therefore, let A and b be the usual $n_p \times n_p$ stiffness matrix and $n_p \times 1$ load vector, respectively, with n_p the total number of nodes in the mesh. The zero mean constraint $\int_\Omega u_h \, dx = \sum_{j=1}^{n_p} \xi_j \int_\Omega \varphi_j \, dx = 0$ can be enforced via the Lagrange multiplier technique. The basic idea is simple, and relies on the fact that the solution ξ (i.e., the n_p nodal values of u_h) to $A\xi = b$ is also the minimizer of the quadratic form $J(\xi) = \frac{1}{2}\xi^T A\xi - \xi^T b$. Indeed, if ξ happens to be subject to a set of $n_c < n_p$ constraints $C\xi = 0$, with C a given $n_c \times n_p$ matrix, then a fundamental result from optimization says that the constrained optimum of J can be found by seeking stationary points to the so-called Lagrangian

$$L(\xi, \mu) = J(\xi) - \mu^T C\xi \tag{4.132}$$

where the $n_c \times 1$ vector μ is called the Lagrange multiplier. Differentiating L with respect to ξ and μ and using the first order optimality conditions $\partial_\xi L = \partial_\mu L = 0$ leads to the augmented linear system

$$\begin{bmatrix} A & C^T \\ C & 0 \end{bmatrix} \begin{bmatrix} \xi \\ \mu \end{bmatrix} = \begin{bmatrix} b \\ 0 \end{bmatrix} \tag{4.133}$$

from which ξ can be obtained. To us, C is just the $1 \times n_p$ matrix with entries $C_{1i} = \int_\Omega \varphi_i \, dx$. The Lagrangian multiplier μ may be though of as a force acting to enforce the constraints. Because the zero mean value on u_h is a constraint, which do not alter the solution to the underlying Neumann problem, the force μ should vanish or, at least, be very small.

4.9 The Eigenvalue Problem

The last of our model problems is the eigenvalue problem: find the function u and the number λ such that

$$-\Delta u = \lambda u, \quad \text{in } \Omega \tag{4.134a}$$

$$n \cdot \nabla u = 0, \quad \text{on } \partial\Omega \tag{4.134b}$$

Here, we have for simplicity assumed a Neumann condition on the boundary, but Dirichlet conditions may also be applied. All the same, the boundary conditions must be homogeneous though.

The significant feature of an eigenvalue problem is that we seek both the function u and the number λ. We call u an eigenfunction, or eigenmode, λ an eigenvalue, and the pair (u, λ) is called an eigenpair.

It follows from spectral theory for operators that there are a countable number of solutions (u_n, λ_n), $n = 1, 2, \dots$ with non-negative increasing eigenvalues. For Neumann conditions the constant function is the only eigenvector with eigenvalue zero, all other eigenvalues are positive and they also tends to infinity as n tends to infinity. For Dirichlet conditions there is no zero eigenvalue. Eigenvectors u_n and u_m associated with different eigenvalues λ_n and λ_m are orthogonal in the sense that

$$\int_\Omega \nabla u_n \cdot \nabla u_m \, dx = \int_\Omega u_n u_m \, dx = 0 \tag{4.135}$$

The set of all eigenvectors associated with one eigenvalue is a vector space called an eigenspace and the dimension of the eigenspace is called the multiplicity of the eigenvalue. We may choose a basis in the eigenspace that satisfies the orthogonality conditions (4.135). Thus the orthogonality condition (4.135) holds for all n and m with $n \neq m$.

To construct a finite element method we proceed as usual. Multiplying $\lambda u = -\Delta u$ by a test function $v \in V$, and integrating using Green's formula, we have

$$\lambda \int_\Omega uv \, dx = -\int_\Omega \Delta uv \, dx \tag{4.136}$$

$$= \int_\Omega \nabla u \cdot \nabla v \, dx - \int_{\partial\Omega} n \cdot \nabla uv ds \tag{4.137}$$

$$= \int_\Omega \nabla u \cdot \nabla v \, dx \tag{4.138}$$

Thus, the variational formulation reads: find $u \in V$ and $\lambda \in \mathbb{R}$ such that

$$\int_\Omega \nabla u \cdot \nabla v \, dx = \lambda \int_\Omega uv \, dx, \quad \forall v \in V \tag{4.139}$$

and the corresponding finite element method reads: find $u_h \in V_h$ and $\lambda_h \in \mathbb{R}$ such that

$$\int_\Omega \nabla u_h \cdot \nabla v \, dx = \lambda_h \int_\Omega u_h v \, dx, \quad \forall v \in V_h \tag{4.140}$$

The finite element method leads, not to a linear system, but to a so-called generalized algebraic eigenvalue problem of the form

$$A\xi = \lambda M \xi \tag{4.141}$$

where A and M are the usual stiffness and mass matrices, and ξ is a vector with the nodal values of u_h. The existence of eigenpairs (ξ_n, λ_n), $n = 1, \dots, n_p$ with n_p the dimension of the finite element space follows from spectral theory for symmetric matrices. The corresponding eigenvectors ξ_i are orthogonal with respect to A and M. Note that if ξ is an eigenvector so is $t\xi$ for any real number t. In practice one often normalizes the eigenvectors so that they have unit L^2-norm.

In MATLAB, generalized sparse eigenvalue problems can be solved using the `eigs` routine. Below we show how to compute the first five eigenmodes with smallest magnitude on the unit disk. Assembly of the matrices A and M is done using the `assema` routine.

```
g = 'circleg'; % built-in geometry of a cricle
[p,e,t] = initmesh(g,'hmax',0.1); % mesh
[A,M] = assema(p,t,1,1,0); % assemble A and M
[Xi,La] = eigs(A,M,5,'SM'); % solve A*Xi=La*M*Xi
pdesurf(p,t,Xi(:,1)) % plot 1:st eigenmode
```

4.10 Adaptive Finite Element Methods

As we have seen a posteriori error estimates are computable error estimates, which can be used to guide adaptive mesh refinement. This iteratively increases the accuracy of the finite element solution, while saving computational resources at the same time. We next formulate adaptive finite elements for Poisson's equation. For simplicity, we restrict attention to the simple model problem (4.5) (i.e., $-\Delta u = f$ in Ω with $u = 0$ on $\partial\Omega$).

4.10.1 A Posteriori Error Estimate

As we know the gradient ∇u_h of the continuous piecewise linear finite element solution u_h is generally a discontinuous piecewise constant vector. Thus, when moving orthogonally across the boundary of one element to a neighbouring element, there is a jump in the normal derivative $n \cdot \nabla u_h$. This jump is denoted $[n \cdot \nabla u_h]$ and plays a key role in the a posteriori error analysis.

Theorem 4.12. *The finite element solution u_h, defined by* (4.13), *satisfies the estimate*

$$|||u - u_h|||^2 \leq C \sum_{K \in \mathcal{K}} \eta_K^2(u_h) \tag{4.142}$$

where the element residual $\eta_K(u_h)$ is defined by

$$\eta_K(u_h) = h_K \|f + \Delta u_h\|_{L^2(K)} + \tfrac{1}{2} h_K^{1/2} \|[n \cdot \nabla u_h]\|_{L^2(\partial K \setminus \partial\Omega)} \tag{4.143}$$

Here, $[n \cdot \nabla u_h]$ denotes the jump in the normal derivative of u_h on the (interior) edges of element K. Also, since u_h is linear on K, $\Delta u_h = 0$.

Proof. Letting $e = u - u_h$ be the error we have

$$|||e|||^2 = \|\nabla e\|^2_{L^2(\Omega)} = \int_\Omega \nabla e \cdot \nabla e dx = \int_\Omega \nabla e \cdot \nabla(e - \pi e)\, dx \tag{4.144}$$

where we have used the Galerkin orthogonality (4.46) to subtract an interpolant $\pi e \in V_{h,0}$ to e. Splitting this into a sum over the elements and using integration by parts we further have

$$\sum_{K\in\mathcal{K}} \int_K \nabla e \cdot \nabla(e - \pi e)\, dx = \sum_{K\in\mathcal{K}} -\int_K \Delta e(e - \pi e)\, dx + \int_{\partial K} n \cdot \nabla e(e - \pi e)\, ds \tag{4.145}$$

$$= \sum_{K\in\mathcal{K}} \int_K (f + \Delta u_h)(e - \pi e)\, dx + \int_{\partial K\setminus\partial\Omega} n \cdot \nabla e(e - \pi e)\, ds \tag{4.146}$$

where we have used that $-\Delta e|_K = f + \Delta u_h|_K$ within the interior of K, and that e and, thus, also πe, vanish on $\partial\Omega$ due to the zero boundary condition. As a consequence we do not need to integrate $n \cdot \nabla e(e - \pi e)$ on the exterior edges, but only on the interior ones. In doing so, we note that there are two contributions from each edge E, namely, one from each element $K^\pm$ sharing E. Summing these contributions we obtain

$$\int_{\partial K^+\cap\partial K^-} n\cdot\nabla e(e-\pi e)\, ds = \int_E (n^+\cdot\nabla e^+(e^+-\pi e^+)+n^-\cdot\nabla e^-(e^--\pi e^-))\, ds \tag{4.147}$$

where we use the notation $v^\pm = v|_{K^\pm}$. Now, the error e is continuous, which implies $(e^+ - \pi e^+)|_E = (e^- - \pi e^-)|_E$, and we can drop the superscripts for ease of notation. Also, the gradient ∇u is continuous, implying $(n^+\cdot\nabla u^+ + n^-\cdot\nabla u^-)|_E = 0$. However, the gradient $\nabla u_h|_E$ is generally not continuous as it is only a piecewise constant vector. Thus, taking into account that $n \cdot \nabla u_h$ might be different on K^+ and K^-, we have

$$\int_E (n^+ \cdot \nabla e^+(e - \pi e) + n^- \cdot \nabla e^-(e - \pi e))\, ds = -\int_E (n^+ \cdot \nabla u_h^+ + n^- \cdot \nabla u_h^-)(e - \pi e)\, ds \tag{4.148}$$

$$= -\int_E [n \cdot \nabla u_h](e - \pi e)\, ds \tag{4.149}$$

From this, we infer

$$\sum_{K\in\mathcal{K}} \int_{\partial K\setminus\partial\Omega} n \cdot \nabla e(e - \pi e)\, ds = -\sum_{E\in\mathcal{E}_I} \int_E [n \cdot \nabla u_h](e - \pi e)\, ds \tag{4.150}$$

Returning to a sum over the elements by simply distributing half of the jump $[n \cdot \nabla u_h]|_E$ on K^+ and half on K^-, we obtain the so-called error representation formula

$$|||e|||^2 = \sum_{K \in \mathcal{K}} \int_K (f + \Delta u_h)(e - \pi e)\, dx - \frac{1}{2} \int_{\partial K \setminus \partial \Omega} [n \cdot \nabla u_h](e - \pi e)\, ds \tag{4.151}$$

Let us estimate the two terms in the right hand side separately.

The interior contribution can be estimated using the Cauchy-Schwarz inequality followed by a standard interpolation error estimate. In doing so, we have

$$\int_K (f + \Delta u_h)(e - \pi e)\, dx \leq \|f + \Delta u_h\|_{L^2(K)} \|e - \pi e\|_{L^2(K)} \tag{4.152}$$

$$\leq \|f + \Delta u_h\|_{L^2(K)} C h_K \|De\|_{L^2(K)} \tag{4.153}$$

For the the edge contributions we need the scaled Trace inequality

$$\|v\|^2_{L^2(\partial K)} \leq C \left(h_K^{-1} \|v\|^2_{L^2(K)} + h_K \|\nabla v\|^2_{L^2(K)} \right) \tag{4.154}$$

Using this and, again, the Cauchy-Schwarz inequality, we have

$$\int_{\partial K} [n \cdot \nabla u_h](e - \pi e)\, ds \leq \|[n \cdot \nabla u_h]\|_{L^2(\partial K)} \|e - \pi e\|_{L^2(\partial K)} \tag{4.155}$$

$$\leq \|[n \cdot \nabla u_h]\|_{L^2(\partial K)} C \left(h_K^{-1} \|e - \pi e\|^2_{L^2(K)} + h_K \|D(e - \pi e)\|^2_{L^2(K)} \right)^{1/2} \tag{4.156}$$

$$\leq \|[n \cdot \nabla u_h]\|_{L^2(\partial K)} C h_K^{1/2} \|De\|_{L^2(K)} \tag{4.157}$$

The estimate now follows from (4.153) and (4.157) together. □

We remark that the a posteriori estimate above also holds in three dimensions on tetrahedral meshes, provided that all integrals over triangle edges are replaced by integrals over tetrahedron faces. Of course, integrals over triangles must also be replaced by integrals over tetrahedrons.

4.10.2 Adaptive Mesh Refinement

Mesh refinement in two and three dimensions is much more complicated than in one dimension. In particular, there are two important issues to consider when constructing a mesh refinement algorithm for a mesh in higher dimension. First, invalid elements (i.e., with hanging nodes) are not allowed, and we wish to refine as few elements as possible, which are not in the list of elements to be refined. Second, it is important that the minimal angle in the mesh is kept as large as possible. Otherwise the quality of the finite element solution might deteriorate as we make successive refinements. In addition, when refining elements on a curved outer domain boundary the curvature should be taken into account, so that the refined mesh represents the domain geometry better than the unrefined mesh. This involves moving newly created nodes on the boundary, which is troublesome as it might lead to elements with poor quality, or even inverted elements.

For triangle and tetrahedral meshes the most popular algorithms used for mesh refinement are:

- Rivara refinement, or, longest edge bisection.
- Regular refinement, or, red-green refinement.

We shall describe these in some detail next.

We remark that the PDE-Toolbox has support for both regular and Rivara refinement.

4.10.2.1 Rivara Refinement

In the Rivara method a triangle is always refined by inserting a new edge from the mid-point of the longest side to the opposite corner. The unspoken hope is that the quality of the mesh can be preserved by always splitting the longest edge within the triangles to be refined. Rivara refinement relies on the concepts of terminal edges and longest edge propagation paths. An edge E is called terminal if E is the longest edge of the triangles that share E. The set of triangles sharing E is called a terminal star. The longest edge propagation path of triangle K, abbreviated LEPP(K), is the sequence of triangles you get if you start with K and successively move to the adjacent triangle with longest edge until you reach a terminal edge. Using these concepts, Rivara and co-workers [53] summarizes longest edge bisection refinement in the following algorithm:

Algorithm 12 Rivara Refinement.

Given a mesh $\mathcal{K}$ and set $\mathcal{S}$ of elements to be refined.
for $K \in \mathcal{S}$ **do**
    while $K \in M$ **do**
        Find LEPP(K) and associated terminal edge.
        Refine the terminal star.
    end while
end for

The refinement of the terminal star is generally done by bisecting the terminal edge and dividing the two triangles making up the star into four smaller subtriangles. If the terminal edge lies on the boundary, then the terminal star only contain one

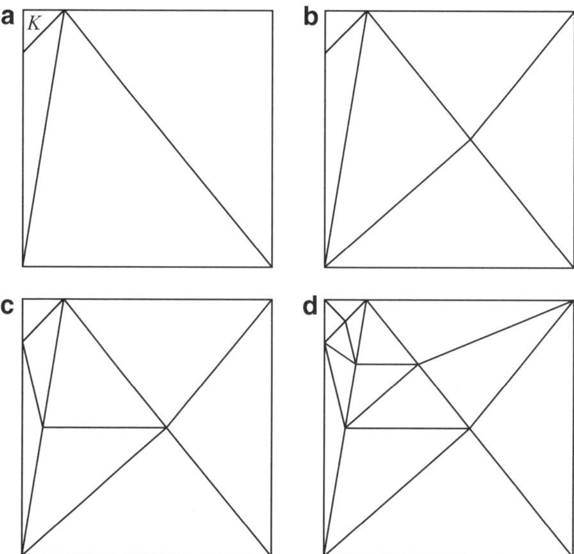

Fig. 4.5 Rivara refinement of target triangle K. (**a**) Initial mesh, (**b**) first refinement step, (**c**) second refinement step, and (**d**) final mesh with K refined

triangle to divide. Note that refinement of the terminal star does not cause additional refinement outside the star. The computational bottleneck of the Rivara algorithm is the repeated computation of LEPP s, which can become expensive for large meshes. Figure 4.5 illustrates the Rivara refinement procedure of a target triangle K on a small mesh.

4.10.2.2 Regular Refinement

Regular refinement consists of splitting each triangle selected for refinement into four smaller ones by inserting a new node on the mid-point of each edge. This gives four child triangles, which are congruent with the original (parent) triangle. The children are therefore of the same quality as their parent. This is called red refinement. See Fig. 4.6a. The insertion of new new nodes on the edges means that triangles adjacent to refined triangles get hanging nodes and must also be refined. This is done by inserting a new edge between the hanging node and the opposite corner. This is called green refinement. See Fig. 4.6b. Green refinement is only applied if there is one hanging node present. Otherwise, extra nodes are inserted so that red refinement is can be applied. This process is iterated until no hanging nodes remain. As a consequence, the refined mesh has regions of red refined triangles surrounded by green refined ones. Naturally, green refined triangles do not have as good quality as red ones. In subsequent refinement loops it is therefore customary not to repeatedly refine previously green marked triangles using green refinement.

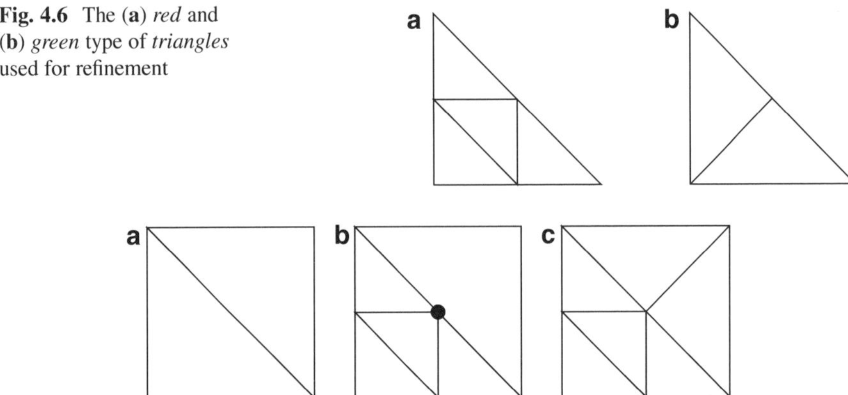

Fig. 4.6 The (**a**) *red* and (**b**) *green* type of *triangles* used for refinement

Fig. 4.7 Illustration of *red green* refinement of a *square* with two *triangles* into one *red* and one *green* type of *triangle*. (**a**) Initial mesh, (**b**) hanging node (•), and (**c**) final mesh

Figure 4.7 illustrates a red green refinement of a square with two triangles into one red and one green type of triangle. The lower triangle is first refined using red refinement. This causes a hanging node, which must be taken care of using green refinement of the upper triangle.

We refer the interested reader to the work by Bank and co-workers [6], and Bey [11] for a more detailed account of red green mesh refinement on triangles and quadrilaterals, and tetrahedra, respectively.

4.10.3 Adaptive Finite Elements Using MATLAB

It is relatively easy to write an adaptive finite element solver in MATLAB.

First we create a (coarse) initial mesh

```
g = 'cardg'; % predefined geometry of a cardioid
[p,e,t] = initmesh(g,'hmax',0.25);
```

Then we compute the finite solution u_h

```
[A,unused,b] = assema(p,t,...);
% .. application of B.C. etc ..
xi = A\b;
```

The next step is to evaluate the element residuals η_K, defined by (4.143). This can be done with the built-in routine `pdejmps`, viz.,

```
eta = pdejmps(p,t,...);
```

The `pdejmps` routine was originally designed for computing the element residuals to $-\nabla \cdot (c\nabla u) + au = f$ and its syntax is therefore

```
eta = pdejmps(p,t,c,a,f,xi,1,1,1);
```

where each of the three inputs c, a, and f can be either a constant or a row vector specifying the values of the coefficients c, a, and f at the centroid of the triangles.

As our refinement criterion we select the 10 % most error prone elements to be refined.

```
tol = 0.9*max(eta);
elements = find(eta > tol);
```

After these calls the vector elements contains the element numbers of the elements to be refined.

The actual refinement is done with the built-in refinemesh routine.

```
[p,e,t] = refinemesh(g,p,e,t,elements,'regular');
```

The mesh refinement algorithm used here is regular refinement. We use the simple stopping criterion that the maximum number of elements in the mesh must not exceed 10, 000.

Below we list a complete routine for adaptively solving Poisson's equation $-\Delta u = 1$ on a domain Ω shaped like a cardioid with $u = 0$ on the boundary $\partial\Omega$.

```
function AdaptivePoissonSolver2D()
% set up geometry
g = 'cardg';
% create initial mesh
[p,e,t] = initmesh(g,'hmax',0.25);
% while not too many elements, do
while size(t,2) < 10000
  % assemble stiffness matrix A, and load vector b
  [A,unused,b] = assema(p,t,1,0,1);
  % get the number of nodes
  np = size(p,2);
  % enforce zero Dirichlet BC
  fixed = unique([e(1,:) e(2,:)]);
  free = setdiff([1:np],fixed);
  A = A(free,free);
  b = b(free);
  % solve for finite element solution
  xi = zeros(np,1);
  xi(free) = A\b;
  figure(1), pdesurf(p,t,xi)
  xlabel('x'), ylabel('y'), title('u_h')
  % compute element residuals
  eta = pdejmps(p,t,1,0,1,xi,1,1,1);
  % choose a selection criteria
  tol = 0.8*max(eta);
  % select elements for refinement
  elements = find(eta > tol)';
  % refine elements using regular refinement
```

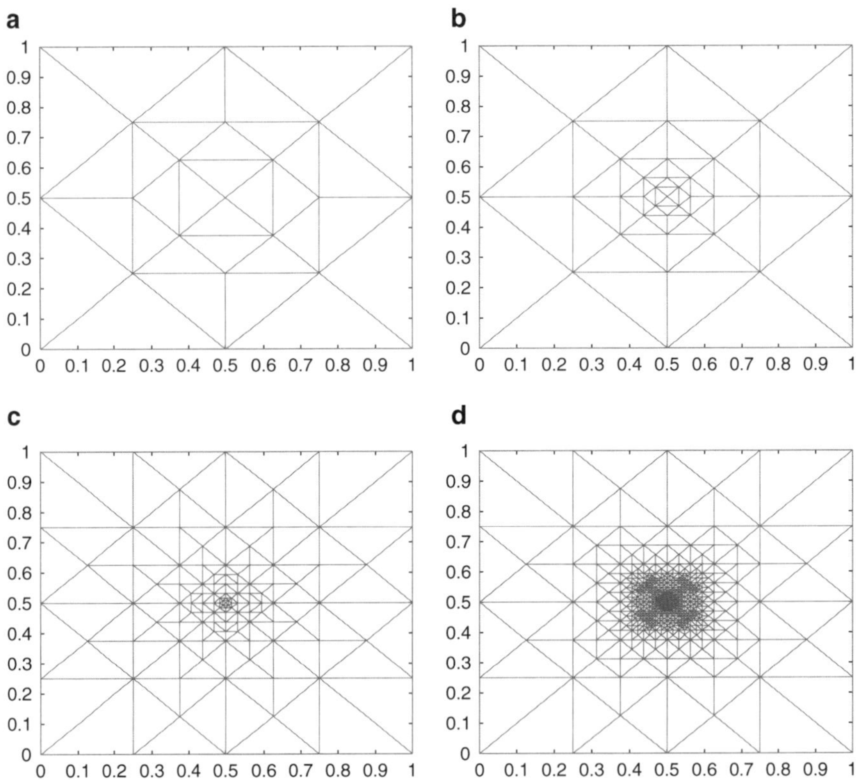

Fig. 4.8 Adaptive meshes for the problem with solution $u = ae^{-ar^2}$. (**a**) Two refinements, (**b**) four refinements, (**c**) six refinements, and (**d**) ten refinements

```
    [p,e,t] = refinemesh(g,p,e,t,elements,'regular');
    figure(2), pdemesh(p,e,t)
end
```

To illustrate adaptive mesh refinement let us solve the problem

$$-\Delta u = 4a^2(1 - ar^2)e^{-ar^2}, \quad \text{in } \Omega = [0, 1]^2 \tag{4.158a}$$

$$u = 0, \quad \text{on } \partial\Omega \tag{4.158b}$$

where a is a parameter and r is the distance from the center of the unitsquare $\Omega = [0, 1]^2$. If a is chosen sufficiently large, say $a = 400$, then the analytical solution is given by $u = ae^{-ar^2}$. This problem is computationally demanding, since the solution is a very narrow pulse, with strong localized gradients, centered at $(0.5, 0.5)$. Thus, to obtain a good finite element approximation we expect to have to resolve the region around this point by placing many triangles there, but maybe we do not need so many triangles in other regions. Figures 4.8 and 4.9 show the

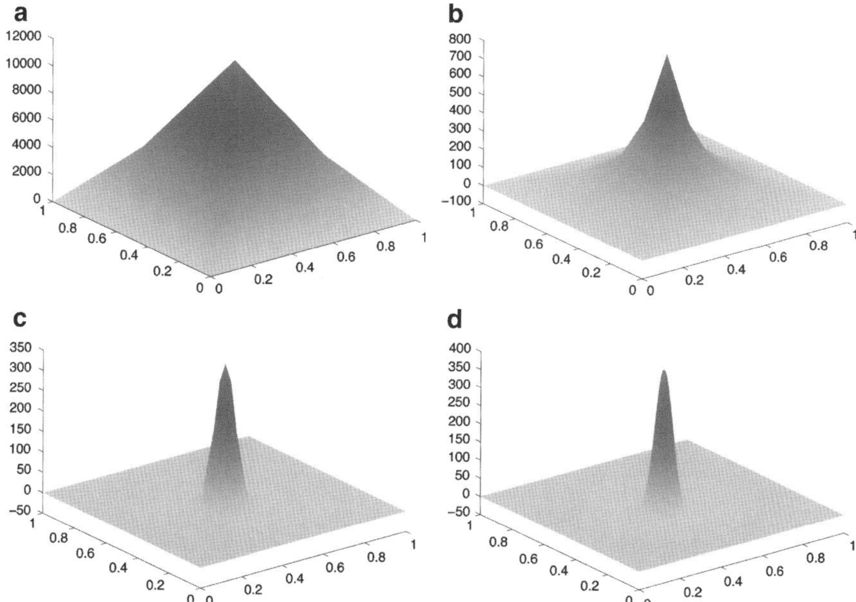

Fig. 4.9 Adaptively computed approximations to $u = ae^{-ar^2}$. (**a**) Two refinements, (**b**) four refinements, (**c**) six refinements, and (**d**) ten refinements

results of running the adaptive code outlined above for ten adaptive loops with a 25 % refinement rule. We see that the expected region is indeed much refined.

4.10.4 Duality Based a Posteriori Error Estimate

In many applications it is not the solution u to Poisson's equation (4.5) itself that is of prime interest, but rather some other quantity involving u. This quantity can often be expressed as a linear functional

$$m(u) = \int_{\Omega} \psi v \, dx \tag{4.159}$$

where ψ is a weighting function to be chosen suitably. For example, choosing $\psi = |\Omega|^{-1}$ gives the average of u in Ω. We say that $m(\cdot)$ expresses the goal of the computation.

To assert that we obtain an accurate value of our quantity of interest, we must find a way of monitoring the error

$$m(e) = m(u) - m(u_h) \tag{4.160}$$

Indeed, we want $m(e)$ to be as small as possible with a minimum of computational effort. More precisely, $m(e)$ should be less than a predefined tolerance using as small mesh as possible. To this end we re-introduce the dual problem: find $\phi \in V$ such that

$$-\Delta\phi = \psi, \quad \text{in } \Omega \tag{4.161a}$$

$$\phi = 0, \quad \text{on } \partial\Omega \tag{4.161b}$$

Recall that this is reminiscent of Nitsche's trick, where $\psi = e$.

Using the dual problem and integration by parts we have

$$m(e) = \int_\Omega e\psi \, dx \tag{4.162}$$

$$= \int_\Omega -e\Delta\phi \, dx \tag{4.163}$$

$$= \int_\Omega \nabla e \cdot \nabla\phi \, dx - \int_{\partial\Omega} en \cdot \nabla\phi \, ds \tag{4.164}$$

$$= \int_\Omega \nabla e \cdot \nabla\phi \, dx \tag{4.165}$$

where we have used that $e = 0$ on $\partial\Omega$ to get rid of the boundary term.

Now, from Galerkin orthogonality (4.46) with v chosen as the piecewise linear interpolant $\pi\phi \in V_{h,0}$ of ϕ, it follows that

$$m(e) = \int_\Omega \nabla e \cdot \nabla(\phi - \pi\phi) \, dx \tag{4.166}$$

$$= \int_\Omega f(\phi - \pi\phi) \, dx - \int_\Omega \nabla u_h \cdot \nabla(\phi - \pi\phi) \, dx \tag{4.167}$$

Integrating by parts on each element K we obtain the error representation formula

$$m(e) = \sum_{K \in \mathcal{K}} \int_K (f + \Delta u_h)(\phi - \pi\phi) \, dx - \frac{1}{2} \int_{\partial K \setminus \partial\Omega} [n \cdot \nabla u_h](\phi - \pi\phi) \, ds \tag{4.168}$$

which provides the exact value of the error in the goal functional $m(\cdot)$ in terms of the finite element solution u_h and the dual ϕ. Of course, this requires knowledge about ϕ, which must generally be replaced by a computed finite element approximation $\phi_h \approx \phi$. In such case, for the term $\phi_h - \pi\phi_h$ to make sense, ϕ_h must be computed using a higher order element than for u_h. This is major concern as it may become computationally costly. However, fortunately, the computation of ϕ_h

can often be afforded. Now, provided that ϕ (or ϕ_h) is sufficiently smooth, we can estimate the terms in the error representation formula using standard interpolation estimates, which yields the following a posteriori error estimate.

Theorem 4.13. *The finite element solution u_h, defined by* (4.13)*, satisfies the estimate*

$$m(e) \leq C \sum_{K \in \mathcal{K}} \eta_K(u_h, \phi) \tag{4.169}$$

where the so-called element indicator $\eta_K(u_h, \phi) = \rho_K(u_h)\omega_K(\phi)$ is the product of the element residual $\rho_K(u_h)$ and the element weight $\omega_K(\phi)$, defined on each element K, by

$$\rho_K(u_h) = \|f + \Delta u_h\|_{L^2(K)} + \tfrac{1}{2} h_K^{-1/2} \|[n \cdot \nabla u_h]\|_{L^2(\partial K \setminus \partial \Omega)} \tag{4.170}$$

$$\omega_K(\phi) = h_K^2 \|D^2 \phi\|_{L^2(K)} \tag{4.171}$$

The element indicator can be used to identify error prone elements, and, thus, drive adaptive mesh refinement. In doing so, only elements contributing significantly to the error in the goal functional $m(\cdot)$ are refined. This is a very nice effect of the dual weight ω_K, which automatically identifies regions of the domain Ω that are have large influence on $m(\cdot)$. Indeed, the dual contains sensitivity information about $m(u)$ with respect to perturbations of u. This is in contrast to energy norm based mesh refinement, which does not take into account the location of the refined elements. Note also that the formal order of convergence is h^2 and, thus, optimal for $m(e)$, but only h for $|||e|||$. We say formally, since it may be dubious to speak about a global mesh size in a refined mesh. However, the use of this kind of dual based a posteriori mesh refinement usually leads to very efficient and cheap adaptive finite element methods.

We mention that not all goal functionals $m(u)$ are computable (e.g., direct point values of u). Indeed, $m(u)$ is usually some kind of average of u. Also, if the quantity of interest is defined on the domain boundary $\partial\Omega$, then it is possible to use a goal functional of the form $m(u) = \int_{\partial\Omega} \psi u \, ds$. In this case, the weight function ψ will enter the error representation formula through the boundary conditions of the dual problem.

By analogy with the dual problem for ϕ, the problem for u is sometimes called the primal problem.

To illustrate goal oriented mesh refinement we consider Poisson's equation $-\Delta u = f$ with f chosen such that the primal is $u = \sin(2\pi x_1/3)\sin(2\pi x_2/3)$ on the square $\Omega = [0, 3]^2$. The weight ψ is chosen such that the dual is $\phi = x_1 x_2 (3 - x_1)^8 (3 - x_2)^8$. This gives the target value $m(u) = \int_\Omega \psi u \, dx = 1.849253113061110 \cdot 10^6$. As element indicator η_K we use the error representation formula (4.168) as is (i.e., we do not estimate the terms). Thus, neglecting errors stemming from insufficient quadrature and from replacing ϕ with ϕ_h we ought to

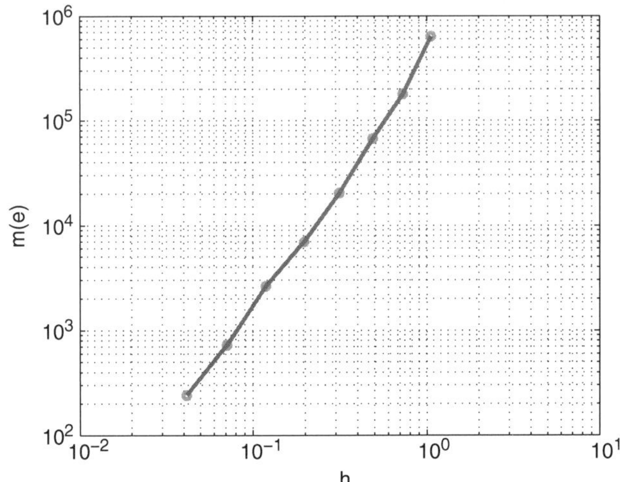

Fig. 4.10 Loglog plot of the mesh size h versus the error $m(e)$

Table 4.1 Convergence of $m(e)$ for a sequence of adaptive meshes

n_t	$10^{-5} \cdot m(e)$	$10^{-5} \cdot \sum_K \eta_K$	I_{eff}
32	6.3789	6.9477	0.9181
68	1.7959	1.7653	1.0173
150	0.6726	0.6673	1.0080
364	0.2034	0.2035	0.9996
920	0.0698	0.0698	1.0002
2,536	0.0263	0.0263	0.9997
7,110	0.0073	0.0073	1.0012
20,800	0.0024	0.0024	1.0006

have $\sum_K \eta_K = m(e)$. The approximate dual ϕ_h is computed using continuous piecewise quadratic finite elements. Making eight refinement loops we obtain the convergence results of Fig. 4.10 and Table 4.1 for the error in the goal functional $m(e)$. Looking at the figure we see that $|m(e)|$ converges asymptotically more or less like h^2, where we choose the mesh size as $h = \sqrt{2|\Omega|/n_t}$ with n_t the number of elements. Moreover, from the table we see that $m(e)$ is accurately predicted by $\sum_{K \in \mathcal{K}} \eta_K$. This is also reflected by the so-called efficiency index I_{eff}, defined by $I_{\text{eff}} = m(e)/\sum_{K \in \mathcal{K}} \eta_K$. Of course, we wish the efficiency index to be close to one.

Even though the global error $e = u - u_h$ is present throughout the whole domain Ω, the dual ϕ is used to automatically identify the relevant region for keeping the goal functional error $m(e)$ small, and, therefore, mesh refinement only takes place in the lower left corner of Ω. This is clearly seen in Fig. 4.11, which shows the final mesh.

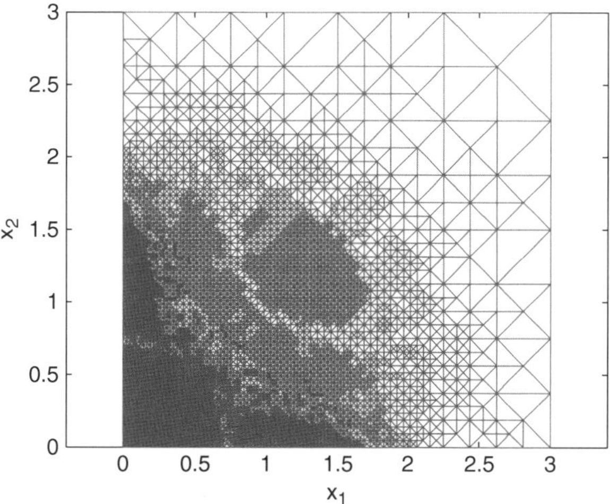

Fig. 4.11 Final mesh for the problem with solution $u = \sin(2\pi x_1/3)\sin(2\pi x_2/3)$

4.11 Further Reading

The theory for elliptic partial differential equations (i.e., Poisson's equation) and their numerical approximation is well understood, and there are very many good books on these topics. We mention the books by Gockenbach [36] and Eriksson et al. [27], which are perhaps best suited for studies at the undergraduate level. Books suitable for the graduate level include those by Johnson [45], Braess [15], Ern and Guermond [37], and Solin [66]. These are all mathematical in style. A book in a more engineering style is the first volume [76] in the the five volumes series overview of the finite element method by Zienkiewicz, Taylor, and Zhu.

Much more information on a posteriori based adaptive mesh refinement can be found in the books by Bangerth and Rannacher [5] and Ainsworth and Oden [2].

4.12 Problems

Exercise 4.1. Prove the Cauchy-Schwarz inequality $|\int_\Omega uv\,dx| \le \|u\|_{L^2(\Omega)} \|v\|_{L^2(\Omega)}$.

Exercise 4.2. Verify that $\|\nabla u\|_{L^2}$ satisfies the requirements of a norm on the space V_0.

Exercise 4.3. Determine f so that $u = x_1(1-x_1)x_2(1-x_2)$ is a solution to $-\Delta u = f$. Compute ∇u, $D^2 u$, $\|u\|_{L^2(\Omega)}$, and $\|\nabla u\|_{L^2(\Omega)}$, with $\Omega = [0,1]^2$.

Exercise 4.4. Write down appropriate test and trial space for the problem

$$-\Delta u = 0, \quad x \in \Omega, \qquad u = 0, \quad x \in \Gamma_D, \qquad n \cdot \nabla u = g_N, \quad x \in \Gamma_N$$

Exercise 4.5. Compute the element mass and stiffness matrices on the triangle $\bar{K}$ with corners at $(0,0)$, $(1,0)$, and $(0,1)$.

Exercise 4.6. Write down the geometry matrix `g` for the domain $\Omega = [-2,3]^2 \setminus [-1,1]^2$ (i.e., a rectangle with a square hole). Use it to make a triangulation of this domain with `initmesh`.

Exercise 4.7. Show, that the solution to

$$-\Delta u = f, \quad x \in \Omega, \qquad u = 0, \quad x \in \partial\Omega$$

satisfies the stability estimate

$$\|\nabla u\|_{L^2(\Omega)} \leq C\|f\|_{L^2(\Omega)}$$

Hint: First, multiply the equation with u and integrate by parts. Then, use the Poincaré inequality.

Also, show that

$$\|\nabla(u - u_h)\|^2_{L^2(\Omega)} = \|\nabla u\|^2_{L^2(\Omega)} - \|\nabla u_h\|^2_{L^2(\Omega)}$$

where u_h is the finite element approximation of u.

Exercise 4.8. Write a MATLAB code to assemble the stiffness matrix A on a mesh of a domain of your choice with `assema`. Use `eigs` to compute the eigenvalues of A and verify that one eigenvalue is zero, while the others are positive. Why?

Exercise 4.9. Write a MATLAB code to solve $-\Delta u = 0$ in $\Omega = [-2,2] \times [-2\pi, 2\pi]$, with Dirichlet boundary conditions $u = \exp(x_1)\arctan(x_2)$ on $\partial\Omega$. Use the routines `StiffnessAssembler2D` and `RobinAssembler2D` to assemble the involved stiffness matrix and load vector.

Exercise 4.10. Consider the problem

$$-\Delta u + u = f, \quad x \in \Omega, \qquad u = 0, \quad x \in \partial\Omega$$

(a) Make a variational formulation of this problem in a suitable space V.
(b) Choose a polynomial subspace V_h of V and write down a finite element method based on the variational formulation.
(c) Deduce the Galerkin orthogonality

$$\int_{\Omega} \nabla(u - u_h) \cdot \nabla v + (u - u_h)v \, dx = 0, \quad \forall v \in V_h$$

(d) Derive the a priori error estimate

$$\|\nabla(u - u_h)\|_{L^2(\Omega)} + \|u - u_h\|_{L^2(\Omega)} \leq Ch\|D^2 u\|_{L^2(\Omega)}$$

where h is the maximum mesh size.

Exercise 4.11. Make a variational formulation of the problem

$$-\frac{\partial^2 u}{\partial x_1^2} - 4\frac{\partial^2 u}{\partial x_2^2} + 2u = 1, \quad x \in \Omega, \qquad u = 0, \quad x \in \partial\Omega$$

Exercise 4.12. Show, that the solution to

$$-\Delta u + u = 0, \quad x \in \Omega, \qquad n \cdot \nabla u = g_N, \quad x \in \partial\Omega$$

satisfies the stability estimate

$$\|u\|^2_{L^2(\Omega)} + \|\nabla u\|^2_{L^2(\Omega)} \leq C\|g_N\|^2_{L^2(\partial\Omega)}$$

Hint: Use the Trace inequality and the arithmetic-geometric mean inequality $2ab \leq a^2 + b^2$.

Exercise 4.13. Derive a posteriori error estimates in the energy norm for the problem

$$-\Delta u = 0, \quad x \in \Omega$$

with boundary conditions

(a) $u = g_D$, $x \in \partial\Omega$.
(b) $u = 0$, $x \in \Gamma_D$, and $n \cdot \nabla u = g_N$, $x \in \Gamma_N$.

where g_D and g_N is given boundary data in, say, $L^2(\partial\Omega)$. For simplicity, assume that g_D can be represented as a continuous piecewise linear.

Chapter 5
Time-Dependent Problems

Abstract Most real-world problems depend on time and in this chapter we shall construct numerical methods for solving time dependent differential equations. We do this by first discretizing in space using finite elements, and then in time using finite differences. Various time stepping methods are presented. As model problems we use two classical equations from mathematical physics, namely, the Heat equation, and the Wave equation. Illustrative numerical examples for both equations are presented. To assert the accuracy of the computed solutions we derive both stability estimates, and a priori error estimates. We also formulate space-time finite elements and use them to derive duality based posteriori error estimates.

5.1 Finite Difference Methods for Systems of ODE

We begin this chapter by reviewing a couple of simple finite difference methods for systems of ordinary differential equations (ODE).

Our basic problem is to find the $n \times 1$ time-dependent solution vector $\xi = \xi(t)$ to the ODE system

$$M\dot{\xi}(t) + A\xi(t) = b(t), \quad t \in J = (0, T] \tag{5.1a}$$

$$\xi(0) = \xi_0 \tag{5.1b}$$

where M and A are given constant $n \times n$ matrices, $b(t)$ is a given time-dependent $n \times 1$ vector, and ξ_0 is a given $n \times 1$ vector with initial data. Further, $\dot{\xi}$ means differentiation of ξ with respect to time t on the time interval J, with final time T.

Assuming that M is invertible we can rewrite (5.1) as

$$\dot{\xi}(t) + \bar{A}\xi(t) = \bar{b}(t) \tag{5.2}$$

M.G. Larson and F. Bengzon, *The Finite Element Method: Theory, Implementation, and Applications*, Texts in Computational Science and Engineering 10,
DOI 10.1007/978-3-642-33287-6_5, © Springer-Verlag Berlin Heidelberg 2013

where $\bar{A} = M^{-1}A$ and $\bar{b}(t) = M^{-1}b(t)$. The analytical solution to this equation can be expressed using matrix exponentials using the so-called Duhamel's formula

$$\xi(t) = \xi(0)e^{-\bar{A}t} + \int_0^t e^{-\bar{A}(t-s)}\bar{b}(s)\,ds \tag{5.3}$$

However, since we can not invert M if its dimension n is large, and since the matrix exponential is complicated to use from the practical point of view, let us look at ways to numerically approximate the solution ξ.

We let $0 = t_0 < t_1 < t_2 < \cdots < t_m = T$ be a partition of the time interval J into m subintervals $J_l = (t_{l-1}, t_l]$ and time steps $k_l = t_l - t_{l-1}$, $l = 1, 2, \ldots, m$. For each time t_l we wish to find an approximation ξ_l to $\xi(t_l)$. We replace the time derivative $\dot{\xi}$ with the backward difference quotient

$$\dot{\xi}(t_l) \approx \frac{\xi_l - \xi_{l-1}}{k_l} \tag{5.4}$$

which yields the so-called Euler backward method

$$M\frac{\xi_l - \xi_{l-1}}{k_l} + A\xi_l = b_l \tag{5.5}$$

or, equivalently,

$$(M + k_l A)\xi_l = M\xi_{l-1} + k_l b_l \tag{5.6}$$

where we have introduced the notation $b_l = b(t_l)$. The initial condition ξ_0 is used to start the time stepping. Indeed, starting with $\xi_0 = \xi(t_0)$, we can successively compute the unknowns $\xi_1, \xi_2, \ldots, \xi_m$ by repeatedly solving the linear system (5.6). This yields the following algorithm for evolving our ODE system in time:

Algorithm 13 The backward Euler method

1: Create a partition $0 = t_0 < t_1 < \cdots < t_m = T$ of the time interval J with m time steps $k_l = t_l - t_{l-1}$.
2: Set $\xi_0 = \xi(0)$.
3: **for** $l = 1, 2, \ldots, m$ **do**
4: Solve

$$(M + k_l A)\xi_l = M\xi_{l-1} + k_l b_l \tag{5.7}$$

5: **end for**

Roughly speaking, by making a Taylor expansion of $\xi(t)$ around t_l it is easy to show that the local truncation error on each subinterval J_l is proportional to k_l^2. Thus, summing up over the the whole time interval J, the total error $\xi(t_m) - \xi_m$

is of size k, with $k = \max_{1 \leq l \leq m} k_l$ the maximum time step. For this reason, the backward Euler method is said to be first order accurate. Moreover, this method is said to be unconditionally stable, since the size of the time step k can be chosen arbitrary large. In particular, we can take large time steps and quickly advance the solution from 0 to T. Of course, the prize we pay for this large time step k is a potentially bad accuracy.

The choice of approximating the time derivative $\dot{\xi}$ with a backward quotient is arbitrary. Using instead a forward quotient, yields

$$M \frac{\xi_l - \xi_{l-1}}{k_l} + A\xi_{l-1} = b_{l-1} \tag{5.8}$$

which is the so-called forward Euler method. Applying M^{-1} from the left and rearranging terms we find

$$\xi_l = \xi_{l-1} + k_l M^{-1}(b_{l-1} - A\xi_{l-1}) \tag{5.9}$$

from which it follows that ξ_l can be obtained at a low cost provided that M is easy to invert (e.g., diagonal). As we shall see, this is, indeed, sometimes the case.

The Euler forward method is also first order accurate. However, it comes with a restriction on the time step k. Indeed, a too large k causes unbounded growth of ξ_l. Therefore, we say that this method is conditionally stable.

Perhaps not so surprising, there is a large range of time stepping methods for the ODE system (5.1). Each of these has its own characteristics regarding accuracy, stability, and computational cost. Let us mention just a few.

The popular so-called Crank-Nicolson (CN) method is defined by

$$M \frac{\xi_l - \xi_{l-1}}{k_l} + A \frac{\xi_l + \xi_{l-1}}{2} = \frac{b_l + b_{l-1}}{2} \tag{5.10}$$

This method is only conditionally stable, but second order accurate.

An alternative to the CN method is the BDF(2) method, defined by

$$M \frac{3\xi_l - 4\xi_{l-1} + \xi_{l-2}}{2k} + A\xi_l = b_l \tag{5.11}$$

The advantages of this method is that it is second order accurate, and simple to implement, in particular for solving time-dependent non-linear problems using fixed-point iteration. The drawbacks are that it is difficult to derive for anything other than a constant time step, and requires two vectors to get started. To construct two starting vectors, ξ_1 is usually found by taking a single Euler step. The acronym BDF stands for backward difference formula. Actually, there is a whole family of BDF methods with BDF(k) denoting a method of k-th order accuracy. BDF methods of order $k \leq 6$ are available.

Finally, we mention the Runge-Kutta (RK) family of methods, the most famous being the RK(4) method, defined by

$$\xi_l = \xi_{l-1} + \frac{\eta_1 + 2\eta_2 + 2\eta_3 + \eta_4}{6} k \tag{5.12}$$

$$\eta_1 = \rho(t_{l-1}, \xi_{l-1}) \tag{5.13}$$

$$\eta_2 = \rho(t_{l-1} + \tfrac{1}{2}k, \xi_{l-1} + \tfrac{1}{2}\eta_1) \tag{5.14}$$

$$\eta_3 = \rho(t_{l-1} + \tfrac{1}{2}k, \xi_{l-1} + \tfrac{1}{2}\eta_2) \tag{5.15}$$

$$\eta_4 = \rho(t_{l-1} + k, \xi_{l-1} + \tfrac{1}{2}\eta_3) \tag{5.16}$$

$$t_l = t_{l-1} + k \tag{5.17}$$

where $\rho(\xi_l) = M^{-1}(b - A\xi_l)$. This method is conditionally stable and fourth order accurate.

Time stepping methods are generally classified as either:

- Implicit, or
- Explicit

Implicit methods are characterized by the fact that the next solution approximation ξ_l must be solved for at time t_l. These methods are therefore often expensive to use. Implicit methods are used for stiff problems, that is, problems with a solution ξ, which contains features with highly different time scales. By contrast, explicit methods are cheap, and the next solution approximation u_l is given by a closed form formula. However, the time step must generally be small in order to guarantee numerical stability. Thus, the low cost of obtaining a single approximation ξ_l is generally counteracted by the large number of levels m needed to reach the final time T. Thus, explicit methods may or may not be more efficient than implicit ones. The Euler forward method is the simplest explicit method, whereas the Euler backward is the simplest implicit method. Also, the CN and BDF methods are implicit, whereas the RK(4) method is explicit.

5.2 The Heat Equation

5.2.1 Derivation of the Transient Heat Equation

In Sect. 2.2.1 we have previously derived the Heat equation by considering heat conduction in a thin rod under the assumption of steady state. We shall now revisit this derivation taking also the dynamics of the heat transfer process into account.

Thus, consider again the thin metal rod of length L, which occupies the interval $I = [0, L]$. Let f be the heat source intensity, q the heat flux along the direction of increasing x, and e the internal energy per unit length in the rod.

The principle of conservation of energy states that the rate of change of internal energy equals the sum of net heat flux and produced heat. In the language of mathematics, this is equivalent to,

$$\int_I \dot{e}\, dx = \int_I f\, dx + A(0)q(0) - A(L)q(L) \tag{5.18}$$

where A is the cross section area of the rod. Dividing by L and letting $L \to 0$ we obtain the differential equation

$$\dot{e} + (Aq)' = f \tag{5.19}$$

Now, under certain physical assumptions it is reasonable to assume that the internal energy e depends linearly on temperature T, that is,

$$e = cT \tag{5.20}$$

where c is the heat conductivity of the rod.

We also assume that Fourier's law, $q = -kT'$, is valid.

Combining (5.19) and (5.20) we infer the time-dependent Heat equation

$$c\dot{T} - (AkT')' = f \tag{5.21}$$

As usual, this differential equation needs to be supplemented by boundary conditions at $x = 0$ and $x = L$ of either Dirichlet, Neumann, or Robin type. The boundary conditions should hold for all times. However, this is not enough to uniquely determine the solution T. An initial condition of the form $T(x, 0) = T_0(x)$, where $T_0(x)$ is a given function at the initial time $t = 0$, is also required for uniqueness of T.

5.2.2 Model Problem

Let us consider the model Heat equation

$$\dot{u} - u'' = f, \quad x \in I = [0, L], \quad t \in J = (0, T] \tag{5.22a}$$

$$u(0, t) = u(L, t) = 0 \tag{5.22b}$$

$$u(x, 0) = u_0(x) \tag{5.22c}$$

where $u = u(x, t)$ is the sought solution, $f = f(x, t)$ a given source function, and $u_0(x)$ a given initial condition.

5.2.3 Variational Formulation

Multiplying $f = \dot{u} - u''$ by a test function $v = v(x, t)$ and integrating by parts we have

$$\int_0^L f v \, dx = \int_0^L \dot{u} v \, dx - \int_0^L u'' v \, dx \tag{5.23}$$

$$= \int_0^L \dot{u} v \, dx - u'(L)v(L) + u'(0)v(0) + \int_0^L u' v' \, dx \tag{5.24}$$

$$= \int_0^L \dot{u} v \, dx + \int_0^L u' v' \, dx \tag{5.25}$$

where we have assumed that $v(0, t) = v(L, t) = 0$.

Next, we introducing the space

$$V_0 = \{v : \|v(\cdot, t)\| + \|v'(\cdot, t)\| < \infty, \ v(0, t) = v(L, t) = 0\} \tag{5.26}$$

where, $\| \cdot \| = \| \cdot \|_{L^2(I)}$ denotes the usual L^2-norm. Note that the norm $\|v\| = \|v(\cdot, t)\|$ is a function of t, but not x.

Using the space V_0 we obtain the following variational formulation of (5.22): find u such that, for every fixed $t \in J$, $u \in V_0$ and

$$\int_I \dot{u} v \, dx + \int_I u' v' \, dx = \int_I f v \, dx, \quad \forall v \in V_0, \quad t \in J \tag{5.27}$$

5.2.4 Spatial Discretization

Let $\mathcal{I} : 0 = x_0 < x_1 < \cdots < x_n = L$ be a mesh of the interval I, and let $V_{h,0} \subset V_0$ be the space of continuous piecewise linears vanishing at $x = 0$ and $x = 1$ on this mesh. The space discrete counterpart of (5.27) takes the form: find u_h such that, for every fixed $t \in J$, $u_h \in V_{h,0}$ and

$$\int_I \dot{u}_h v \, dx + \int_I u_h' v' \, dx = \int_I f v \, dx, \quad \forall v \in V_{h,0}, \quad t \in J \tag{5.28}$$

We observe that (5.28) is equivalent to

$$\int_I \dot{u}_h \varphi_i \, dx + \int_I u_h' \varphi_i' \, dx = \int_I f \varphi_i \, dx, \quad i = 1, 2, \ldots, n-1, \quad t \in J \tag{5.29}$$

where φ_i are the interior hat functions.

Next, we seek a solution u_h to (5.29), expressed for every fixed t, as a linear combination of hat functions $\varphi_j(x)$, $j = 1, 2, \dots, n-1$, and time-dependent coefficients $\xi_j(t)$. That is, we make the ansatz

$$u_h(x,t) = \sum_{j=1}^{n-1} \xi_j(t)\varphi_j(x) \tag{5.30}$$

and seek to determine the vector

$$\xi(t) = \begin{bmatrix} \xi_1(t) \\ \xi_2(t) \\ \vdots \\ \xi_{n-1}(t) \end{bmatrix} = \begin{bmatrix} u_h(x_1,t) \\ u_h(x_2,t) \\ \vdots \\ u_h(x_{n-1},t) \end{bmatrix} \tag{5.31}$$

so that (5.29) is satisfied.

Consider carefully the construction of u_h. For every fixed time t, u_h is a continuous piecewise linear function with time-dependent nodal values $\xi_j(t)$.

Now, substituting (5.30) into (5.29) we have

$$\int_I f\varphi_i \, dx = \sum_{j=1}^{n-1} \dot{\xi}_j(t) \int_I \varphi_j \varphi_i \, dx \tag{5.32}$$

$$+ \sum_{j=1}^{n-1} \xi_j(t) \int_I \varphi_j' \varphi_i' \, dx, \quad i = 1, 2, \dots, n-1, \quad t \in J$$

Introducing the notation

$$M_{ij} = \int_I \varphi_j \varphi_i \, dx \tag{5.33}$$

$$A_{ij} = \int_I \varphi_j' \varphi_i' \, dx \tag{5.34}$$

$$b_i(t) = \int_I f(t)\varphi_i \, dx \tag{5.35}$$

where $i, j = 1, 2, \dots, n-1$, we further have

$$b_i(t) = \sum_{j=1}^{n-1} M_{ij}\dot{\xi}_j(t) + \sum_{j=1}^{n-1} A_{ij}\xi_j(t), \quad i = 1, 2, \dots, n-1, \quad t \in J \tag{5.36}$$

which is nothing but a system of $n-1$ ODEs for the $n-1$ coefficients $\xi_j(t)$. In matrix form, we write this

$$M\dot{\xi}(t) + A\xi(t) = b(t), \quad t \in J \tag{5.37}$$

where the entries of the $(n-1) \times (n-1)$ matrices M and A, and the $(n-1) \times 1$ vector $b(t)$ are defined by (5.33)–(5.35), respectively. We recognize M, as the mass matrix, A as the stiffness matrix, and $b(t)$ as a time-dependent load vector.

The ODE system (5.37) is sometimes called a spatial semi-discretization of the Heat equation, since the dependence on the space coordinate x has been eliminated.

We, thus, conclude that the coefficients $\xi_j(t)$, $j = 1, 2, \dots, n-1$, in the ansatz (5.30) satisfy a system of ODE s, which must be solved in order to obtain the space discrete solution $u_h(t)$.

5.2.5 *Time Discretization*

Applying the backward Euler method to the ODE system (5.37) we immediately obtain the following algorithm for the numerical solution of the Heat equation:

Algorithm 14 The backward Euler method for solving the Heat equation

1: Create a mesh with n elements on the interval I.
2: Create a partition $0 = t_0 < t_1 < \dots < t_m = T$ of the time interval J with m time steps $k_l = t_l - t_{l-1}$.
3: Choose ξ_0.
4: **for** $l = 1, 2, \dots, m$ **do**
5: Compute the $(n-1) \times (n-1)$ mass and stiffness matrices M and A, and the $(n-1) \times 1$ load vector $b_l = b(t_l)$ with entries

$$M_{ij} = \int_I \varphi_j \varphi_i \, dx, \quad A_{ij} = \int_I \varphi_j' \varphi_i' \, dx, \quad b(t_l)_i = \int_I f(t_l) \varphi_i \, dx \tag{5.38}$$

6: Solve

$$(M + k_l A)\xi_l = M\xi_{l-1} + k_l b_l \tag{5.39}$$

7: **end for**

Regarding the starting vector ξ_0 there are a few different choices. One choice, and perhaps the simplest, is to let ξ_0 be the nodal values of the interpolant πu_0 of the initial data u_0. Another choice is to let ξ_0 be the nodal values of the L^2-projection $P_h u_0$ of u_0, but this is, of course, more computationally expensive. Yet another choice is to let ξ_0 be the nodal values of the so-called Ritz projection Ru_h of u_0, to be defined shortly.

5.3 Stability Estimates

For time-dependent problems it is generally of interest to know something about the long term behavior of the solution. In particular, one would like to know if it grows with time and if it can be bounded by the coefficients (e.g., the initial condition and the right hand side). To this end, stability estimates are used.

5.3.1 *A Space Discrete Estimate*

We first derive a stability estimate for the space discrete solution u_h. Recall that $u_h = u_h(x,t)$ is a continuous smooth function if viewed as a function of time t, but only a continuous piecewise linear function if viewed as a function of space x.

Choosing $v = u_h$ in the variational formulation (5.27) we have

$$\int_I \dot{u}_h u_h + (u_h')^2 \, dx = \int_I f u_h \, dx \tag{5.40}$$

Noting that the first term can be written

$$\int_I \dot{u}_h u_h \, dx = \int_I \tfrac{1}{2}\partial_t (u_h^2) \, dx = \tfrac{1}{2}\partial_t \|u_h\|^2 = \|u_h\| \partial_t \|u_h\| \tag{5.41}$$

and using the Cauchy Schwarz inequality we further have

$$\|u_h\| \partial_t \|u_h\| + \|u_h'\|^2 \leq \|f\| \|u_h\| \tag{5.42}$$

which implies

$$\partial_t \|u_h\| \leq \|f\| \tag{5.43}$$

Integrating this result with respect to time from 0 to t we obtain the stability estimate

$$\|u_h(\cdot,t)\| = \|u_h(\cdot,0)\| + \int_0^t \|f(\cdot,s)\| \, ds \tag{5.44}$$

This shows that the size of u_h is bounded in time by the initial condition $u_h(\cdot,0)$ and the source function f.

5.3.2 A Fully Discrete Estimate

Let $u_{h,l}$ denote the continuous piecewise linear function $u_h(t_l)$, that is, the sum $\sum_{i=1}^{n-1}(\xi_l)_i\varphi_i$ with $(\xi_l)_i$ component i of the vector ξ_l. We wish to derive a stability estimate for $u_{h,l}$. To do so, we multiply the backward Euler method with the vector ξ_l, which yields

$$\xi_l^T(M + k_l A)\xi_l = \xi_l^T(M\xi_{l-1} + b_l) \tag{5.45}$$

Clearly, this is the matrix form of the equation

$$\|u_{h,l}\|^2 + k_l\|u_{h,l}'\|^2 = \int_I u_{h,l-1}u_{h,l}\,dx + k_l\int_I f_l u_{h,l}\,dx \tag{5.46}$$

Using the Cauchy-Schwarz inequality we have

$$\|u_{h,l}\|^2 + k_l\|u_{h,l}'\|^2 \leq \|u_{h,l-1}\|\,\|u_{h,l}\| + k_l\|f_l\|\,\|u_{h,l}\| \tag{5.47}$$

which implies

$$\|u_{h,l}\| \leq \|u_{h,l-1}\| + k_l\|f_l\| \tag{5.48}$$

Iterated use of this result gives us

$$\|u_{h,l}\| \leq \|u_{h,0}\| + \sum_{i=1}^{l} k_i\|f_i\| \tag{5.49}$$

which is our stability estimate.

This shows that the size of $u_{h,l}$ is bounded for all times by the timestep k_l, the initial condition $u_{h,0}$, and the source function f.

5.4 A Priori Error Estimates

Error estimates for time-dependent problems can be derived by combining error estimates for the corresponding stationary problem with stability estimates. We use this approach below to derive a priori error estimates for both the space discrete solution u_h and the fully discrete solution $u_{h,l}$.

5.4.1 Ritz Projection

Ritz projection is a technique for approximating a given function u, and is very similar to L^2-projection. Indeed, both L^2- and Ritz projection compute the orthogonal projection of u onto a finite dimensional subspace with respect to a certain scalar product. For L^2-projection the subspace is V_h and the scalar product the usual L^2-product $\int uv\,dx$. However, for Ritz projection the subspace is $V_{h,0}$ and the scalar product $\int u'v'\,dx$. More precisely, the Ritz projection $R_h u \in V_{h,0}$ of a given function $u \in V_0$ is defined by

$$\int_I (u - R_h u)' v'\,dx = 0, \quad \forall v \in V_{h,0} \tag{5.50}$$

With this definition we have the following approximation result.

Proposition 5.1. *The Ritz projection $R_h u$, defined by* (5.50)*, satisfies the estimate*

$$\|u - R_h u\| \le C h^2 \|u''\| \tag{5.51}$$

Proof. The proof follows from Nitsche's trick with dual problem $-\phi'' = u - R_h u$ with boundary conditions $\phi(0) = \phi(L) = 0$. □

5.4.2 A Space Discrete Estimate

Theorem 5.1. *The space discrete solution u_h, defined by* (5.28)*, satisfies the estimate*

$$\|u(t) - u_h(t)\| \le C h^2 \left(\|u_0''\| + \int_0^t \|\dot{u}''(\cdot, s)\|\,ds \right) \tag{5.52}$$

Proof. Following standard procedure (cf., Larsson and Thomeé [70]), let us write the error $u - u_h$ as

$$u - u_h = (u - R_h u) + (R_h u - u_h) = \rho + \theta \tag{5.53}$$

We can bound ρ at once using the approximation result of the Ritz projector $R_h u$.

$$\|\rho(\cdot, t)\| \le C h^2 \|u''(\cdot, t)\| \tag{5.54}$$

$$\le C h^2 \left\| u''(\cdot, 0) + \int_0^t \dot{u}''(\cdot, s)\,ds \right\| \tag{5.55}$$

$$\le C h^2 \left(\|u_0''\| + \int_0^t \|\dot{u}''(\cdot, s)\|\,ds \right) \tag{5.56}$$

Continuing, to bound also θ we subtract (5.28) from (5.27), which yields

$$\int_I (\dot{u} - \dot{u}_h)v\,dx + \int_I (u - u_h)'v'\,dx = 0, \quad \forall v \in V_{0,h}, \quad t \in J \tag{5.57}$$

or

$$\int_I \dot{\theta} v\,dx + \int_I \theta' v'\,dx = -\int_I \dot{\rho} v\,dx - \int_I \rho' v'\,dx \tag{5.58}$$

$$= -\int_I \dot{\rho} v\,dx \tag{5.59}$$

Here, we have used the definition of the Ritz projector to get rid of the last term. From this we see that θ satisfies the space discrete variational equation with $-\dot{\rho}$ as source function. Consequently, we can apply the stability estimate (5.44) to obtain

$$\|\theta(\cdot,t)\| \leq \|\theta(\cdot,0)\| + \int_0^t \|\dot{\rho}(\cdot,s)\|\,ds \tag{5.60}$$

Further, choosing $u_h(x,0) = R_h u(x,0) = R_h u_0(x)$, we have $\theta(x,0) = 0$ and the first term drops out. For the second term, we obtain

$$\|\dot{\rho}(\cdot,t)\| = \|\partial_t(u(\cdot,t) - R_h u(\cdot,t))\| \leq Ch^2\|\partial_t u''(\cdot,t)\| = Ch^2\|\dot{u}''(\cdot,t)\| \tag{5.61}$$

Thus, we have

$$\|\theta(\cdot,t)\| \leq Ch^2 \int_0^t \|\dot{u}''(\cdot,s)\|\,ds \tag{5.62}$$

Combining (5.56) and (5.62) concludes the proof. □

From this we see that the error at any given time t in the space discrete solution $u_h(t)$ is proportional to the mesh size squared h^2, which is to be expected when using a piecewise linear ansatz in space.

5.4.3 A Fully Discrete Estimate

Theorem 5.2. *The fully discrete solution $u_{h,l}$ satisfies the estimate*

$$\|u(t_l) - u_{h,l}\| \leq Ch^2\left(\|u_0''\| + \int_0^{t_l} \|\dot{u}''(\cdot,s)\|\,ds\right) + Ck\int_0^{t_l} \|\ddot{u}''(\cdot,s)\|\,ds \tag{5.63}$$

Proof. For simplicity, let us assume that the time interval J has a uniform partition with time step k. As before, we follow the standard procedure and write the error

$u(t_l) - u_{h,l} = (u(t_l) - R_h u(t_l)) + (R_h u(t_l) - u_{h,l}) = \rho_l + \theta_l$. Also, as before, ρ_l can be bounded by

$$\|\rho_l\| \le Ch^2 \left(\|u_0''\| + \int_0^{t_l} \|\dot{u}''(\cdot, s)\| \, ds \right) \tag{5.64}$$

To bound also θ_l we use the fact that it satisfies the backward Euler method in the sense that

$$\int_I \frac{\theta_l - \theta_{l-1}}{k} v \, dx + \int_I \theta_l' v' \, dx = - \int_I \chi_l v \, dx \tag{5.65}$$

where

$$\chi_l = \dot{u}(t_l) - \frac{R_h u(t_l) - R_h u(t_{l-1})}{k} \tag{5.66}$$

Adding and subtracting $k^{-1}(u(t_l) - u(t_{l-1}))$ from χ_l yields

$$\chi_l = \left(\dot{u}(t_l) - \frac{u(t_l) - u(t_{l-1})}{k} \right) \tag{5.67}$$

$$+ \left(\frac{u(t_l) - R_h u(t_l)}{k} - \frac{u(t_{l-1}) - R_h u(t_{l-1})}{k} \right)$$

$$= \chi_l^1 + \chi_l^2 \tag{5.68}$$

Application of the stability estimate (5.49) further yields

$$\|\theta_l\| \le \|\theta_0\| + k \sum_{i=1}^{l} \|\chi_i^1\| + k \sum_{i=1}^{l} \|\chi_i^2\| \tag{5.69}$$

where, as before, θ_0 can be eliminated by choosing $u_{h,0} = R_h u_0$. Now, from Taylor's formula, we have

$$\dot{u}(t_l) - \frac{u(t_l) - u(t_{l-1})}{k} = \frac{1}{k} \int_{t_{l-1}}^{t_l} (s - t_{l-1}) \ddot{u}(\cdot, s) \, ds \tag{5.70}$$

and it follows that

$$k \sum_{i=1}^{l} \|\chi_i^1\| \le \sum_{i=1}^{l} \left\| \int_{t_{n-1}}^{t_l} (s - t_{l-1}) \ddot{u}(\cdot, s) \, ds \right\| \le k \int_0^{t_l} \|\ddot{u}(\cdot, s)\| \, ds \tag{5.71}$$

Further, since

$$\frac{u(t_l) - R_h u(t_l)}{k} - \frac{u(t_{l-1}) - R_h u(t_{l-1})}{k} = \frac{1}{k}\int_{t_{l-1}}^{t_l} (1 - R_h)\dot{u}(x,s)\, ds \qquad (5.72)$$

it also follows that

$$k\sum_{i=1}^{l} \|\chi_i^2\| \le \sum_{i=1}^{l}\int_{t_{l-1}}^{t_l} Ch^2\|\dot{u}''(\cdot,s)\|\, ds \le Ch^2\int_0^{t_l} \|\dot{u}''(\cdot,s)\|\, ds \qquad (5.73)$$

Combining (5.64), (5.71), and (5.73) concludes the proof. □

From this we see that the error at a given time t_l in the fully discrete solution $u_{h,l}$ consists of two parts. First, there is one part that stems from the space discretization and the piecewise linear finite elements. It is of size h^2. Then, there is another part that stems from the time discretization and the time stepping method. This contribution to the error depends linearly on the time step k, which is to be expected since it reflects the first order accuracy of the Euler backward method.

5.5 Computer Implementation

To implement our numerical method based on the Euler backward method we can reuse the assembly routines `MassAssembler1D`, `StiffnessAssembler1D`, and `LoadAssembler1D` to assemble the mass and stiffness matrices M, and A, and the load vector $b = b(t)$. Let the load be given by $f = 2x$, and let the initial condition be given by $u_0 = 0.5 - |x - 0.5|$. u_0 looks like a triangle with its peak at $x = 0.5$. After roughly 0.5 time units the solution u assumes a steady state given by $u(x,\infty) = \frac{3}{2}x(x^2 - x)$. The Dirichlet boundary condition $u(0,t) = u(L,t) = 0$ is approximated by a Robin ditto. The code for this solver is listed below. Note that the assembly of A, M, and b is done outside the time loop to save computer resources, since neither of these are time-dependent.

```
function BackwardEulerHeatSolver1D()
h = 0.01; % mesh size
x = 0:h:1; % mesh
m = 100; % number of time levels
T = 0.5; % final time
t = linspace(0,T,m+1); % time grid
xi = 0.5-abs(0.5-x)'; % inital condition
kappa = [1.e+6 1.e+6]; % Robin BC parameters
g = [0 0];
one = inline('1','x'); % dummy
twox = inline('2*x','x'); % load f=2x
```

```
A = StiffnessAssembler1D(x,one,kappa); % stiffness matrix
M = MassAssembler1D(x); % mass matrix
b = LoadAssembler1D(x,twox,kappa,g); % load vector
for l = 1:m % time loop
  k = t(l+1) - t(l); % time step;
  xi = (M + k*A)\(M*xi + k*b); % backward Euler method
  plot(x,xi), axis([0 1 0 0.5]), pause(0.1) % plot
end
```

Figure 5.1 shows a series of snapshots of the computed solution $u_{h,l}$, for $l = 0, 2, 4, 8$ etc., as it evolves towards steady state. From the figure we see that the triangle peak in the initial condition u_0 quickly diffuses and disappears. Indeed, the smoothing of any sharp features in the solution is a characteristic feature of the Heat equation, which is therefore said to have smoothing properties. This is a characteristic feature of transient so-called parabolic type differential equations.

5.5.1 Mass Lumping

Had we used the forward, instead of backward, Euler method to time step the ODE system (5.30), the involved time loop for doing so would have taken a form similar to

```
for l = 1:m % time loop
  xi = xi + k*M\(b - A*xi); % forward Euler method
end
```

Here, we see that a linear system involving the mass matrix M must be solved at each iteration. To speed up this computation a common trick is to replace M with a diagonal matrix D with the row sums of M on the diagonal. This is called mass lumping and D is called the lumped mass matrix. Thus, the diagonal entries D_{ii} are given by

$$D_{ii} = \sum_{j=1}^{n-1} M_{ij}, \quad i = 1, 2, \dots, n-1 \tag{5.74}$$

The lumped mass matrix can be viewed as a result of under integration, which means to use a quadrature rule with too low precision. Indeed, D results if we use the Trapezoidal rule, which has precision 1 instead of the necessary 2, to compute the matrix entries M_{ij}.

The use of mass lumping speeds up the Euler forward method considerably. However, a careful analysis shows that the restriction on the maximal time step is of the form $k = Ch^2$ for the Heat equation, which may be prohibitively small if h is small.

We remark that mass lumping can only be used for low order approximations, such as, piecewise linear, for instance. Also, the use of lumping on any other matrix

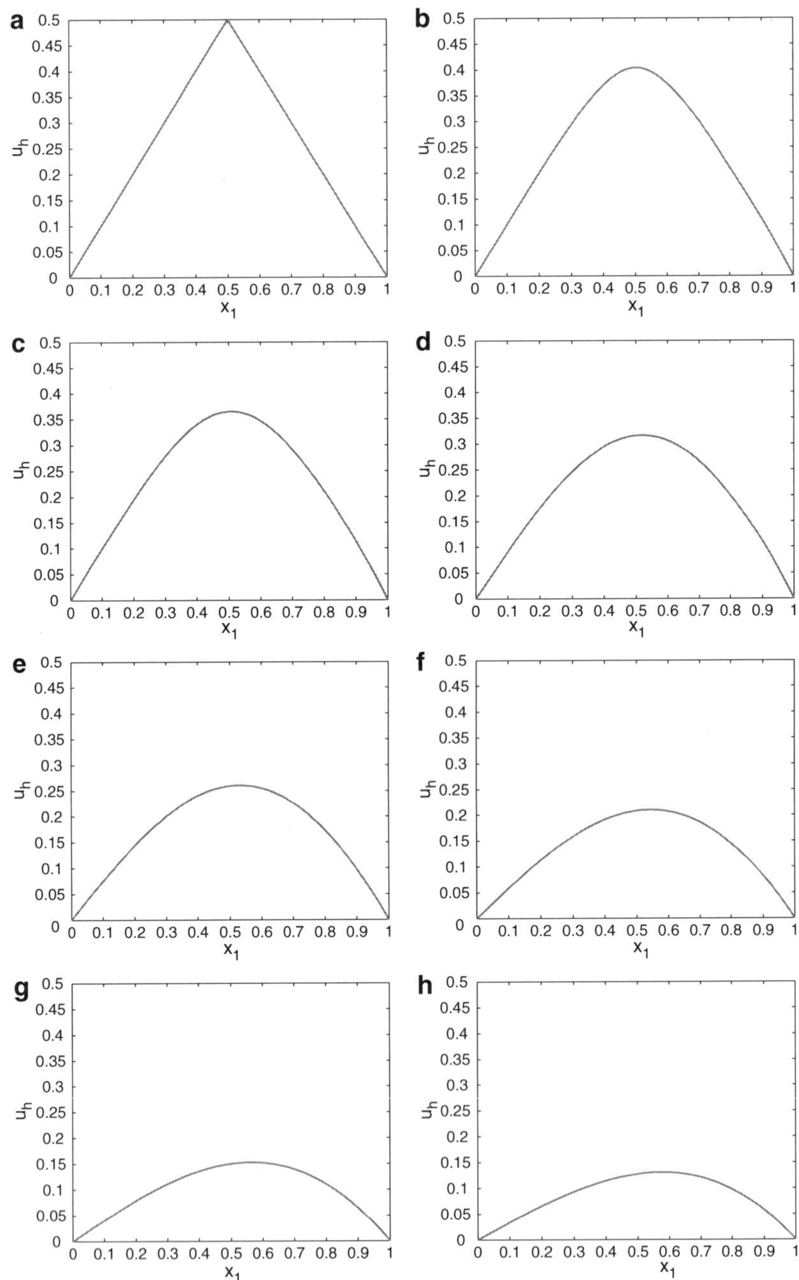

Fig. 5.1 Snapshots of transient solution $u_{h,l}$ evolving to steady state. (**a**) $t = 0$. (**b**) $t = 0.01$. (**c**) $t = 0.02$. (**d**) $t = 0.04$. (**e**) $t = 0.075$. (**f**) $t = 0.125$. (**g**) $t = 0.25$. (**h**) $t = 0.5$

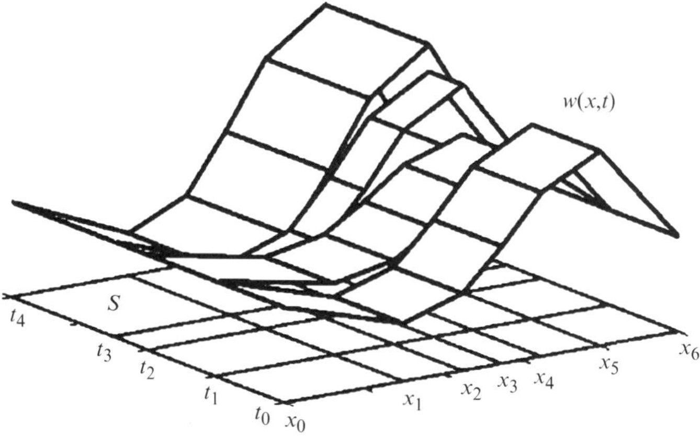

Fig. 5.2 A function w in W_h on the space-time slab S

is generally not recommended, since the matrix entries may be both positive and negative and cancel out on summation. The stiffness matrix, for example, can never be lumped.

5.6 Space-Time Finite Element Approximation

It is also possible to use a more finite element oriented approach for discretizing time-dependent partial differential equations than semi-discretization. The characteristic feature of such discretizations is that no difference is made between the space and time variables. Indeed, consider the so-called space-time slab $S = I \times J$. By partitioning this slab it is possible to define piecewise polynomials that are functions of both space and time. In the simplest case, since S is a rectangular domain, we can divide it into a regular grid. The space-time grid is the tensor product of the spatial and temporal mesh $0 = x_0 < x_1 < \cdots < x_n = 1$ and $0 = t_0 < t_1 < \ldots < t_m = T$, respectively. A small space-time grid is shown in Fig. 5.2.

Let $S_l = I \times J_l$ be a strip of the slab S, and let W_h be the space of functions that are continuous piecewise linear in space, and constant in time on S_l, that is,

$$W_{hl} = \{v(x,t) : v(\cdot,t)|_I \in V_h,\ t \in J_l,\ v(x,\cdot)|_{J_l} \in P_0(J_l),\ x \in I\} \tag{5.75}$$

where V_h is the usual space of continuous piecewise linears on I, and $P_0(J_l)$ the space of constants on J_l. A space on the whole slab S can then trivially be constructed by the direct sum $W_h = \oplus_{l=1}^m W_{hl}$. In other words, W_h is the space of functions w, such that $w|_{S_l}$ in W_{hl}, for $l = 1, 2, \ldots, m$.

The functions in W_h are generally discontinuous in time across adjacent subintervals J_l and J_{l+1}. At time t_l the jump of $w \in W_h$ is denoted $[w]_l = w_l^+ - w_l^-$ with $w_l^{\pm} = \lim_{\epsilon \to 0} w(t_l \pm \epsilon)$.

Now, by definition, the so-called cG(1)dG(0) method amounts to the time stepping method: find $u_h \in W_h$, such that, for $l = 1, 2, \dots, m$,

$$\int_{J_l} \int_I u_h' v' \, dx dt + \int_I [u_h]_{l-1} v_{l-1}^+ \, dx = \int_{J_l} \int_I f v \, dx dt, \quad \forall v \in W_{hl} \tag{5.76}$$

or, upon writing out the jump,

$$\int_{J_l} \int_I u_h' v' \, dx dt + \int_I (u_h^+ - u_h^-)_{l-1} v_{l-1}^+ \, dx = \int_{J_l} \int_I f v \, dx dt, \quad \forall v \in W_{hl} \tag{5.77}$$

Setting $u_{h,l} = (u_h)_l^-$, we can rewrite this as

$$\int_{J_l} \int_I u_{h,l}' v' \, dx dt + \int_I (u_{h,l} - u_{h,l-1}) v \, dx = \int_{J_l} \int_I f v \, dx dt, \quad \forall v \in W_{hl} \tag{5.78}$$

Because both the trial and test function $u_{h,l}$ and v are constant in time on J_l, it is easy to integrate (5.78) in time, yielding

$$k_l \int_I u_{h,l}' v' \, dx + \int_I (u_{h,l} - u_{h,l-1}) v \, dx = k_l \int_I f v \, dx, \quad \forall v \in V_h \tag{5.79}$$

We recognize this as the Euler backward method. Thus, the nodal values of $u_{h,l}$ at time t_l is obtained by a simple time stepping method.

The very close connection between the cG(1)dG(0) method and the Euler backward method might raise questions why we should use finite elements, and not just finite differences to discretize the Heat equation. The cG(1)dG(0) method is after all more complicated to define and understand, compared to the Euler backward method.

Although semi-discretization in combination with finite differences is both easy to comprehend and implement, the finite element approach has certain advantages since it provide a systematic framework for analysis compared to finite differences. For example, in the finite element setting we know that the solution $u_{h,l}$ is constant in time between time t_{l-1} and t_l, whereas we know nothing about this behavior of $u_{h,l}$ in the finite difference setting. This is due to the fact that $u_{h,l}$ lies in a function space W_h in the former case, but is only defined by a set of point values in the latter case. In the next section we shall use the finite element framework to derive an posteriori error estimate.

In this context we remark that the acronym cG(1) stands for continuous Galerkin of order 1, which is the space approximation. By analogy, dG(0) means discontinuous Galerkin of order 0, which is the time approximation. The cG(1)dG(0) method can be generalized to a cG(1)dG(q) method, which implies using discontinuous piecewise polynomials of order $q \geq 0$ in time.

5.6.1 A Posteriori Error Estimate

The space W_h is a subspace of $W = L^2(J; V)$, which contains all functions $v = v(x,t)$ that are square integrable in time on J, and for any given time belongs to the space $V = \{v : \|v'\|_{L^2(I)} + \|v\|_{L^2(I)} \leq \infty,\ v(0) = v(L) = 0\}$. Thus, we can state the following variational formulation of the Heat equation (5.22): find $u \in W$ such that

$$\int_J \int_I \dot{u} v \, dx dt + \int_J \int_I u' v' \, dx dt = \int_J \int_I f v \, dx dt, \quad \forall v \in W \tag{5.80}$$

Subtracting the cG(1)dG(0) method (5.76) from the variational formulation (5.80) we obtain the Galerkin orthogonality

$$\sum_{l=1}^{m} \int_{J_l} \int_I (\dot{e} v + e' v') \, dx dt + \int_I [e]_{l-1} v_{l-1}^+ \, dx = 0, \quad \forall v \in W_h \tag{5.81}$$

where $e = u - u_h$, as usual, denotes the error.

To derive a goal oriented a posteriori error estimate we next introduce the dual problem

$$-\dot{\phi} - \phi'' = \psi, \quad x \in I, \quad t \in J \tag{5.82a}$$

$$\phi(0,t) = \phi(L,t) = 0 \tag{5.82b}$$

$$\phi(x,T) = 0 \tag{5.82c}$$

This is an unusual problem since the it runs backwards in time, starting at $t = T$ and ending at $t = 0$. This is visible from the reversed sign on the time derivative, and the initial condition, which is given at the final time T. The load for the dual problem is a given weight function $\psi = \psi(x,t)$, which expresses the goal of the computation. Indeed, we let $m(\cdot)$ be the goal functional

$$m(v) = \int_J \int_I \psi v \, dx dt \tag{5.83}$$

Now, integrating by parts in time and space over each interval J_l we have

$$m(e) = \int_J \int_I e\psi \, dxdt \tag{5.84}$$

$$= \sum_{l=1}^{m} \int_{J_l} \int_I e(-\dot{\phi} - \phi'') \, dxdt \tag{5.85}$$

$$= \sum_{l=1}^{m} \int_{J_l} \int_I (\dot{e}\phi + e'\phi') \, dxdt + \int_I [e]_{l-1} \phi_{l-1}^{+} \, dx \tag{5.86}$$

where the jump $[e]_{l-1}$ is due to the partial integration in time. Because u is a continuous function in time $[u]_{l-1} = 0$, and the jump can therefore be simplified to $[e]_{l-1} = [u - u_h]_{l-1} = -[u_h]_{l-1}$. Using the Galerkin orthogonality, we further have

$$m(e) = \sum_{l=1}^{m} \int_{J_l} \int_I (\dot{e}(\phi - \pi\phi) + e'(\phi - \pi\phi)') \, dxdt + \int_I [e]_{l-1} (\phi - \pi\phi)_{l-1}^{+} \, dx \tag{5.87}$$

where $\pi\phi \in W_h$ is the interpolant to ϕ.

For any $v \in W$ it is natural to use the interpolant $\pi v \in W_h$ defined by

$$\pi v = \overline{P_h v} \tag{5.88}$$

where $P_h v$ is the L^2-projection onto V_h, and $\overline{v}|_{J_l}$ is the time average of v on J_l. This interpolant satisfies the interpolation error estimates

$$\|v - \pi v\|_{L^2(J_l)} \leq C k_l \|\dot{v}\|_{L^2(J_l)}, \quad \|\pi v - v\|_{L^2(I_i)} \leq C h_i^2 \|v''\|_{L^2(I_i)} \tag{5.89}$$

and, in particular,

$$\|v - \pi v\|_{L^2(S_{il})} = \left(\int_{J_l} \int_{I_i} (v - \pi v)^2 \, dxdt \right)^{1/2} \leq C \left(k_l \|\dot{v}\|_{L^2(S_{il})} + h_i^2 \|v''\|_{L^2(S_{il})} \right) \tag{5.90}$$

where S_{il} is the space-time element $S_{il} = I_i \times J_l$.

Now, breaking (5.87) into a sum over the n elements I_i, $i = 1, 2, \ldots, n$, and integrating by parts on each of these, we obtain the error representation formula

$$m(e) = \sum_{l=1}^{m} \sum_{i=1}^{n} \int_{J_l} \int_{I_i} (f - \dot{u}_h + u_h'')(\phi - \pi\phi) \, dxdt - \int_I [u_h]_{l-1} (\phi - \pi\phi)_{l-1}^{+} \, dx \tag{5.91}$$

Note that the integration by parts does not give rise to additional jump terms, since $\phi - \pi\phi$ is vanish at the nodes x_i.

Finally, using the Cauchy-Schwarz inequality on each term, and the interpolation error estimates (5.89) and (5.90) we have

$$m(e) \leq C \sum_{l=1}^{m} \sum_{i=1}^{n} \|f - \dot{u}_h + u_h''\|_{L^2(S_{il})} \|\phi - \pi\phi\|_{L^2(S_{il})} \tag{5.92}$$

$$+ \|[u_h]_{l-1}\|_{L^2(I_i)} \|(\phi - \pi\phi)_{l-1}^{+}\|_{L^2(I_i)}$$

$$\leq C \sum_{l=1}^{m} \sum_{i=1}^{n} \|f - \dot{u}_h + u_h''\|_{L^2(S_{il})} \left(k_l \|\dot{\phi}\|_{L^2(S_{il})} + h_i^2 \|\phi''\|_{L^2(S_{il})}\right) \tag{5.93}$$

$$+ \|[u_h]_{l-1}\|_{L^2(I_i)} h_i^2 \|\phi_{l-1}^{+\,\prime\prime}\|_{L^2(I_i)}$$

Thus, we have the following a posteriori error estimate.

Theorem 5.3. *The space-time finite element solution u_h, defined by* (5.76)*, satisfies the estimate*

$$m(e) \leq C \sum_{l=1}^{m} \sum_{i=1}^{n} \eta_{il}(u_h, \phi) \tag{5.94}$$

where the element indicator $\eta_{il}(u_h, \phi)$ is the dot product of the element residual $\rho_i(u_h)$ and the element weight $\omega_l(\phi)$, defined on each space-time element S_{il}, by

$$\rho_i(u_h) = \begin{bmatrix} \|f - \dot{u}_h + u_h''\|_{L^2(S_{il})} \\ \|[u_h]_{l-1}\|_{L^2(I_i)} \end{bmatrix} \tag{5.95}$$

$$\omega_l(\phi) = \begin{bmatrix} k_l \|\dot{\phi}\|_{L^2(S_{il})} + h_i^2 \|\phi''\|_{L^2(S_{il})} \\ h_i^2 \|\phi_{l-1}^{+\,\prime\prime}\|_{L^2(I_i)} \end{bmatrix} \tag{5.96}$$

Note that $\dot{u}_h = u_h'' = 0$ on S_{il} due to the low order of u_h.

Loosely speaking we may interpret $\|f\|_{L^2(S_{il})}$ and $\|[u_h]_{l-1}\|_{L^2(I_i)}$as a spatial and temporal residual, respectively. These can be used to adaptively adjust the mesh size and time step, which gives rise to so-called space-time adaptive finite element methods. In the simplest case, a basic such adaptive method first solves the primal problem, then the dual, and finally the element indicators are computed and the mesh refined. This is repeated a couple of times until satisfactory results have been obtained.

The element indicator η_{il} depends on the dual solution ϕ, which is generally not available but has to be computed using finite elements. Indeed, both the primal and the dual problem must be solved in order to be able to compute the element indicators. Moreover, the size of η_{il} depends on the size of ϕ and its derivatives. The dual is therefore said to contain sensitivity information about the error $m(e)$. That is, how the error at a specific time affects the error at another, later, time. For time dependent problems $m(e)$ depends on the integrated behavior of ϕ during the time interval J. Thus, the error is in a sense dependent on the history of ϕ. Naturally, ϕ can either increase or decrease with time, which signifies that the error amplifies or dampens the longer we run the simulation.

The fact that the error depends on its history is troublesome from the point of view of adaptive mesh refinement. Indeed, if the error in one subdomain $\omega \subset \Omega$ depends on the error in many other subdomains, we may need to refine large portions of Ω to obtain good control over the error. Moreover, if the error spreads rapidly, the mesh may have to be heavily adjusted from one time level to the next. This is very complicated to implement on unstructured meshes as it generally involves refinement and derefinement of already refined elements. Recall that the mesh quality must always be preserved under mesh refinement. A related trouble, which arises when refining the mesh from time t_{l-1} to t_l, is how to interpolate the nodal values of the previous solution $u_h(t_{l-1})$ onto the new mesh. This is often hard to do efficiently.

Obviously, there are much more to say about this topic. However, suffice perhaps for now to say that space-time adaption is still very much an active area of research.

5.7 The Wave Equation

As we have seen the Heat equation quickly diffuses any high gradients in the initial conditions to produce smooth solutions at steady state. This is a typical feature for equations involving just one time derivative. We shall now study what happens if the number of time derivatives in the equation is increased from one to two. It turns out that this seemingly small change dramatically alters the behavior of the solution. Indeed, this modified equation allows for oscillating solutions, and does not have a steady state. It is called the Wave equation.

5.7.1 Derivation of the Acoustic Wave Equation

The Wave equation is a frequently occurring partial differential equation in science and engineering and can be derived in many ways. We derive it from the point of view of acoustics. In this context, the Wave equation describes sound waves propagating in a medium, such as a liquid or a gas.

Let Ω be a domain occupied by a liquid or gas with density ρ, pressure p, and velocity u. From the physical point of view sound is a perturbation of the pressure caused by movement of the particles in the medium. In particular, it is assumed that the force required to move a small volume of matter is balanced out by the build up of a pressure gradient. Under this assumption, Newton's second law (i.e., net force equals mass times acceleration) in differential form yields the equilibrium equation

$$\rho \dot{u} = -\nabla p \tag{5.97}$$

Further, the principle of conservation of energy states that any expansion of the small volume of matter due to the movement will cause a drop in pressure, whereas any contraction will cause a pressure rise. Now, a local measure of volume expansion or contraction is the divergence of the velocity, which suggests the constitutive relation

$$\dot{p} = -k\nabla \cdot u \tag{5.98}$$

where k is a material parameter. Differentiating (5.98) with respect to time t and using (5.97) yields

$$\ddot{p} = -k\nabla \cdot \dot{u} = k\nabla \cdot \frac{\nabla p}{\rho} \tag{5.99}$$

which, assuming ρ and k are constant, simplifies to

$$\ddot{p} = c^2 \Delta p \tag{5.100}$$

where $c^2 = k/\rho$. This is the acoustic Wave equation.

The boundary conditions for the Wave equation (5.100) are the same as for any equation involving the Laplacian Δp, and can be of either Dirichlet, Neumann, or Robin type. However, since this equation also involves the second order time derivative $\ddot{p}$, two initial conditions are required, namely, one for p and one for $\dot{p}$.

Perhaps needless to say, the solution p to the Wave equation looks like a wave. Hence, its name.

5.7.2 Model Problem

Let us consider the model Wave equation

$$\ddot{u} - c^2 \Delta u = f, \quad \text{in } \Omega \times J \tag{5.101a}$$

$$u = 0, \quad \text{on } \partial\Omega \times J \tag{5.101b}$$

$$u = u_0, \quad \text{in } \Omega, \text{ for } t = 0 \tag{5.101c}$$

$$\dot{u} = v_0, \quad \text{in } \Omega, \text{ for } t = 0 \tag{5.101d}$$

where $J = (0, T]$ is the time interval, c^2 is a parameter, f is a given source function, and u_0 and v_0 given initial conditions.

5.7.3 Variational Formulation

Multiplying $f = \ddot{u} - c^2 \Delta u$ by a test function $v \in V_0 = \{v : \|v\| + \|\nabla v\| < \infty,\ v|_{\partial\Omega} = 0\}$, and integrating using Green's formula we obtain the variational formulation of (5.101): find u such that, for every fixed $t \in J$, $u \in V_0$ and

$$\int_\Omega \ddot{u} v \, dx + c^2 \int_\Omega \nabla u \cdot \nabla v \, dx = \int_\Omega f v \, dx, \quad \forall v \in V_0, \quad t \in J \tag{5.102}$$

5.7.4 Spatial Discretization

Let $V_{h,0} \subset V_0$ be the usual space of continuous piecewise linears on a mesh of Ω. The space discrete counterpart of (5.102) reads: find u_h such that, for every fixed $t \in J$, $u_h \in V_{h,0}$ and

$$\int_\Omega \ddot{u}_h v \, dx + c^2 \int_\Omega \nabla u_h \cdot \nabla v \, dx = \int_\Omega f v \, dx, \quad \forall v \in V_{h,0}, \quad t \in J \tag{5.103}$$

We note that (5.103) is equivalent to

$$\int_\Omega \ddot{u}_h \varphi_i \, dx + c^2 \int_\Omega \nabla u_h \cdot \nabla \varphi_i \, dx = \int_\Omega f \varphi_i \, dx, \quad i = 1, 2, \dots, n_i, \quad t \in J \tag{5.104}$$

where φ_i, $i = 1, 2, \dots, n_i$ are the usual hat basis functions for $V_{h,0}$ and n_i the number of interior nodes.

As before, we make the space discrete ansatz

$$u_h = \sum_{j=1}^{n_i} \xi_j(t) \varphi_j \tag{5.105}$$

where $\xi_j(t)$ are n_i time-dependent coefficients to be determined.

Substituting (5.105) into (5.104) we have

$$\sum_{j=1}^{n_i} \ddot{\xi}_j(t) \int_\Omega \varphi_j \varphi_i \, dx + c^2 \sum_{j=1}^{n_i} \xi_j(t) \int_\Omega \nabla \varphi_j \cdot \nabla \varphi_i \, dx$$
$$= \int_\Omega f \varphi_i \, dx, \quad i = 1, 2, \dots, n_i, \quad t \in J \tag{5.106}$$

We recognize this as an $n_i \times n_i$ system of ODE s

$$M \ddot{\xi}(t) + c^2 A \xi(t) = b(t), \quad t \in J \tag{5.107}$$

where M, A, and $b(t)$ are the usual mass matrix, stiffness matrix, and load vector, respectively.

5.7.5 Time Discretization

Looking at the ODE system (5.107) we see that it is of second order, which poses a new problem as all our finite difference time stepping methods are designed for first order systems only. The solution for this problem is to introduce a new variable $\eta = \dot{\xi}$ and rewrite the second order system as two first order systems. In doing so, we end up with the new ODE system

$$M\dot{\xi}(t) = M\eta(t) \tag{5.108}$$

$$M\dot{\eta}(t) + c^2 A\xi(t) = b(t) \tag{5.109}$$

Application of, say, the Crank-Nicolson method to each of these two ODE s gives us

$$M\frac{\xi_l - \xi_{l-1}}{k_l} = M\frac{\eta_l + \eta_{l-1}}{2} \tag{5.110}$$

$$M\frac{\eta_l - \eta_{l-1}}{k_l} + c^2 A\frac{\xi_l + \xi_{l-1}}{2} = \frac{b_l + b_{l-1}}{2} \tag{5.111}$$

In block matrix form, we write this

$$\begin{bmatrix} M & -\frac{k_l}{2}M \\ \frac{c^2 k_l}{2}A & M \end{bmatrix} \begin{bmatrix} \xi_l \\ \eta_l \end{bmatrix} = \begin{bmatrix} M & \frac{k_l}{2}M \\ -\frac{c^2 k_l}{2}A & M \end{bmatrix} \begin{bmatrix} \xi_{l-1} \\ \eta_{l-1} \end{bmatrix} + \begin{bmatrix} 0 \\ \frac{k_l}{2}(b_l + b_{l-1}) \end{bmatrix} \tag{5.112}$$

Here, ξ_0 and η_0 can be chosen as the nodal interpolants of u_0 and v_0, for example.

The reason for choosing the Crank-Nicolson time stepping method is that it is more accurate than the Euler methods, and that it has the property of conserving energy, which loosely speaking means that the computed solution will not get numerically smeared out. Thus, it is a suitable method for the Wave equation.

We summarize the Crank-Nicolson method for solving the Wave equation in the following algorithm:

5.7.6 Conservation of Energy

In the absence of external forces, or damping, the solution u to the Wave equation (5.101) is a traveling wave, which move back and forth over the domain eternally, and although the wave may disperse, the energy content (i.e., the sum of kinetic and potential energy) of the initial condition is not diminished. This is the content of the following stability estimate.

Algorithm 15 The Crank-Nicolson method for solving the Wave equation

Create a mesh of Ω with n_i interior nodes, and use it to define the space of all continuous piecewise linear functions $V_{h,0}$, with hat function basis $\{\varphi_i\}_{i=1}^{n_i}$.

Create a partition $0 = t_0 < t_1 < \cdots < t_m = T$ of the time interval J with m time steps $k_l = t_l - t_{l-1}$.

Choose ξ_0 and η_0.

for $l = 1, 2, \dots, m$ **do**

  Compute the $n_i \times n_i$ mass and stiffness matrices M and A, and the $n_i \times 1$ load vector $b_l = b_l(t)$, with entries

$$M_{ij} = \int_\Omega \varphi_j \varphi_i \, dx, \quad A_{ij} = \int_\Omega \nabla\varphi_j \cdot \nabla\varphi_i \, dx, \quad b_i = \int_\Omega f(t_l)\varphi_i \, dx \tag{5.113}$$

  Solve the linear system

$$\begin{bmatrix} M & -\frac{k_l}{2}M \\ \frac{c^2 k_l}{2}A & M \end{bmatrix} \begin{bmatrix} \xi_l \\ \eta_l \end{bmatrix} = \begin{bmatrix} M & \frac{k_l}{2}M \\ -\frac{c^2 k_l}{2}A & M \end{bmatrix} \begin{bmatrix} \xi_{l-1} \\ \eta_{l-1} \end{bmatrix} + \begin{bmatrix} 0 \\ \frac{k_l}{2}(b_l + b_{l-1}) \end{bmatrix} \tag{5.114}$$

end for

Theorem 5.4. *With* $f = 0$ *the solution* u *to the Wave equation* (5.101) *satisfies the estimate*

$$\|\dot{u}(\cdot,t)\|^2 + \|\nabla u(\cdot,t)\|^2 = C \tag{5.115}$$

with constant C *independent of time* t.

Proof. Choosing $v = \dot{u}$ in the variational formulation (5.102) we have

$$0 = \int_\Omega \ddot{u}\dot{u} \, dx + \int_\Omega \nabla u \cdot \nabla \dot{u} \, dx \tag{5.116}$$

$$= \int_\Omega \tfrac{1}{2}\partial_t(\dot{u})^2 \, dx + \int_\Omega \tfrac{1}{2}\partial_t(\nabla u)^2 \, dx \tag{5.117}$$

$$= \tfrac{1}{2}\partial_t \left(\|\dot{u}\|^2 + \|\nabla u\|^2\right) \tag{5.118}$$

Integrating this result with respect to to time t from 0 to T we further have

$$\|\dot{u}(\cdot,T)\|^2 + \|\nabla u(\cdot,T)\|^2 = \|v_0\|^2 + \|\nabla u_0\|^2 \tag{5.119}$$

The proof ends by noting that the right hand side is independent of time t. □

The fact that the solution is wave like is a characteristic feature of transient so-called hyperbolic type differential equations.

5.8 Computer Implementation

A MATLAB code for solving the Wave equation is given below. The problem under consideration is $\ddot{u} - \Delta u = 0$ in a domain Ω composed of a square with two smaller rectangular strips added on one side. This domain is shown in Fig. 5.3. The boundary conditions are $u = 0.1\sin(8\pi t)$ on the line segments $\Gamma_D = \{x : x_1 = -0.25\}$, and $n \cdot \nabla u = 0$ on the rest of the boundary Γ_N. Thus, we have both Dirichlet and Neumann boundary conditions. The initial condition u_0 is zero.

This problem set up corresponds to a situation where coherent light in the form of a sine wave impinges on a screen with two narrow slits. This creates so-called interference on the other side of the screen. This is due to the fact that the distance traveled by the wave from the two slits is different. Indeed, as the light passes the screen the waves from the two sources are in phase. However, as we move away from the screen, the path traveled by the light from one slit is larger than that traveled by the light from the other slit. When this difference in path is equal to half a wavelength, the waves extinguish each other and the amplitude of their sum vanish. Similarly, when the difference in path length is equal to a wavelength, the waves interact to enhance each other.

```
function CNWaveSolver2D()
g = Dslitg(); % double slit geometry
h = 0.025; % mesh size
k = 0.005; % time step
T = 2; % final time
[p,e,t] = initmesh(g,'hmax',h);
np = size(p,2); % number of nodes
x = p(1,:)'; y = p(2,:)'; % node coordinates
fixed = find(x < -0.24999); % Dirichlet nodes
xi = zeros(np,1); % set zero IC
eta = zeros(np,1);
[A,M,b] = assema(p,t,1,1,0); % assemble A, M, and b
for l = 1:round(T/k) % time loop
  time = l*k;
  LHS = [M -0.5*k*M;  0.5*k*A M]; % Crank-Nicholson
  rhs = [M  0.5*k*M; -0.5*k*A M]*[xi; eta] ...
        + [zeros(np,1); k*b];
  sol = LHS\rhs;
  xi = sol(1:np);
  eta = sol(np+1:end);
  xi(fixed) = 0.1*sin(8*pi*time); % set BC the ugly way
  pdesurf(p,t,xi), axis([-1 1 -1 1 -0.25 0.5]) % plot
  pause(0.1)
end
```

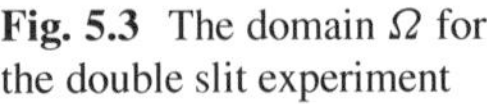

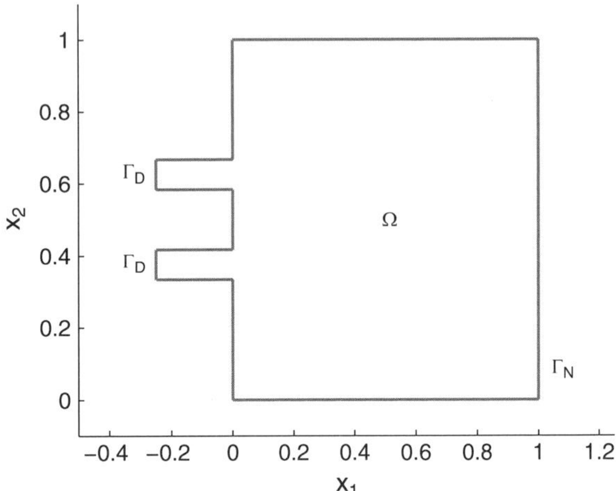

Fig. 5.3 The domain Ω for the double slit experiment

Here, the enforcement of the Dirichlet boundary condition perhaps requires some commenting. Because these are time-dependent they have to be enforced inside the time loop at every time level. We choose to do this in a rather quick and dirty way. Indeed, we first solve the linear system resulting from the CN method using Neumann boundary conditions (i.e., no boundary conditions). Then we apply the Dirichlet boundary conditions to the resulting solution. This is done for each time level.

The geometry matrix for the double slit domain is given by the routine `Dslit` listed in the Appendix.

Figure 5.4 shows a few snapshots of the amplitude of the light wave. The evolution of an interference pattern is clearly seen.

5.9 Further Reading

For more information on numerical methods for ODE s and their properties we refer to the introductory book by Heath [41] or the advanced books by Hairer et al. [40] Regarding time dependent partial differential equations a nice and mathematically accessable text is Strikwerda [67], which uses finite differences in both space and time. Mathematically more advanced books with emphasis on using finite elements in space and finite differences in time include those of Larsson and Thomeé [70], Valli and Quarteroni [72], and Knabner and Angermann [47].

Adaptive space-time finite elements are described and discussed by Bangerth and Rannacher [5].

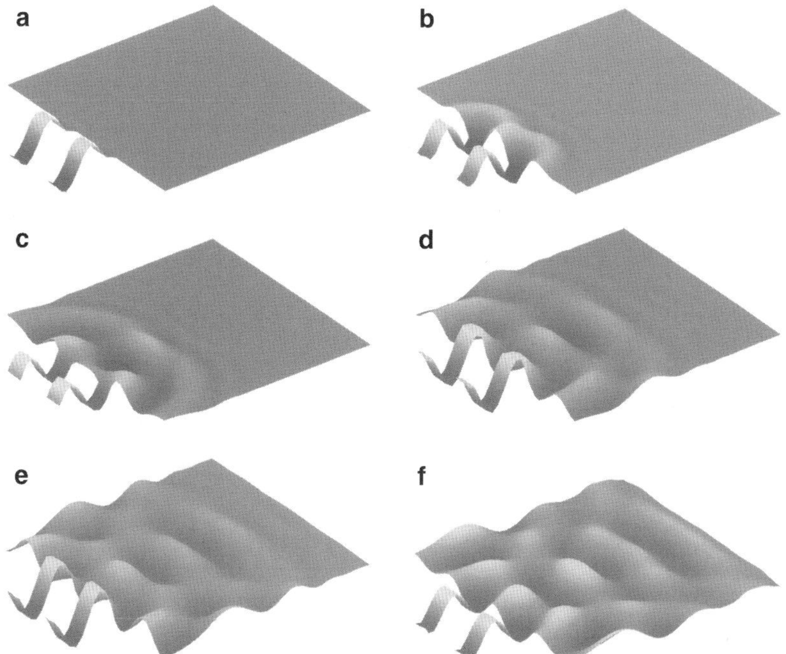

Fig. 5.4 Amplitude of finite element solution u_h at a various times. The interference pattern is clearly visible. (**a**) $t = 0.1$. (**b**) $t = 0.4$. (**c**) $t = 0.7$. (**d**) $t = 1.0$. (**e**) $t = 1.3$. (**f**) $t = 1.6$

5.10 Problems

Exercise 5.1. Explain the difference between implicit and explicit time stepping.

Exercise 5.2. Make two iterations using backward Euler on the ODE system

$$\dot{\xi}(t) + A\xi(t) = f, \quad t > 0, \quad \xi(0) = \xi_0,$$

where

$$A = \begin{bmatrix} 1 & 0 \\ 0 & 2 \end{bmatrix}, \quad f = \begin{bmatrix} 0 \\ -1 \end{bmatrix}, \quad \xi_0 = \begin{bmatrix} 1 \\ 1 \end{bmatrix}$$

Assume time step $k = 1/2$.

Exercise 5.3. Show that a space discretization of the problem

$$\begin{aligned} \dot{u} - \Delta u + u &= f, \quad x \in \Omega,\ t > 0 \\ u &= 0, \quad x \in \partial\Omega,\ t > 0 \\ u &= u_0, \quad x \in \Omega,\ t = 0 \end{aligned}$$

leads to a system of ODE s of the form

$$M\dot{\xi}(t) + A\xi(t) + M\xi(t) = b(t)$$

Identify the entries of the involved matrices and vectors.

Exercise 5.4. Show that the homogeneous Heat equation $\dot{u} - \Delta u = 0$, with homogeneous Dirichlet boundary conditions, and initial condition $u(x,0) = u_0(x)$ satisfies the stability estimate

$$\|u(\cdot,t)\| \le C\|u_0\|$$

Interpret this result. *Hint:* Multiply by u and integrate.

Exercise 5.5. Modify `BackwardEulerHeatSolver1D` to solve the heat problem

$$\dot{u} = \tfrac{1}{10}u'', \quad 0 < x < 1, \quad t > 0$$
$$u(0) = u(1) = 0$$
$$u(x,0) = x(1-x)$$

Use a mesh with 100 elements, final time 0.1 and timestep 0.001. Plot the finite element solution at each time step. Compare with the exact solution, given by the infinite sum

$$u(x,t) = \frac{4}{\pi^3}\sum_{n=1}^{\infty}\frac{(-1)^n - 1}{n^2}e^{-n^2\pi^2 t/10}\sin(n\pi x)$$

You can truncate the sum after, say, 25 terms.

Exercise 5.6. Show that the Ritz projector $R_h u$ satisfies the estimate $\|e'\| \le Ch\|u''\|$, where $e = u - R_h u$. *Hint:* Start from $\|e'\|^2 = \int_I e'^2\,dx$ and write $e = u - \pi u - \pi u - R_h u$, where $\pi u \in V_{h,0}$ is the usual node interpolant of u. Then use the definition of the Ritz projector, the Cauchy-Schwarz inequality $\int_I e'(u-\pi u)'\,dx \le \|e'\|\|(u-\pi u)'\|$, and a standard interpolation estimate.

Chapter 6
Solving Large Sparse Linear Systems

Abstract In the previous chapters we have seen how finite element discretization gives rise to linear systems, which must be solved in order to obtain the discrete solution. The size of these linear systems is generally large, as it is direct proportional to the number of nodes in the mesh. Indeed, it is not unusual to have millions of nodes in large meshes. This puts high demands on the linear algebra algorithms and software that is used to solve the linear systems in terms of computational complexity (i.e., number of floating point operations), memory requirements, and time consumption. To cope with these problems it is necessary and important to exploit the fact that these linear systems are sparse, which means that they have very few non-zero entries as compared to their size. This is due to the fact that the finite element basis functions have very limited support and only interact with their nearest neighbours. In this chapter we review some of the most common direct and iterative methods for solving large sparse linear systems. We emphasize that the aim is not to present and analyze these methods rigorously in any way, but only to give an overview of them and their connection to finite elements.

6.1 Linear Systems

We consider the problem of solving the linear system

$$Ax = b \tag{6.1}$$

where A is a given $n \times n$ matrix, b is a given $n \times 1$ vector, and x the sought $n \times 1$ solution vector. The assumption is that n is large, say, 10^6, and that A is sparse. A sparse matrix is somewhat vaguely defined as one with very few non-zero entries A_{ij}. A prime example of such a matrix is the stiffness matrix resulting from finite element discretization of the Laplace operator $-\Delta$. If A is invertible, which is the case when the underlying differential equation is well posed, the solution x to (6.1) can formally be found by first computing the inverse A^{-1} of A, and then multiplying

M.G. Larson and F. Bengzon, *The Finite Element Method: Theory, Implementation, and Applications*, Texts in Computational Science and Engineering 10,
DOI 10.1007/978-3-642-33287-6_6,

it with b to yield $x = A^{-1}b$. However, this requires the computation of the $n \times n$ matrix A^{-1}, which is usually impossible and wasteful, since our goal is to find the vector x and not the inverse of A. Indeed, it is almost never necessary to compute any matrix inverse to solve a linear system.

There are two main classes of solution methods for linear systems, namely:

- Direct methods.
- Iterative methods.

In addition to this, there are special methods, such as multigrid, for instance.

In the following we shall quickly describe these methods.

6.2 Direct Methods

6.2.1 Triangular Systems, Backward and Forward Substitution

Direct methods refers to Gaussian elimination, or, LU factorization, and variants hereof. The common feature of direct methods is that the solution x is obtained after a fixed number of floating point operations. The basic idea exploits the fact that triangular linear systems are easy to solve and that almost any square matrix can be written as the product of two triangular matrices. Indeed, given A the LU factorization factors it into

$$A = LU \tag{6.2}$$

where L and U are lower and upper triangular $n \times n$ matrices, respectively. If we allow permutation of A, then it can be shown that the LU factorization exist for all square invertible matrices.

For a symmetric and positive definite (SPD) A, U and L can be made to coincide, which reduces (6.2) to

$$A = LL^T \tag{6.3}$$

This is the so-called Cholesky factorization.

Once the left hand side matrix A has been factorized the solution x can be determined from the right hand side b by first solving

$$Ly = b \tag{6.4}$$

for y, and then

$$Ux = y \tag{6.5}$$

for x. This can be cheaply and efficiently done by forward and backward substitution, respectively.

6.2.2 Fill-ins, Graphs, and Symbolic and Numeric Factorization

The solution procedure described above is the same for both dense and sparse matrices. However, the LU factorization for sparse matrices is much more complicated than for dense matrices. The main complication is due to the fact that the structure of the L and U factors, that is, the locations of any non-zero entries, is different as compared to A. Matrix entries which are zero in A, but non-zero in L or U are generally called fill-ins. Because there are usually many fill-ins, they are a major concern from the computational point of view. Indeed, fill-in requires a larger amount of memory and cause a more expensive factorization. In this context a key observation is that the amount of fill-in depends on the structure of A. By permuting the rows and columns of A, L and U can be made more or less sparse. It is therefore important to find fill-in reducing permutations or reorderings. Row permutation means reordering the equations, while column permutation means reordering the unknowns x_i, $i = 1, 2, \ldots, n$. As there are $2n!$ ways to permute a general $n \times n$ matrix, finding the optimal reordering is practically hopeless for large n, and various heuristic methods must be used. The universal tool for predicting and reducing the amount of fill-in are graphs.

A graph $G(A)$ of a sparse symmetric matrix A consists of a set of vertices V and a set of edges E connecting a pair of vertices. The vertex set corresponds to the column or row indices, whereas the edge set corresponds to the non-zero matrix entries in the sense that there is an edge between vertex i and j if $A_{ij} \neq 0$. For example, the matrix

$$A = \begin{bmatrix} 1 & x & & & & & x \\ & 2 & x & x & & & \\ x & x & 3 & & x & & \\ & x & & 4 & & & \\ & & x & & 5 & x & x \\ & & & & x & 6 & x \\ x & & & & x & x & 7 \end{bmatrix} \tag{6.6}$$

has the graph shown in Fig. 6.1. A cross or a number indicate a non-zero matrix entry.

Let us try to figure out how the graph of a sparse matrix can be of help during Cholesky factorization. For simplicity, let us restrict attention to SPD matrices, since this enables us to treat the structural properties of matrices separate from their numerical ones. Indeed, LU factorization of indefinite matrices is complicated by the fact that pivoting (i.e., reordering of equations and unknowns) is needed for numerical stability. This mixes the structure of the matrix with the actual numerical values of its entries. Besides, as we have seen, SPD matrices occur in many important applications.

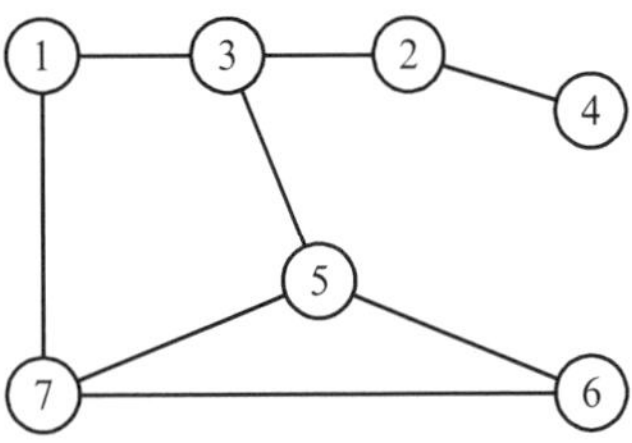

Fig. 6.1 Graph of A, given by (6.6), with vertices (◯) and edges (-). An edge between vertex i and j means that matrix entry A_{ij} is non-zero

Cholesky factorization is perhaps best described step by step. Let A be an $n \times n$ SPD matrix, set $A = A_0 = C_0$, and write C_0 as

$$C_0 = \begin{bmatrix} d_1 & c_1^T \\ c_1 & \bar{C}_1 \end{bmatrix} \tag{6.7}$$

where the scalar d_1 is the first diagonal entry of C_0, the $n \times 1$ vector c_1 is the first column of C_0, except d_1, and the $(n-1) \times (n-1)$ submatrix $\bar{C}_1$ the rest of C_0. By direct multiplication it is straight forward to show that C_0 affords the factorization

$$C_0 = \begin{bmatrix} \sqrt{d_1} & 0 \\ c_1/\sqrt{d_1} & I_{n-1} \end{bmatrix} \begin{bmatrix} 1 & 0 \\ 0 & \bar{C}_1 - c_1 c_1^T/d_1 \end{bmatrix} \begin{bmatrix} \sqrt{d_1} & c_1^T/\sqrt{d_1} \\ 0 & I_{n-1} \end{bmatrix} \tag{6.8}$$

$$= L_1 \begin{bmatrix} 1 & 0 \\ 0 & C_1 \end{bmatrix} L_1^T \tag{6.9}$$

$$= L_1 A_1 L_1^T \tag{6.10}$$

where I_m is the $m \times m$ identity matrix. Continuing the factorization with C_1, which is also SPD, we find

$$A_1 = \begin{bmatrix} 1 & 0 \\ 0 & C_1 \end{bmatrix} \tag{6.11}$$

$$= \begin{bmatrix} 1 & 0 & 0 \\ 0 & d_2 & c_2^T \\ 0 & c_2 & \bar{C}_2 \end{bmatrix} \tag{6.12}$$

$$= \begin{bmatrix} 1 & & \\ 0 & \sqrt{d_2} & \\ 0 & c_2/\sqrt{d_2} & I_{n-2} \end{bmatrix} \begin{bmatrix} 1 & 0 & 0 \\ 0 & 1 & 0 \\ 0 & 0 & \bar{C}_2 - c_2 c_2^T/d_2 \end{bmatrix} \begin{bmatrix} 1 & 0 & 0 \\ & \sqrt{d_2} & c_2^T/\sqrt{d_2} \\ & & I_{n-2} \end{bmatrix} \tag{6.13}$$

$$= L_2 A_2 L_2^T \tag{6.14}$$

$$\vdots$$

$$A_{n-1} = L_n I_n L_n^T \tag{6.15}$$

Hence, after n steps we obtain the Cholesky factorization

$$A = (L_1 L_2 \dots L_n) I_n (L_n^T \dots L_2^T L_1^T) = L L^T \tag{6.16}$$

Here, it can be shown that column $k = 1, 2, \dots, n$ of L is precisely column k of L_k.

From the point of view of fill-in the structure of A_k is important, since this is the matrix we have to store at stage k of the Cholesky factorization process. (In principle, the memory space for A is reused so that the vectors c_k are stored in the first k columns, and the matrix A_k in the remaining $n-k$ columns.) Because $\bar{C}_k$ is a submatrix of A_{k-1}, it can not contribute with any new fill-ins. However, the update $c_k c_k^T$ might do so. Structurally, the non-zeros of c_k are the neighbours of vertex k in the graph of A_k, which implies that the non-zeros of $c_k c_k^T$ is the combination of all these neighbours. If any of these non-zeros do not already exist new fill-ins will be created.

Computing L_k is equivalent to eliminating vertex k in the graph of A. This amounts to first deleting vertex k and all its adjacent edges and then adding edges to the graph between any two vertices that were adjacent to vertex k. Starting with the first vertex and eliminating all subsequent vertices one-by-one gives rise to a sequence of smaller and smaller graphs called elimination graphs. This is illustrated in Fig. 6.2, which shows the elimination graphs $G(A_k)$, $k = 0, 1, 2, 3$, for our example matrix A, defined by (6.6).

Clearly, the elimination graphs $G(A_k)$, $k = 4, 5, 6, 7$, can not give rise to any fill-ins.

Collecting all fill-ins from the elimination graphs we obtain the location of the non-zero entries L_{ij} of the Cholesky factor L. For our example matrix, we have

$$L = \begin{bmatrix} L_{11} & & & & & & \\ & L_{22} & & & & & \\ L_{31} & L_{32} & L_{33} & & & & \\ & L_{42} & L_{43} & L_{44} & & & \\ & & L_{53} & L_{54} & L_{55} & & \\ L_{61} & & & & L_{65} & L_{66} & \\ L_{71} & & L_{73} & L_{74} & L_{75} & L_{76} & L_{77} \end{bmatrix} \tag{6.17}$$

where the fill-ins are colored red.

The procedure performed above for finding fill-ins can be described more theoretically using the concept of a fill-path. A fill-path is a path (i.e., a sequence of consecutive edges) between two vertices i and j in the graph of A, such that all vertices in the path, except i and j, are numbered less than i and j. The fill-path is closely connected to the fill-in through the next theorem.

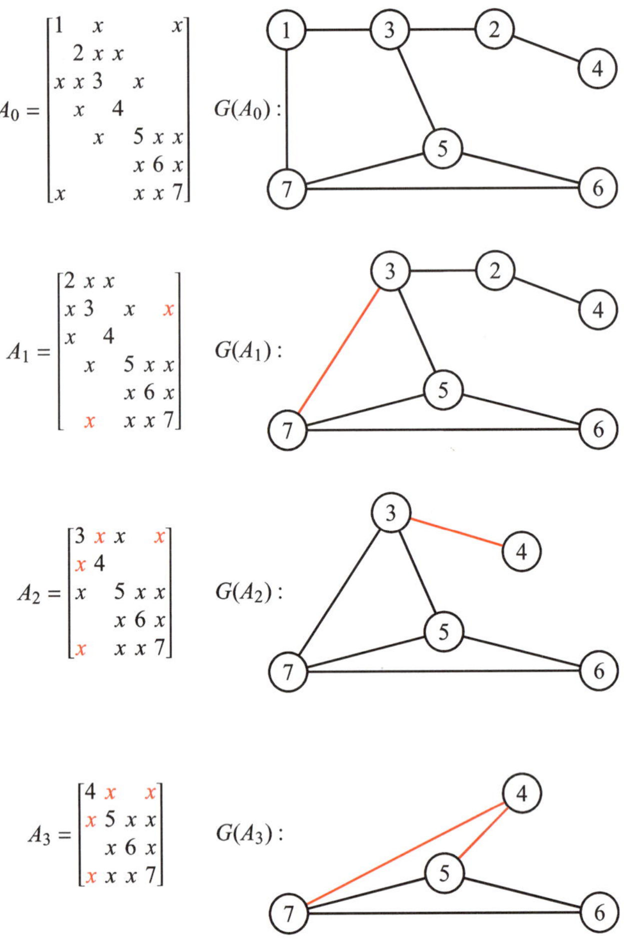

Fig. 6.2 Elimination graphs for A given by (6.6). New edges and corresponding fill-ins are colored *red*

Theorem 6.1 (Rose-Tarjan). *There will be fill-in in entry L_{ij} of the Cholesky factor of the matrix A if and only if there is a fill-path between vertex i and j in the graph of A.*

Thus, for instance, inspecting Fig. 6.1 we see that there is a path $p = \{7, 1, 3, 2, 4\}$ in $G(A)$ from vertex 7 to 4 via vertices 1, 3, and 2. Because 1, 3, and 2 are less than both 7 and 4, p is a fill-path. Consequently, there should be a fill-in at A_{74} or L_{74}. Indeed, this fill-in is created at stage 3 in the elimination process. By contrast, there is no fill-path from vertices 6 to 4, say.

Finding the structure of the Cholesky factor L is commonly referred to as symbolic factorization, since it only involves symbolic operations and no numerical computations. Once the symbolic factorization is complete the numerical factorization (i.e., the actual computation of the non-zero entries L_{ij}) can begin. The numerical factorization is done either by the right-looking, the left-looking, or the multifrontal method. The left and right-looking methods compute one column of L at a time using the other columns to either the left or right, whereas the multifrontal method compute submatrices of L that have the same sparsity pattern. Because it is more efficient to operate on matrices than on vectors, the multifrontal method performs better than the left- and right-looking methods. However, it is harder to implement. For simplicity, we shall focus on the right-looking method. Actually, we have already defined this method when computing the Cholesky factor L in the step by step way we did, which lead up to (6.16). However, the right-looking method can also be recast as a nested triple loop. Indeed, for a dense $n \times n$ matrix A we have the following algorithm.

Algorithm 16 Dense right-looking Cholesky factorization

for $k = 1$ to n **do**
  $A_{kk} = \sqrt{A_{kk}}$
  for $i = k + 1$ to n **do**
    $A_{ik} = A_{ik}/A_{kk}$
    for $j = k + 1$ to n **do**

$$A_{ij} = A_{ij} - A_{ik}A_{kj} \tag{6.18}$$

    end for
  end for
end for

Here, upon completion L has overwritten the lower triangle of A.

Let us next introduce the two macros div(j) and add(j,k), meaning

- div(j) – divide column j by the square root of its diagonal.
- Add(j,k) – add into column j a multiple of column k, with $k < j$.

Using these, we have the sparse counterpart of Algorithm 16.

Algorithm 17 Sparse right-looking Cholesky factorization

for $k = 1$ to n **do**
  div(k)
  for j such that $L_{jk} \neq 0$ **do**
    add(j,k)
  end for
end for

Looking at Algorithm 17, it is clearly important to have access to the structure of L to do the inner loop efficiently. However, as we know, this is precisely the information available from the elimination graphs $G(A_k)$. Indeed, it is the justification for setting up these.

6.2.3 *Matrix Reorderings*

In order to save computer memory and execution time it is highly desirable to minimize the amount of fill-in created during the Cholesky factorization. To this end, the graph $G(A)$ is analyzed for obtaining a good permutation of the matrix A. In the language of graphs this amounts to relabeling the vertices of $G(A)$. In doing so, two of the most common algorithms are the so-called Minimum Degree, and Nested Dissection orderings.

6.2.3.1 Minimum Degree

The degree of a vertex in a graph is the number of connecting edges (i.e., neighbouring vertices).

The minimum degree ordering originates from the observation that the largest possible fill-in when eliminating vertex i is the product of the non-diagonal non-zeros within row and column i. That is, the square of the degree of vertex i. This naturally leads to the idea that by always eliminating the vertex, or one of the vertices, with minimum degree in the elimination graph, we will get a small fill-in.

In practice, it turns out that the performance of the minimum degree ordering is often best for problems of moderate size. However, it is not guaranteed to minimize fill-in. This has to do with the fact that the vertex to eliminate is chosen based only on the information available at a specific step of the elimination process. Indeed, in step k we search only elimination graph $G(A_k)$ for the this next vertex. Because of this, we say that minimum degree is a local reordering strategy.

Also, minimum degree ordering is quite hard to implement efficiently in software.

Application of the minimum degree reordering strategy to our example matrix A in (6.6) implies that vertex 4 should be eliminated in the first elimination step, since this vertex has degree one, whereas all other vertices have degree two, in the graph $G(A)$. Indeed, eliminating the vertices in the order, say, 4, 2, 1, 3, 5, 6, and 6, the only fill-in in the Cholesky factor L occurs at L_{73}. This is easy to see by drawing the elimination graphs. Hence, with this minimum degree ordering we get a single fill-in, instead of four as with the natural ordering 1 to 7.

MATLAB supports a variant of the minimum degree reordering through the command `symamd`. Using this we can permute a random symmetric sparse matrix by typing

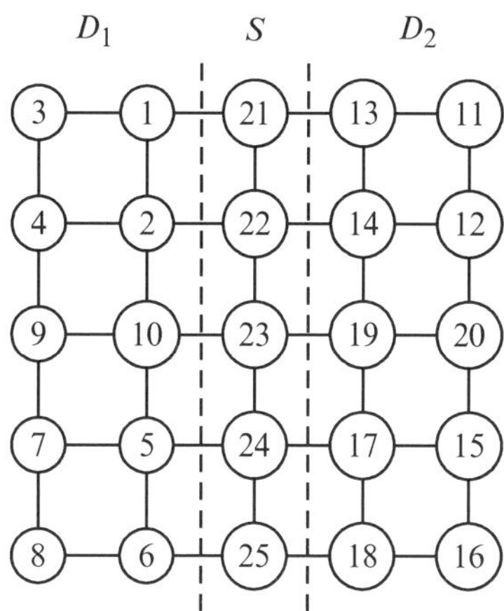

Fig. 6.3 Matrix graph dissected using nested dissection into two subgraphs $D_i, i = 1, 2$, with separator S

```
n=100;
A0=sprandsym(n,0.01); % random symmetric matrix
perm=symamd(A0); % minimum degree permutation
Ap=A0(perm,perm); % permuted matrix
```

We can use `spy` to view the structure of the matrix before and after reordering.

6.2.3.2 Nested Dissection

The nested dissection matrix reordering strategy tries to reorder the matrix A, so that the fill-in is kept within certain matrix blocks in the Cholesky factor L.

To this end, the graph of A is dissected into smaller subgraphs by so-called separators. A separator S between two subgraphs D_1 and D_2 is the set of vertices containing all paths between D_1 and D_2. The rationale for making this dissection is that there can not be any fill-in L_{ij} with vertex i in D_1 and j in D_2. For example, consider the graph $G(A)$ shown in Fig. 6.3.

Here, the vertices has been labeled with nested dissection. The vertices in the center column are the separator S, and the two columns of vertices to the left and right of S are the subgraph D_1, and D_2, respectively. In this case it is easy to find a separator due to the grid like stricture of the graph. Note that the numbering of the vertices is such that those of D_1 are less than those of D_2, which in turn are less than those of S. This is a characteristic feature of nested dissection reorderings.

Making a symbolic Cholesky factorization of the matrix A we obtain a factor L with the structure shown in Fig. 6.4.

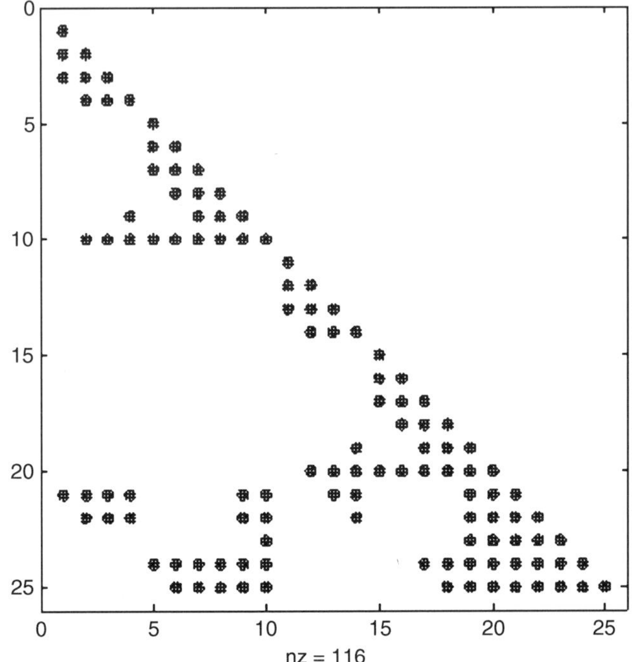

Fig. 6.4 Structure of Cholesky factor L

From this figure we see that L has the block structure

$$L = \begin{bmatrix} L_{11} & 0 & 0 \\ 0 & L_{22} & 0 \\ L_{S1} & L_{S2} & L_{SS} \end{bmatrix} \tag{6.19}$$

where the diagonal blocks L_{ii}, $i = 1, 2$, and L_{SS} stems from the vertex sets D_i and S, and the off-diagonal blocks L_{Si} stems from the edge intersection of these sets. Thus, due to the nested dissection any fill-in must occur in L_{ii} or L_{SS}. Clearly, this limits the number of fill-ins.

The nested dissection procedure exemplified above is usually repeated recursively, yielding a Cholesky factor L with a more fine grained block structure than that of (6.19).

In the example above it was kind of obvious how to choose the separator S due to the grid like nature of the graph. Indeed, this graph stems from a stiffness matrix assembled on a quadrilateral mesh of a square. However, on unstructured (e.g., triangular) meshes, it is generally very difficult to choose good (i.e., small) separators. In fact, this is the most complicated task when implementing nested dissection in software. Also, for finite element applications, instead of partitioning the graph of the stiffness matrix, it is common to partition and renumber the elements and nodes of the mesh. This is done using a so-called mesh partitioner, such as the software METIS by Karypis and Kumar [46], for example.

Nested dissection often performs at its best for large problems. Due to its recursive substructuring of a matrix into submatrices it is usually combined with the multifrontal factorization method. Loosely speaking, a primitive multifrontal method uses five dense Cholesky factorizations to compute the matrix blocks L_{ii}, L_{Si} and L_{SS} of (6.19) as follows:

1. Compute L_{ii} from $L_{ii}L_{ii}^T = A_{ii}$, for $i = 1, 2$.
2. Compute L_{Si} from $L_{ii}L_{Si}^T = A_{Si}^T$, for $i = 1, 2$.
3. Update $A_{SS} = A_{SS} - \sum_{i=1}^{2} L_{Si}L_{Si}^T$.
4. Compute L_{SS} from $L_{SS}L_{SS}^T = A_{SS}$.

We remark that nested dissection is a global reordering strategy, unlike the minimum degree, which, as we know, is a local ditto.

6.3 Iterative Methods

Loosely speaking, discretization of one- and two-dimensional problems yields linear systems that are small enough to be efficiently solved with direct methods. However, three-dimensional problems usually leads to linear systems that are too large and expensive for direct methods. Instead, cheaper iterative methods must be utilized. Unlike direct methods, iterative methods do not have a fixed number of floating point operations attached to them for computing the solution x to a linear system. Instead, a sequence of approximations $x^{(k)}$ is sought successively, such that $x^{(k)} \to x$ in the limit $k \to \infty$. Of course, the unspoken hope is that convergence in one way or another will be reached with only a small number of iterations.

6.3.1 *Basic Iterative Methods*

It is actually quite simple to create a framework for a set of basic iterative methods. To this end consider, again, the linear system $Ax = b$, and let us split A into

$$A = M - K \tag{6.20}$$

where M is any non-singular matrix, and K the remainder $K = M - A$. This gives

$$(M - K)x = b \tag{6.21}$$

$$Mx = Kx + b \tag{6.22}$$

$$x = M^{-1}Kx + M^{-1}b \tag{6.23}$$

Now, if we have a starting guess $x^{(0)}$ for x, this suggests the iteration

$$x^{(k+1)} = M^{-1}Kx^{(k)} + M^{-1}b \tag{6.24}$$

Although we do not know for which linear systems the iteration scheme (6.24) converges, if any, let us tacitly take it as our basic iterative method for solving linear systems. Thus, we have the following algorithm.

Algorithm 18 Basic iterative method for solving a linear system

---

1: Choose a staring guess $x^{(0)}$.
2: **for** $k = 0, 1, 2, \ldots$ until convergence **do**
3:

$$x^{(k+1)} = M^{-1}Kx^{(k)} + M^{-1}b \tag{6.25}$$

4: **end for**

For this algorithm to be computationally practical, it is important that the splitting of A is chosen such that $M^{-1}K$ and $M^{-1}b$ are easy to calculate, or at least their action on any given vector. Recall that we do not want to compute any inverses.

Choosing, for example, $M = \alpha^{-1}I$ and $K = \alpha^{-1}I - A$, with $\alpha > 0$ a constant, we obtain the simple so-called Richardson iteration

$$x^{(k+1)} = (I - \alpha A)x^{(k)} + \alpha b \tag{6.26}$$

which can be written as

$$x^{(k+1)} = x^{(k)} + \alpha r^{(k)} \tag{6.27}$$

where $r^{(k)} = b - Ax^{(k)}$ is the residual at iteration k. The residual $r^{(k)}$ can be viewed as a correction to $x^{(k)}$ and α as a relaxation parameter. This is a typical construction of the next iterate $x^{(k+1)}$.

A naive MATLAB implementation of Richardson iteration takes only a few lines of code.

```
function x = Richardson(A,x,b,maxit,alpha)
for k=1:maxit
    x=x+alpha*(b-A*x);
end
```

Here, `maxit` denotes the maximum number of iterations.

In the following we shall study splittings of A of the form

$$A = D - U - L \tag{6.28}$$

where D is the diagonal of A, and $-U$ and $-L$ the strictly upper and lower triangular part of A, respectively. This leads to two classical iterative methods, known as the Jacobi and the Gauss-Seidel methods.

6.3.2 The Jacobi Method

Jacobi iteration is defined by choosing $M = D$ and $K = L + U$, which gives the iteration scheme

$$x^{(k+1)} = D^{-1}(L + U)x^{(k)} + D^{-1}b \tag{6.29}$$

We observe that D is easy to invert, since it is a diagonal matrix.

6.3.3 The Gauss-Seidel Method

In the Gauss-Seidel method $M = D - L$ and $K = U$, which gives the iteration scheme

$$x^{(k+1)} = (D - L)^{-1}(Ux^{(k)} + b) \tag{6.30}$$

We observe that since $D-L$ has a lower triangular structure, the effect of $(D-L)^{-1}$ can be computed by forward elimination.

6.3.4 Convergence Analysis

We now return to the question of convergence of our basic iterative method. By inspection, we see that it can be rewritten as

$$x^{(k+1)} = Rx^{(k)} + c \tag{6.31}$$

where $R = M^{-1}K$ is called the relaxation matrix, and $c = M^{-1}b$.

Let $e^{(k)} = x - x^{(k)}$ be the error after k iterations. A relation between the errors in successive approximations can be be derived by subtracting $x = Rx + c$ from (6.31). Indeed, we have

$$e^{(k+1)} = x^{(k+1)} - x = R(x^{(k)} - x) = \ldots = R^{k+1}(x^{(0)} - x) = R^{k+1}e^{(0)} \tag{6.32}$$

Here, for convergence, we require $R^{k+1}e \to 0$ as $k \to \infty$ for any e. It turns out that this requirement is equivalent to $\rho(R) < 1$, where $\rho(R) = \max_{1\le k\le n} |\lambda_k(R)|$ is the so-called spectral radius of R, that is, the magnitude of the extremal eigenvalue of R.

Moreover, $\rho(R) < 1$ implies that there exist a consistent matrix norm $\|R\|$, such that $\|R\| < 1$. Conversely, if $\|R\| < 1$ in any such matrix norm, then $\rho(R) < 1$.

Now, taking norms and using the Cauchy-Schwarz inequality, we have

$$\|e^{(k+1)}\| \leq \|R^{k+1}\| \, \|e^{(0)}\| \leq \|R\|^{k+1} \|e^{(0)}\| \tag{6.33}$$

From this it readily follows that $\|R\|$ should be as small as possible, since it is the amplification factor for the error in each iteration. Hence, the splitting of A should preferably be chosen such that

- $Rx = M^{-1}Kx$ and $x = M^{-1}b$ are easy to evaluate.
- $\rho(R)$ is small.

Unfortunately, these goals are contradictory, and a balance has to be struck. For example, choosing

- $M = I$ makes M^{-1} trivial, but we probably do not have $\rho(A - I) < 1$.
- $M = A$ gives $K = 0$, and, thus, $\rho(R) = \rho(M^{-1}K) = 0$, but then $M^{-1} = A^{-1}$, which is very expensive to compute.

For the Jacobi and Gauss-Seidel methods it is possible to state more general convergence criterions than just $\rho(R) < 1$, but before we do that, let us pause for a moment to introduce the concept of a diagonally dominant matrix.

An $n \times n$ matrix A is said to be (strictly) diagonally dominant if, for each row, the absolute value of the diagonal entry is larger than the sum of the absolute values of the other entries. That is, if

$$|A_{ii}| > \sum_{j=1, j\neq i}^{n} |A_{ij}|, \quad \forall i = 1, 2, \ldots, n \tag{6.34}$$

For example, the matrix

$$A = \begin{bmatrix} 4 & 1 & 0 \\ -2 & -5 & 1 \\ 6 & 0 & -7 \end{bmatrix} \tag{6.35}$$

is diagonally dominant.

We have following convergence criteria.

Theorem 6.2.

- *Jacobi's method converges if A is strictly diagonally dominant.*
- *The Gauss-Seidel method converges if A is SPD.*

Proof. Let us prove the first part of the theorem as the second part is somewhat technical.

In Jacobi's method, the relaxation matrix R has the entries

$$R_{ij} = \frac{A_{ij}}{A_{ii}}, \quad i \neq j, \quad R_{ii} = 0 \tag{6.36}$$

Taking the infinity norm of R we have, by definition,

$$\|R\|_\infty = \max_{1 \leq i \leq n} \sum_{j=1,\ j \neq i}^{n} \frac{|A_{ij}|}{|A_{ii}|} \tag{6.37}$$

which shows that $\|R\|_\infty < 1$, and, thus, also $\rho(R) < 1$, provided that A is strictly diagonally dominant. □

We remark that if A is non-singular, but unsymmetric or indefinite, then it is possible to apply the Gauss-Seidel method to the normal equations

$$A^T A x = A^T b \tag{6.38}$$

In this case, since $A^T A$ is SPD, the iteration scheme will converge. However, the rate of convergence may be very slow.

6.3.5 Projection Methods

The basic iterative methods are cheap and easy to implement, but generally slow to converge. This has lead to the development of faster and smarter iterative methods, which are based on the requirement that the residual $r = b - Ax$ should be orthogonal to subspaces of $\mathbb{R}^n$. This is analogous to the finite element method, which requires the residual of a partial differential equation to be orthogonal to the finite element space V_h. Indeed, modern iterative methods for linear systems share many features with finite elements.

Suppose we seek a solution approximation $\tilde{x}$ to $Ax = b$ from a (small) m-dimensional subspace $K \subset \mathbb{R}^n$, such that the residual

$$r = b - A\tilde{x} \tag{6.39}$$

is orthogonal to another m-dimensional subspace $L \subset \mathbb{R}^n$, that is,

$$w^T r = w^T (b - A\tilde{x}) = 0, \quad \forall w \in L \tag{6.40}$$

By analogy with finite elements the subspace K is called trial space, and the subspace L is called test space.

In case we have a starting guess x_0 for x, the solution must be sought in the space

$$x_0 + K \tag{6.41}$$

Thus, we have

$$\tilde{x} = x_0 + d \tag{6.42}$$

where d is some vector in K. Our problem is then to find $\tilde{x} \in x_0 + K$ such that

$$w^T r = w^T (b - A(x_0 + d)) = w^T (r_0 - Ad) = 0, \quad \forall w \in L \tag{6.43}$$

where we have introduced the initial residual $r_0 = b - Ax_0$.

Suppose that $V = [v_1, v_2, \dots, v_m]$ and $W = [w_1, w_2, \dots, w_m]$ are two $n \times m$ matrices whose columns $\{v_i\}_{i=1}^m$ and $\{w_i\}_{i=1}^m$ form a basis for K and L, respectively. Then, we can write

$$\tilde{x} = x_0 + d = x_0 + Vy \tag{6.44}$$

for some $m \times 1$ vector y to be determined. The orthogonality of r against L means that

$$w^T (r_0 - AVy) = 0, \quad \forall w \in L \tag{6.45}$$

and since W is a basis for L this is equivalent to

$$W^T (r_0 - AVy) = 0 \tag{6.46}$$

or

$$W^T AVy = W^T r_0 \tag{6.47}$$

Hence, if the $m \times m$ matrix $W^T AV$ can be inverted then we end up with

$$\tilde{x} = x_0 + Vy = x_0 + V(W^T AV)^{-1} W^T r_0 \tag{6.48}$$

Now, there are two instances when it is certain that $W^T AV$ can be inverted, namely:

- If A is SPD, and $L = K$.
- If A is non-singular, and $L = AK$.

From the point of view of finite elements this is reminiscent of the Galerkin method, in which the trial and test space is the same, and the so-called GLS stabilization technique, in which the test space is the trial space plus a piece of the differential operator applied to the trial space.

In practice, (6.48) is iterated. In each iteration a new pair of subspaces K and L is used, with the initial guess x_0 equal to the solution approximation from the previous iterate.

6.3.6 One-Dimensional Projection Methods

The simplest choice is to let the trial and test space K and L to be one-dimensional.

$$K = \text{span}\,\{v\}, \qquad W = \text{span}\,\{w\} \tag{6.49}$$

where v and w are two given $n \times 1$ vectors. This yields

$$\tilde{x} = x_0 + \alpha v \tag{6.50}$$

where the scalar α, by virtue of (6.48), equals

$$\alpha = \frac{w^T r_0}{w^T A v} \tag{6.51}$$

Further, choosing v and w to be equal to the current residual r in an iteration yields the so-called Steepest Descent algorithm.

Algorithm 19 Steepest descent

```
1: Choose a starting guess x^(0).
2: for k = 0, 1, 2, ... until convergence do
3:     r^(k) = b - A x^(k)
4:     α^(k) = r^(k)^T r^(k) / r^(k)^T A r^(k)
5:     x^(k+1) = x^(k) + α^(k) r^(k)
6: end for
```

Since $L = K$ steepest decent works in case A is SPD.

Other choices of v and w include $v = r$ and $w = Ar$, which leads to the so-called Minimal Residual algorithm (MINRES).

One-dimensional projection methods are simple, but generally not very efficient.

6.3.7 Krylov Subspaces

The most important iterative methods for sparse linear systems use projection onto so-called Krylov subspaces.

The m-th Krylov subspace $K_m(A; v) \subset \mathbb{R}^n$ is defined by

$$K_m(A; v) = \text{span}\{v, Av, A^2 v, \ldots, A^{m-1} v\} \tag{6.52}$$

where A is a given $n \times n$ matrix and v is a given $n \times 1$ vector. We say that A and v generates K_m.

Let us try to motivate why the Krylov subspaces are defined as they are. Consider a linear system with

$$A = \begin{bmatrix} 5 & 1 \\ 0 & 2 \end{bmatrix}, \qquad b = \begin{bmatrix} 20 \\ 10 \end{bmatrix} \tag{6.53}$$

The so-called characteristic polynomial $p(\lambda)$ of A is defined by

$$p(\lambda) = \det(A - \lambda I) = \lambda^2 - 7\lambda + 10 \tag{6.54}$$

The roots of the polynomial $p(\lambda)$ are the eigenvalues of A. Now, according to a result called the Cayley-Hamilton theorem, a matrix satisfies its characteristic equation. That is, $p(A) = 0$, or in our case

$$0 = A^2 - 7A + 10I \tag{6.55}$$

Thus, multiplying with A^{-1} and rearranging terms, we end up with

$$A^{-1} = \tfrac{7}{10}I - \tfrac{1}{10}A \tag{6.56}$$

Hence, we have

$$x = A^{-1}b = (\tfrac{7}{10}I - \tfrac{1}{10}A)b = \tfrac{7}{10}b - \tfrac{1}{10}Ab = \begin{bmatrix} 3 \\ 5 \end{bmatrix} \tag{6.57}$$

Here, the key observation is that the solution x to $Ax = b$ is a linear combination of the vectors b and Ab, which make up the Krylov subspace $K_2(A; b)$. In other words, the solution to $Ax = b$ has a natural representation as a member of a Krylov space, and therefore we may understand why one would construct approximations to x from this space. The hope is that a good approximation to x can be found in K_m with a space dimension m small compared to the matrix dimension n.

Because the Krylov vectors $\{A^j v\}_{j=0}^{m-1}$ tend very quickly to become almost linearly dependent, methods relying on Krylov subspaces frequently involve some orthogonalization procedure for numerical stability. The most general of these is the Arnoldi procedure, which is an algorithm for building an orthonormal basis $\{q_j\}_{j=1}^{m}$ to $K_m(A; v)$.

Algorithm 20 Arnoldi's Orthogonalization procedure

Choose a vector v and set $q_1 = v/\|v\|$
for $j = 1, 2, \dots, m$ **do**
    Compute $z = Aq_j$
    for $i = 1, 2, \dots, j$ **do**
        $H_{ij} = q_i^T z$
        $z = z - H_{ij} q_i$
    end for
    $H_{j+1j} = \|z\|$
    if $H_{j+1j} = 0$ **then**
        break
    end if
    $q_{j+1} = z/H_{j+1j}$
end for

Here, at stage j, the basic idea is to first multiply the last Arnoldi vector q_j by A and then orthogonalize $z = Aq_j$ against all the other Arnoldi vectors q_i, $i = 1, 2, \ldots, j$ using the well-known Gram-Schmidt method. Indeed, inspecting the algorithm, we see that z is a linear combination of the Arnoldi vectors q_i, $i = 1, 2, \ldots j + 1$. The coefficients in this linear combination are the numbers H_{ij}.

A MATLAB realization of the Arnoldi algorithm is given below.

```
function [Q,H] = Arnoldi(A,q,m)
n=size(A,1);
Q=zeros(n,m+1);
H=zeros(m+1,m);
Q(:,1)=q/norm(q);
for j=1:m
    z=A*Q(:,j);
    for i=1:j
        H(i,j)=dot(z,Q(:,i));
        z=z-H(i,j)*Q(:,i);
    end
    H(j+1,j)=norm(z);
    if H(j+1,j)==0, break, end
    Q(:,j+1)=z/H(j+1,j);
end
```

Loosely speaking, the Arnoldi procedure factorizes the matrix A. At stage m of the Arnoldi algorithm, it computes the factorization

$$AQ_m = Q_{m+1}\bar{H}_m \tag{6.58}$$

where

$$Q_m = \begin{bmatrix} q_1 \; q_2 \; \ldots \; q_m \end{bmatrix} \tag{6.59}$$

is the $n \times m$ orthonormal matrix containing the Arnoldi vectors q_j, $j = 1, 2, \ldots, m$, as columns, and $\bar{H}_m$ is the $(m + 1) \times m$ upper Hessenberg matrix

$$\bar{H}_m = \begin{bmatrix} H_{11} & H_{12} & H_{13} & H_{14} & \ldots & H_{1m} \\ H_{21} & H_{22} & H_{23} & H_{24} & & H_{2m} \\ 0 & H_{32} & H_{33} & & & H_{3m} \\ 0 & 0 & H_{43} & & & H_{4m} \\ \vdots & & & \ddots & & \vdots \\ 0 & 0 & \ldots & 0 & H_{m+1m-1} & H_{m+1m} \end{bmatrix} \tag{6.60}$$

By definition, upper Hessenberg matrices have all zero entries below their first subdiagonal. The factorization (6.58) is sometimes written

$$AQ_m = Q_m H_m + H_{m+1m} e_m q_{m+1}^T \tag{6.61}$$

where H_m is the $m \times m$ matrix obtained by deleting the last row from $\bar{H}_m$, and e_m is column m of the $m \times m$ identity matrix I_m. Because the columns of Q_m are orthonormal it is easy to confirm that

$$Q_m^T A Q_m = H_m \tag{6.62}$$

In case A is symmetric Arnoldi's algorithm simplifies, and is then called Lanczos algorithm.

6.3.8 *The Conjugate Gradient Method*

Combining the prototype projection method (6.48) with the Krylov space $K_m(A; v)$ we can derive the Conjugate Gradient (CG) method, which is the most famous Krylov method.

Given a linear system $Ax = b$ with A SPD, and a starting guess x_0 for its solution, let us consider a projection method with similar test and trial space $L = K = K_m(A, r_0)$, where r_0 is the initial residual. First, to generate a basis Q_m for $K_m(A, r_0)$ we do m steps of the Arnoldi procedure with the first Arnoldi vector chosen as $q_1 = r_0/\|r_0\|$. Then, substituting $V = W = Q_m$ into the left hand side of (6.47) we have, by virtue of (6.62),

$$W^T A V = Q_m^T A Q_m = H_m \tag{6.63}$$

where H_m is the $m \times m$ Hessenberg matrix. Further, setting $\beta = \|r_0\|$ and substituting $W = Q_m$ into the right hand side of (6.47) we have, since all columns of Q_m except the first are orthogonal against r_0,

$$Q_m^T r_0 = Q_m^T(\beta q_1) = \beta e_1 \tag{6.64}$$

where the $m \times 1$ vector $e_1 = [1, 0, \ldots, 0]^T$. Thus, (6.47) reduces to

$$H_m y_m = \beta e_1 \tag{6.65}$$

and, as a result, the approximate solution x_m in (6.48) is given by

$$x_m = x_0 + Q_m y_m = x_0 + Q_m H_m^{-1} \beta e_1 \tag{6.66}$$

Now, the accuracy of x_m depends on the dimension m of the Krylov space, and, of course, A and r_0. In practice, we would like to be able to improve x_m by choosing m dynamically during the iteration process. This is possible and leads to the following algorithm called the Full Orthogonalization Method (FOM), which is mathematically equivalent to the CG method.

Algorithm 21 The full Orthogonalization method

```
Choose a starting guess x_0.
for m = 1, 2, 3, ... until convergence do
    β = ||r_0||
    Compute Q_m by doing m steps of Arnoldi's procedure.
    Solve H_m y_m = β e_1.
    x_m = x_0 + Q_m y_m
end for
```

In FOM all m Arnoldi vectors q_j, $j = 1, 2, \ldots, m$ are needed for computing x_m, and after m iterations the cost of storing Q_m is therefore mn, which may be too expensive for large n. A remedy for this is to periodically restart FOM after a given number of iterations with the previous solution approximation x_m as initial guess x_0. The restarting of a Krylov method is a common technique for saving memory.

As a by-product of the Arnoldi process we obtain a way of computing the residual residual r_m at a low cost. Indeed, from (6.61) we have

$$
\begin{aligned}
r_m &= b - Ax_m && (6.67)\\
&= r_0 - AQ_m y_m && (6.68)\\
&= \beta e_1 - Q_m H_m y_m - H_{m+1m} e_m^T y_m q_{m+1} && (6.69)\\
&= -H_{m+1m} e_m^T y_m q_{m+1} && (6.70)
\end{aligned}
$$

Thus, the old residual r_m is a multiple of the new Arnoldi vector q_{m+1}. This yields

$$\|r_m\| = H_{m+1m} |e_m^T y_m| \tag{6.71}$$

which is a simple and important result, since the norm of r_m is used to decide when to terminate the iteration.

Because A is symmetric the Hessenberg matrix H_m becomes tridiagonal, and the linear system $H_m y_m = \beta e_1$ can be solved efficiently using LU factorization. This can be further exploited by noting that the difference in the L_m and U_m factors of H_m between successive iterates $m-1$ and m lies only in the last rows and columns to derive short recurrences between the involved vectors and matrices. In particular, the current solution can be updated as $x_m = x_{m-1} + \alpha_m p_m$, where α_m is a scalar and p_m a so-called search direction. Similar three term recursions hold for the residuals r_m and search directions p_m. The search directions are A-conjugate, which means that $p_i^T A p_j = 0, i \neq j$. Hence, the name conjugate gradients. All in all, we get the very concise and elegant CG algorithm.

Algorithm 22 The Conjugate Gradient method

Choose a starting guess $x^{(0)}$.
Compute the initial residual $r^{(0)} = b - Ax^{(0)}$, and set $p^{(1)} = r^{(0)}$.
for $m = 1, 2, 3, \dots$ until convergence **do**
  $t^{(m)} = Ap^{(m)}$
  $\alpha^{(m)} = r^{(m-1)^T} r^{(m-1)} / p^{(m)^T} t^{(m)}$
  $x^{(m)} = x^{(m-1)} + \alpha^{(m)} p^{(m)}$
  $r^{(m)} = r^{(m-1)} - \alpha^{(m)} t^{(m)}$
  $\beta^{(m+1)} = r^{(m)^T} r^{(m)} / r^{(m-1)^T} r^{(m-1)}$
  $p^{(m+1)} = r^{(m)} + \beta^{(m+1)} p^{(m)}$
end for

This is a cheap algorithm both regarding computational complexity and memory requirements, since it only requires:

- One matrix-vector multiplication $t_m = Ap_m$ per iteration.
- Storage of four vectors, and not all the m Arnoldi vectors.

In theory, the CG method should converge to the solution after at most n iterations, since the Krylov space then span all of $\mathbb{R}^n$. In practice, however, this is not so due to the finite precision of computer arithmetic. In fact, rounding errors can quickly destroy the orthogonality of the involved Arnoldi vectors. The simple remedy for this is to make more iterations.

The rate of convergence of the CG method depends on the condition number $\kappa(A)$ of A. It is a tedious task to show that the error $e_m = x - x_m$ decreases as

$$\frac{\|e_m\|_A}{\|e_0\|_A} \leq 2 \left(\frac{\sqrt{\kappa(A)} - 1}{\sqrt{\kappa(A)} + 1} \right)^m \tag{6.72}$$

where $\|\cdot\|_A$ denotes the energy norm $\|v\|_A = \sqrt{v^T A v}$. Now, this may look like bad news considering the fact that, as we might recall, $\kappa(A) \leq Ch^{-2}$, with h the minimum mesh size, and A the usual stiffness matrix. With h small we may end up with virtually no convergence as the quotient $(\sqrt{\kappa} - 1)/(\sqrt{\kappa} + 1)$ is close to unity. However, this is often too pessimistic as the problem with slow convergence can, at least partially, be overcome by proper so-called preconditioning. Also, from (6.72) it follows that the number of iterations m needed to get $\|e_m\|_A / \|e_0\|_A$ within a prescribed tolerance $\epsilon > 0$ is given by

$$m \leq \log(2/\epsilon) \sqrt{\kappa(A)} \leq Ch^{-1} \tag{6.73}$$

In MATLAB the CG method is implemented as a black box solver with syntax

```
x=pcg(A,b,tol,maxit)
```

Input is the system matrix `A`, the right hand side vector `b`, a desired tolerance `tol`, and the maximum number of iterations `maxit`. Output in the simplest case is the solution approximation `x`.

6.3.9 The Generalized Minimal Residual Method

Conjugate gradients only works for linear systems with A SPD. This is somewhat limiting and we shall therefore next study another Krylov method, which works also for A indefinite and unsymmetric.

The Generalized Minimum Residual method (GMRES) is a projection method based on taking $K = K_m(A; r_0)$ and $L = AK_m(A; r_0)$. This choice of trial and test space implies that the solution approximation x_m minimizes the residual norm $\|r_m\|$ over $x_0 + K_m(A; r_0)$. Thus, GMRES is a least squares method.

To derive GMRES, first recall that any vector x in $x_0 + K_m(A; r_0)$ can be written as

$$x = x_0 + Q_m y \tag{6.74}$$

for some $m \times 1$ vector y to be determined. Then, defining the least squares functional

$$J(y) = \|b - Ax_m\|^2 = \|b - A(x_0 + Q_m y_m)\|^2 \tag{6.75}$$

the property (6.58) of $\bar{H}_m$ implies

$$\begin{aligned} b - Ax &= b - A(x_0 + Q_m y) && (6.76)\\ &= r_0 - AQ_m y && (6.77)\\ &= \beta q_1 - Q_{m+1}\bar{H}_m y && (6.78)\\ &= Q_{m+1}(\beta e_1 - \bar{H}_m y) && (6.79) \end{aligned}$$

Also, the orthonormality of Q_m implies

$$J(y) = \|b - A(x_0 + Q_m y)\|^2 = \|\beta e_1 - \bar{H}_m y\|^2 \tag{6.80}$$

Now, the GMRES solution approximation x_m is defined as the $n \times 1$ vector $x_m = x_0 + Q_m y_m$, where the $m \times 1$ vector y_m minimizes the least squares functional $J(y)$ over $x_0 + K_m(A; r_0)$. This minimizer is inexpensive to compute, since it only requires the solution of an $(m+1)\times m$ linear least squares problem. Indeed, y_m is the solution to the normal equations $\bar{H}_m^T \bar{H}_m y_m = \bar{H}_m^T \beta e_1$. However, as it is inefficient to compute the matrix product $\bar{H}_m^T \bar{H}_m$ other methods from dense linear algebra are used to efficiently compute y_m. A common method is QR decomposition involving so-called Givens rotations.

Summarizing, we have the following algorithm for GMRES.

Algorithm 23 The generalized minimum residual method

Compute $r_0 = b - Ax_0$ and set $q_1 = r_0/\|r_0\|$.
for $m = 1, 2, 3, \ldots$ until convergence **do**
    Compute Q_m with Arnoldi's process.
    Compute y_m, the minimizer of $J(y) = \|\beta e_1 - \bar{H}_m y\|$.
    $x_m = x_0 + Q_m y_m$
end for

6.3.10 Other Krylov Methods

There are many Krylov methods. For example, there is the CG on the Normal Equations (CGNE) method, which solves $A^T Ax = A^T b$ using conjugate gradients. The rationale for doing so is that the matrix A need not be SPD, but only invertible. However, CGNE has poor convergence due to the squared condition number $\kappa(A^T A) = \kappa(A)^2$. Also, there is the Bi-Conjugate Gradient (BiCG) method, which stems from choosing the test space L as $K_m(A^T; \cdot)$. This method is cheaper than GMRES, but its convergence is erratic and break down often occurs. To remedy these shortcomings there is the CG Squared (CGS) method, which avoids multiplication with A^T in BiCG, and the Quasi Minimal Residual method (QMR), which has a smoother convergence behavior than BiCG.

6.4 Preconditioning

As we have seen the convergence rate of many iterative methods depend on the spectrum (i.e., eigenvalues) of the system matrix A. Thus, to accelerate this rate, it is tempting to try to transform the linear system $Ax = b$ into one that has better spectral properties and can be solved with fewer number of iterations. This is generally accomplished through so-called preconditioning. A preconditioner M is a matrix that approximates A in some sense, but is easier to construct and invert.

Multiplying $Ax = b$ by M^{-1} from the left we have the transformed linear system

$$M^{-1}Ax = M^{-1}b \tag{6.81}$$

which, obviously, has the same solution as $Ax = b$, but whose matrix $M^{-1}A$ may be better from the numeric point of view (e.g., a smaller spectral radius). Loosely speaking, if M is a good preconditioner, then $M^{-1}A$ is approximately the identity I, with eigenvalues clustered around one, or at least away from zero, and a condition number $\kappa(M^{-1}A)$ close to one.

In constructing a good preconditioner a balance has to be struck between the cost of computing it, and its efficiency per iteration. For example, $M = I$ is a useless preconditioner, while $M = A$ is the most expensive as it means inverting A. For linear systems of equations with A constant throughout the iteration process it usually pays off to use a sophisticated and expensive preconditioner, whereas cheaper and maybe more primitive preconditioning is most economic to use for non-linear systems with $A = A(x)$, since A changes in each iteration. The cost of computing a preconditioner must always be amortized by frequent application.

Preconditioners for finite element applications are generally of two types. First, there are problem specific preconditioners, which are constructed based on information about the particular differential equation under consideration. For example, M may be defined as the stiffness matrix A assembled on a coarse mesh or using finite elements of lower order (e.g., piecewise constants). Second, there are black box preconditioners, which need only the matrix entries of A for their construction. Examples of such preconditioners include the Jacobi and the incomplete LU preconditioner.

We remark that it is generally not necessary to form the matrix $M^{-1}A$ explicitly, since this would be too expensive and lead to loss of symmetry. Instead, the iterative methods are rewritten to incorporate the effect of the preconditioner M through matrix-vector multiplications with A and solutions of linear systems of the form $Mz = r$.

6.4.1 Jacobi Preconditioning

The simplest black box preconditioner consists of just the diagonal of the matrix A, that is, $M = \text{diag}(A)$. This is known as Jacobi preconditioning and can successfully be applied to linear systems with A SPD. The Jacobi preconditioner needs no additional storage, and is very easy to implement. However, it is usually quite inefficient on real-world problems.

6.4.2 Polynomial Preconditioners

If B is a matrix with spectral radius $\rho(B) < 1$, then the series

$$(I - B)^{-1} = \sum_{j=0}^{\infty} B^j \tag{6.82}$$

converges and gives a simple representation of the inverse of $I - B$. By choosing

$$B = I - A/\omega \tag{6.83}$$

with $\omega > 0$ a big enough constant we can not only keep $\rho(B) < 1$, but also get $A^{-1} = \omega(I - (I - A/\omega))^{-1} = (I - B)^{-1}$. Further, by truncating the sum (6.82) after just a few terms we obtain a simple approximate inverse of A that can be used as the preconditioner M^{-1}. This is called polynomial preconditioning, and works for definite matrices A.

6.4.3 Incomplete Factorizations

An advanced and modern black box preconditioner is the so-called incomplete LU factorization (ILU). The idea underlying ILU factorization is surprisingly simple. When computing the ordinary LU factorization $A = LU$, entries of L and U that are deemed too small are discarded to save memory. Hence, the name incomplete LU factorization. The preconditioner M is then defined by $M = LU$. This type of preconditioning has proven to be efficient in combination with GMRES. The difficulty is to choose a good drop tolerance, that is, the level below which the matrix entries of L and U or M are ignored. A high drop tolerance yields a dense M, while a low drop tolerance makes M inefficient. A trial and error approach is usually used to decide on a good drop tolerance. However, there is no guarantee that a high drop tolerance per say leads to a more efficient preconditioner. In fact, the preconditioner may get worse before getting better. This has to do with the fact that the fill-in of L and U occur in a more or less random fashion, and it may happen that $M = LU$ is a too crude approximation of A. Because of this ILU preconditioners are sometimes known to be unstable. Various strategies have been proposed to automatically choose the drop tolerance, sparsity pattern, and, thus, fill-in of M.

For SPD matrices ILU is known as incomplete Cholesky factorization (IC). It is commonly used in combination with the CG method. For low values of the the drop tolerance it can be shown that the condition number of $M^{-1}A$ is $\kappa(M^{-1}A) = Ch^{-2}$, that is, similar to that of A, but with a smaller constant C. However, for a particular factorization known as modified incomplete Cholesky (MIC) this condition number can be improved to $\kappa(M^{-1}A) = Ch^{-1}$. This dramatically speeds up the CG method. The basic idea behind MIC is that the row sums of A and M should be the same. Loosely speaking this amounts to adding the discarded fill-in of M to the diagonal of M.

MATLAB has two built-in routines called `luinc` and `cholinc` for computing incomplete LU and Cholesky factorizations, respectively. For example, to solve a linear system using GMRES with ILU preconditioning we type

```
[L,U]=luinc(A,1.e-3); % drop tolerance=0.001
x=gmres(A,b,[],tol,m,L,U);
```

6.5 Towards Multigrid

6.5.1 The Smoothing Property

Many basic iterative methods have a so-called smoothing property, which loosely speaking means that fast frequencies of the error is damped faster than slow frequencies. In order to study this property in more detail we consider again the simple Richardson iteration

$$x^{(k+1)} = (I - \Lambda^{-1}A)x^{(k)} + \Lambda^{-1}b \tag{6.84}$$

where Λ is an upper bound on the spectral radius of A. That is, $\lambda_i \leq \Lambda$ for all eigenvalues λ_i, $i = 1, 2, \dots, n$, of A. For simplicity, let us assume that A is SPD.

Then, following our convergence analysis for basic iterative methods, we have the error equation for $e^{(k)} = x - x^{(k)}$

$$e^{(k)} = R^k e^{(0)} \tag{6.85}$$

where R is the relaxation matrix

$$R = I - \frac{1}{\Lambda}A \tag{6.86}$$

Now, letting η_i denote the eigenvectors of A with corresponding eigenvalues λ_i and writing

$$e^{(0)} = \sum_{i=1}^{n} e_i^{(0)} \eta_i \tag{6.87}$$

we get

$$e^{(k)} = R^k \sum_{i=1}^{n} e_i^{(0)} \eta_i \tag{6.88}$$

$$= \sum_{i=1}^{n} e_i^{(0)} R^k \eta_i \tag{6.89}$$

$$= \sum_{i=1}^{n} e_i^{(0)} \left(1 - \frac{\lambda_i}{\Lambda}\right)^k \eta_i \tag{6.90}$$

Thus, we arrive at the identity

$$e_i^{(k)} = e_i^{(0)} \left(1 - \frac{\lambda_i}{\Lambda}\right)^k, \quad i = 1, 2, \dots, n \tag{6.91}$$

Since $\lambda_i \leq \Lambda$ for all $i = 1, 2, \dots, n$ we have

$$0 \leq 1 - \frac{\lambda_i}{\Lambda} < 1 \tag{6.92}$$

and we note that we have faster convergence for larger eigenvalues since the larger λ_i is the smaller is $1 - \frac{\lambda_i}{\Lambda}$ and, thus, $\left(1 - \frac{\lambda_i}{\Lambda}\right)^k \to 0$ faster as $k \to \infty$.

6.5.2 A Two-Grid Method

Based on the observation that a simple iterative solver like the Richardson iteration quickly decreases the high frequency components of the error but does a poor job on the low frequency components we propose to first use Richardson iteration and then use another solver to correct the remaining low frequency part of the error. This can be done by using a direct solver on a coarser grid. Since the coarser mesh has fewer degrees of freedom it is cheaper to solve on that grid and therefore we can afford to use a more expensive solver on the coarse mesh.

To make this idea precise, let V_h be our finite element space consisting of continuous piecewise linear functions on a fine mesh $\mathcal{K}_h$, and let $V_H \subset V_h$ be a subspace on a coarser mesh $\mathcal{K}_H$. We assume that $\mathcal{K}_h$ is obtained by a finite number of uniform refinements of $\mathcal{K}_H$. In order to transfer data between the grids we define a prolongation operator $P_{h,H}$ that takes a function v_H on the coarse mesh and represents it as another function v_h on the fine mesh. Since V_H is a subspace of V_h we let $P_{h,H}$ be the matrix of the natural embedding (i.e., linear interpolation) of V_H into V_h. That is, given the nodal values ξ_H of v_H we have $\xi_h = P_{h,H}\xi_H$ with ξ_h the nodal values of v_h. By analogy, we let the restriction operator $R_{H,h}$ take a function in V_h and represent it in V_H. In doing so, a common choice is to take the restriction operator to be more or less the transpose of the prolongation operator $R_{H,h} = P_{h,H}^T$. We let A_h and b_h denote the stiffness matrix and load vector on the fine grid and we let A_H denote the stiffness matrix on the coarse grid. With this notation a two-grid method can be summarized in the following algorithm.

Algorithm 24 A Two-Grid method

Given $u_h^{(0)} \in V_h$ with nodal values $\xi_h^{(0)}$.
for $k = 0, 1, \dots,$ until convergence **do**
  Compute $\xi_h^{(k)}$ by doing m Richardson iterations on the linear system $A_h \xi_h = b_h$.
  Compute the residual $r_h^{(k)} = b_h - A_h \xi_h^{(k)}$.
  Restrict the residual to the coarse grid $r_H^{(k)} = R_{H,h} r_h^{(k)}$.
  Solve the linear system $A_H \delta_H^{(k)} = r_H^{(k)}$ for the nodal values $\delta_H^{(k)}$ of the correction $e_H^{(k)} \in V_H$.
  Prolongate the correction to the fine grid $\delta_h^{(k)} = P_{h,H} \delta_H^{(k)}$.
  Update the solution $\xi_h^{(k+1)} = \xi_h^{(k)} + \delta_h^{(k)}$.
end for

As a small example, consider $-u'' = 1$ with $u(0) = u(L) = 0$ on a uniform grid with nine uniformly spaced nodes $0 = x_0 < x_1 < \dots < x_8 = L$. The fine grid consists of all the nodes $x_1, x_2, \dots, x_7$, with $h = 1/8$, whereas the coarse grid consists of the three nodes x_2, x_4, and x_6, with $H = 2h = 1/4$. The nodes x_0 and x_8 are not considered since they are associated with boundary conditions. Thus, the unknowns are the 7 non-zero nodal values $\xi_1, \dots, \xi_7$ of $u_h \in V_h$. The 7×7 tridiagonal stiffness matrix A_h, and the 7×1 load vector b_h are given by

$$A_h = \frac{1}{h} \begin{bmatrix} 2 & -1 & & & & \\ -1 & 2 & -1 & & & \\ & -1 & 2 & -1 & & \\ & & \ddots & \ddots & \ddots & \\ & & & -1 & 2 & -1 \\ & & & & -1 & 2 \end{bmatrix}, \qquad b_h = h \begin{bmatrix} 1 \\ 1 \\ 1 \\ \vdots \\ 1 \\ 1 \end{bmatrix} \tag{6.93}$$

Now, given a function on the coarse grid we choose to transfer it to the fine grid using linear interpolation. In doing so, we let the coarse nodes that are also in the fine grid keep their value. However, the fine nodes that are not in the coarse grid are given the average value of their left and right neighbour nodes. This gives the 7×3 prolongation matrix

$$P_{h,H} = \begin{bmatrix} \frac{1}{2} & & \\ 1 & & \\ \frac{1}{2} & \frac{1}{2} & \\ & 1 & \\ & \frac{1}{2} & \frac{1}{2} \\ & & 1 \\ & & \frac{1}{2} \end{bmatrix} \tag{6.94}$$

Figure 6.5a shows the prolongation v_h with nodal values $\xi_h = P_{h,H} \xi_H$ of a function v_H with nodal values $\xi_H = [2, 3, 2]^T$.

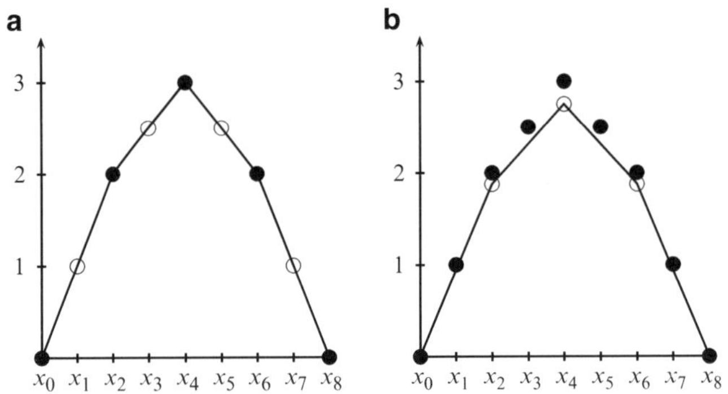

Fig. 6.5 Prolongation and restriction (-) with given (•) and computed (○) node values. (**a**) Prolongation. (**b**) Restriction

Utilizing the prolongation, the 3×7 restriction matrix is given by

$$R_{H,h} = P_{h,H}^T/2 \tag{6.95}$$

where we have divided by a factor two since we want the rows to sum up to unity. This is natural for any interpolation or averaging operator. Figure 6.5b shows the restriction v_H with nodal values $\xi_h = P_{h,H}\xi_H$ of a function v_h with nodal values $\xi_H = [1, 2, 2.5, 3, 2.5, 2, 1]^T$.

In MATLAB, our two-grid method can be coded as shown below.

```
function TwoGrid()
nf=2*25-1; % number of fine nodes
nc=(nf-1)/2; %           coarse nodes
h=1/(nf+1); % mesh size
x=0:h:1; % mesh
e=ones(nf,1);
A=spdiags([-e 2*e -e],-1:1,nf,nf)/h; % fine stiffness matrix
b=ones(nf,1)*h; % load vector
u=zeros(nf,1); % solution guess
P=sparse(nf,nc); % prolongation matrix
for i=1:nc
    P(2*i-1,i)=0.5;
    P(2*i,i)=1;
    P(2*i+1,i)=0.5;
end
R=0.5*P'; % prolongation matrix
RAP=R*A*P; % coarse stiffness matrix
for k=1:5 % outer iteration loop
    u=Richardson(A,u,b,4,0.25*h);
    r=R*(b-A*u); % residual
    e=RAP\r; % correction
    u=u+P*e; % solution update
end
plot(x,[0 u' 0]), xlabel('x'), ylabel('u_h') % plot
```

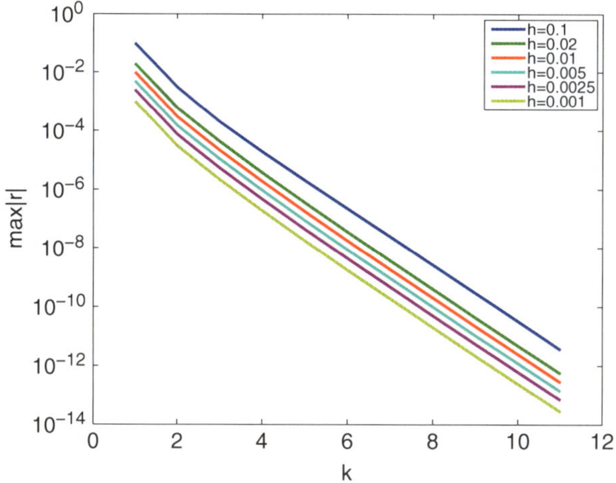

Fig. 6.6 Semilog plot of max $|r_H^{(k)}|$ for the iterates $k = 1, 2, \ldots, 11$

Here, we make 5 outer iterations with 4 Richardson smoothing steps per iteration. The involved relaxation parameter is heuristically chosen as $\alpha = h/4$.

By design, the two-grid method is good at decaying the low frequency content of the error. In fact, it turns out that the error reduction per iteration is practically independent of the mesh size h. That is, the convergence of the two-grid method does not deteriorate significantly as $h \to 0$. This is clearly seen in Fig. 6.6, which shows the maximum absolute value of the coarse grid residual $r_H^{(k)}$ for 11 iterations $k = 1, 2, \ldots, 11$.

In the two-grid method, we have used a direct solver to make the coarse grid correction, which is generally too expensive for large problems. To overcome this difficulty it is customary to replace this solve by recursive smoothing on a sequence of successively coarser grids with the correction being solved for only on the coarsest grid. To obtain a symmetric method it is common to apply smoothing not just before but also after the coarse grid correction. This is called a V-cycle and is the basic idea behind the so-called multigrid method.

6.6 Further Reading

The basic numerical algorithms, such as Gaussian elimination, and forward and backward substitution, for instance, for solving dense linear systems, can be found in almost any text book on numerical linear algebra. We refer to Heath [41] or Demmel [23].

An in-depth treatment of direct methods for large sparse SPD matrices is given by George and Liu [32]. The generalization of these methods to unsymmetric and indefinite matrices is given by Duff et al. [25]. Iterative methods for large sparse linear systems is comprehensively treated by Saad [57], and Barret and co-authors [7]. A pedagogical and painless introduction to the CG method is given by Shewchuk [63]. A survey of preconditioning techniques is given by Benzi [10].

An introduction to multigrid is given by Demmel [23]. A more advanced treatment is given by Hackbush [39].

6.7 Problems

Exercise 6.1. Plot the structure, using `spy`, of the usual stiffness matrix `A`, given by

```
[p,e,t]=initmesh('circleg','hmax',0.1);
A=assema(p,t,1,0,0);
```

Plot also the structure of its inverse. Use `nnz` to find out how many non-zeros there are in each matrix.

Exercise 6.2. Reorder the stiffness matrix `A` from the previous exercise using the approximate minimum degree ordering `symamd`. Plot the matrix structure before and after reordering.

Exercise 6.3. On the L-shaped domain `lshapeg` with mesh size one tenth, assemble the stiffness and mass matrix, viz.,

```
[A,M]=assema(p,t,1,1,0);
```

Compute the Cholesky factorization of `A+M` using `chol`. Reorder the matrix using `symamd` and repeat. How much sparser is the Cholesky factor after reordering?

Exercise 6.4. Draw the graph $G(A)$ of the symmetric matrix

$$A = \begin{bmatrix} 10 & 1.2 & 0 & 0 \\ 1.2 & 3 & -5.3 & 2.4 \\ 0 & -5.3 & 1.6 & 7 \\ 0 & 2.4 & 7 & 8 \end{bmatrix}$$

(a) How many vertices and edges are there in the graph?
(b) What is the degree of each vertex?

Exercise 6.5. Suppose A is SPD with structure

$$A = \begin{bmatrix} 1 & x & & x & & x \\ x & 2 & & & x & \\ & & 3 & & x & x \\ x & & & 4 & & \\ & x & x & & 5 & \\ x & & x & & & 6 \end{bmatrix}$$

(a) Draw the graph of A.
(b) Draw the sequence of elimination graphs corresponding to the Cholesky factorization $A = LL^T$.
(c) Draw the structure of L.

Exercise 6.6. Write two routines `Jacobi` and `GaussSeidel` implementing Jacobi and Gauss-Seidel iteration. Let the syntax for calling the routines be given by

```
[x,k]=Jacobi(A,b,tol)
[x,k]=GaussSeidel(A,b,tol)
```

where `tol` is a number specifying the desired relative residual $\|r^{(k)}\|/\|r^{(0)}\|$, and `k` is the number of iterations performed. Test your codes by solving the linear system with

$$A = \begin{bmatrix} 12 & 1 & 0 & 0 & 0 & -1 \\ 1 & 10 & 1 & 0 & 0 & 0 \\ 2 & 0 & 20 & 2 & 0 & 0 \\ 0 & 0 & 1 & 12 & -1 & 0 \\ 0 & 3 & 0 & 0 & 30 & 3 \\ 0 & 0 & 0 & 2 & -2 & 24 \end{bmatrix}, \qquad b = \begin{bmatrix} 8 \\ 24 \\ 70 \\ 46 \\ 174 \\ 142 \end{bmatrix}$$

Exercise 6.7. Use `Jacobi` and `GaussSeidel` to compare the number of iterations required by these methods to converge to a given accuracy from a zero starting guess. Let A and b be defined by

```
e=ones(n,1);
A=spdiags([-e 2*e -e], -1:1, n, n);
b=rand(n,1);
```

Record the number of iterations needed to achieve the tolerance 0.1, 0.01, 0.001, and 0.0001 for a few different values of n, say 10, and 100. How many times faster is Gauss-Seidel than Jacobi?

Exercise 6.8. Show that Jacobi iteration may take the form

$$x^{(k+1)} = x^{(k)} + Hr^{(k)}$$

where H is a matrix to be defined by you and $r^{(k)} = b - Ax^{(k)}$ is the residual at stage k. Can you interpret this result from the point of view of one-dimensional projection methods for $Ax = b$.

Exercise 6.9. Consider the m-th Krylov space $K_m(A; b)$, and the corresponding Krylov matrix

$$K_m = \left[b \; Ab \; A^2b \; \dots \; A^{m-1}b\right]$$

Let `A = diag([1 2 3 4])` and `b = [1 1 1 1]'`.

(a) Compute the Krylov matrix K_4. Then express the vector $x = A^{-1}b = [1, \frac{1}{2}, \frac{1}{3}, \frac{1}{4}]^T$ as a linear combination of the columns of K_4.
(b) Use **`Arnoldi`** to compute the 4×4 matrices Q and H in the Arnoldi factorization of A, (i.e., such that $AQ = QH$). Use $q_1 = b/\|b\|$ as starting vector. (Note that since the Arnoldi algorithm stops at stage 3, the last column of H is not actually computed. It comes from a final command `H(:,4) = Q'*A*Q(:,4)`.)
(c) Assume that we have run Arnoldi's algorithm for 2 steps so that we have access to the orthogonal basis $Q_2 = [q_1, q_2]$ that span the Krylov subspace $K_2(A; b)$. Show how the matrix H_2 can be used to get a Galerkin solution x_2, that is, such that the residual $r_2 = b - Ax_2$ is orthogonal to the span of the basis vectors q_1 and q_2. Compute x_2. What is the residual r_2?

Exercise 6.10. Consider a linear system with coefficient matrix and right hand side vector, `A` and `b`, defined by

```
[p,e,t]=initmesh('lshapeg','hmax',0.1); x=p(1,:); y=p(2,:);
 [A,M]=assema(p,t,2,1,0); b=M*(x+y)'; A=A+M;
```

Solve `A*x=b` using `pcg` and `cholinc`. Use the parameters `tol=1e-6`, `maxit=10000`, and `droptol=1e-3`. How many iterations are needed to obtain a solution within the desired tolerance with and without the preconditioner?

Exercise 6.11. Use `TwoGrid` to verify the results of Fig. 6.6.

Chapter 7
Abstract Finite Element Analysis

Abstract In this chapter we study the mathematical theory of finite element methods from a broader perspective by introducing a general theory for linear second order elliptic partial differential equations. This allows us to handle a large class of problems with the same analytical techniques. We do this by first introducing a general elliptic problem and its abstract weak form posed on a so-called Hilbert space. We show that this weak problem has a solution by proving the Lax-Milgram lemma, and that this solution is unique. Knowing that the solution exists we then show how to approximate it by finite elements. Finally, we prove basic a priori and a posteriori error estimates for the finite element approximation.

7.1 Function Spaces

In this section we introduce spaces of functions that are useful in the analysis of partial differential equations and the finite element method. We begin by recalling some basic concepts from linear algebra.

7.1.1 Normed and Inner Product Vector Spaces

A real vector space is V is a set with operations $+ : V \times V \to V$ and $\cdot : \mathbb{R} \times V \to V$ such that

- $u + v = v + u$
- $(u + v) + w = u + (v + w)$
- $\lambda \cdot (u + v) = \lambda \cdot u + \lambda \cdot v$
- $(\lambda + \mu) \cdot u = \lambda \cdot u + \mu \cdot u$

for all vectors $u, v, w \in V$ and scalars $\lambda, \mu \in \mathbb{R}$. In addition, there should be a zero vector 0 such that $u + 0 = u$, and a negative vector $-u$ such that $u + (-u) = 0$ for

M.G. Larson and F. Bengzon, *The Finite Element Method: Theory, Implementation, and Applications*, Texts in Computational Science and Engineering 10, DOI 10.1007/978-3-642-33287-6_7, © Springer-Verlag Berlin Heidelberg 2013

any vector $u \in V$. Examples of real vector spaces include Euclidean space $\mathbb{R}^n$, the space $P_k(I)$ of polynomials of order k on an interval I, and the space of continuous functions $C^0(I)$ on I. In general, we simplify the notation for the product between a scalar $\lambda \in \mathbb{R}$ and a vector $v \in V$ and write $\lambda \cdot v = \lambda v$.

A norm on a vector space V is a mapping $\|\cdot\| : V \to \mathbb{R}$ that satisfies

- $\|u + v\| \leq \|u\| + \|v\|$
- $\|\lambda u\| = |\lambda| \, \|u\|$
- $\|u\| \geq 0$, with equality if and only if $u = 0$

for all $u, v \in V$ and $\lambda \in \mathbb{R}$. A semi-norm $|\cdot| : V \to \mathbb{R}$ is a mapping that satisfies the requirements of a norm, except that $|u| = 0$ does not need to imply $u = 0$. A normed vector space is a vector space equipped with a norm. Perhaps needless to say there are many norms. For example, we may equip $\mathbb{R}^n$ with the Euclidean 2-norm or the max norm

$$\|v\|_2 = \left(\sum_{i=1}^{n} |v_i|^2 \right)^{1/2}, \quad \|v\|_\infty = \max_{1 \leq i \leq n} |v_i| \tag{7.1}$$

We may also equip the function spaces $P_k(I)$ and $C^0(I)$ with the corresponding norms

$$\|v\|_{L^2(I)} = \left(\int_I |v|^2 \, dx \right)^{1/2}, \quad \|v\|_{L^\infty(I)} = \sup_{x \in I} |v(x)| \tag{7.2}$$

A basis for the vector space V is a minimal set of vectors $\{\phi_i\}_{i=1}^n$ that span V. This means that any vector v in V can be written as the linear combination $v = c_1\phi_1 + c_2\phi_2 + \cdots + c_n\phi_n$ for some coefficients c_i, $i = 1, 2, \ldots, n$. The bases ϕ_i must be linearly independent in the sense that $v = 0$ if and only if $c_i = 0$ for all i. The number of basis vectors n is called the dimension of V. For example, the set $\{1, x, x^2, \ldots, x^k\}$ of monomials forms a basis for the polynomial space $P_k(I)$. Since there are $k + 1$ monomials, the dimension of $P_k(I)$ is $k + 1$. If $k < \infty$ we say that $P_k(I)$ is of finite dimension. If $k = \infty$ the set $\{x^i\}_{i=0}^\infty$ forms a basis for the space of square integrable functions $L^2(I)$. In this case, since the number of basis functions are infinite, $L^2(I)$ is said to be of infinite dimension.

If a subset V_0 of V is a vector space itself under the operations of V, then it is a subspace of V.

A functional or linear form on a vector space V is a mapping $l(\cdot) : V \to \mathbb{R}$ that satisfies

- $l(u + v) = l(u) + l(v)$
- $l(\lambda u) = \lambda l(u)$

for all $u, v \in V$ and $\lambda \in \mathbb{R}$. A linear form is continuous, or bounded, if there is a constant C such that

$$|l(v)| \leq C \|v\| \tag{7.3}$$

for all $v \in V$. The norm of $l(\cdot)$ is defined by

$$\|l\| = \sup_{v \in V} \frac{|l(v)|}{\|v\|} \tag{7.4}$$

The set of all continuous functionals on V is called the dual space V^* of V. It can be shown that V^* is a linear vector space under the $+$ and $\cdot$ operations, $(\lambda \cdot l + \mu \cdot m)(v) = \lambda \cdot l(v) + \mu \cdot m(v)$ for any functionals $l, m \in V^*$ and scalars $\lambda, \mu \in \mathbb{R}$. The dual space V^* can be normed by setting $\|l\|_{V^*} = \|l\|$.

A bilinear form on a vector space V is a mapping $a(\cdot,\cdot) : V \times V \to \mathbb{R}$ such that

- $a(u + v, w) = a(u, w) + a(v, w)$
- $a(u, v + w) = a(u, v) + a(u, w)$
- $a(\lambda u, v) = \lambda a(u, v)$
- $a(u, \lambda v) = \lambda a(u, v)$

for all $u, v, w \in V$ and $\lambda \in \mathbb{R}$. The bilinear form is symmetric if

$$a(u, v) = a(v, u) \tag{7.5}$$

for all $u, v \in V$ and continuous, or bounded, if there is a constant C such that

$$a(u, v) \leq C \|u\| \, \|v\| \tag{7.6}$$

for all $u, v \in V$.

A symmetric bilinear form $a(\cdot,\cdot)$ is called an inner product if

- $a(u, u) \geq 0$, with equality if and only if $u = 0$

for all $u \in V$. Inner products are often also denoted $(\cdot,\cdot)$. An inner product defines a so-called induced norm by

$$\|u\|^2 = (u, u) \tag{7.7}$$

on V. In particular, $a(\cdot,\cdot)$ defines the induced energy norm $|||u|||^2 = a(u, u)$.

A vector space equipped with an inner product is called an inner product vector space. In such spaces the Cauchy-Schwarz inequality

$$(u, v) \leq \|u\| \, \|v\| \tag{7.8}$$

holds for all $u, v \in V$. In order to prove Cauchy-Schwarz inequality we note that it is trivially satisfied if $\|v\| = 0$. Assuming $\|v\| \neq 0$ we have

$$0 \leq \|u + \lambda v\|^2 = (u + \lambda v, u + \lambda v) = \|u\|^2 + 2\lambda(u, v) + \lambda^2\|v\|^2 = g(\lambda) \tag{7.9}$$

for all $u, v \in V$ and $\lambda \in \mathbb{R}$. Computing the minimum of the function $g(\lambda)$ with respect to the scalar variable λ we eventually end up with

$$0 \leq \|u\|^2 - \frac{|(u, v)|^2}{\|v\|^2} \tag{7.10}$$

and the inequality follows. Using the Cauchy-Schwarz inequality we can easily verify that the norm defined by the inner product satisfies the Triangle inequality and therefore indeed is a norm.

We define the angle θ between two vectors $u, v \in V$ by

$$\theta = \arccos \frac{(u, v)}{\|u\| \, \|v\|} \tag{7.11}$$

and we say that u and v are orthogonal if $(u, v) = 0$. If $u, v \in V$ are orthogonal, then the Pythagorean theorem

$$\|u + v\|^2 = \|u\|^2 + \|v\|^2 \tag{7.12}$$

holds. Finally, we recall that the Parallelogram law

$$\|u + v\|^2 + \|u - v\|^2 = 2\|u\|^2 + 2\|v\|^2 \tag{7.13}$$

holds for all $u, v \in V$ in an inner product space.

7.1.2 Banach and Hilbert Spaces

In order to define Banach and Hilbert spaces we first need to define the concept of completeness.

A Cauchy sequence in a normed vector space V is a sequence $\{v_i\}_{i=1}^{\infty}$ of elements $v_i \in V, i = 1, 2, \ldots$, such that for all $\epsilon > 0$ there is a positive integer n such that

$$\|v_i - v_j\| \leq \epsilon, \quad \text{for } i, j \geq n \tag{7.14}$$

A sequence $\{v_i\}_{i=1}^{\infty}$ is convergent if there exists $v \in V$ such that for all $\epsilon > 0$ there is a positive integer n such that

$$\|v - v_i\| \leq \epsilon, \quad \text{for } i \geq n \tag{7.15}$$

If a sequence is convergent, then it is also a Cauchy sequence. A vector space is said to be complete if every Cauchy sequence is also convergent.

A Banach space is a complete normed vector space and a Hilbert space is a complete inner product vector space. We shall see below that completeness plays an important role when one seeks to prove the existence of solutions to certain problems. Then a common strategy is to construct a sequence of approximate solutions to the problem, show that the sequence is indeed a Cauchy sequence and finally use completeness to show that the sequence converges.

A linear subspace $V_0 \subset V$ of a Hilbert space V is said to be closed if all sequences in V_0 has a limit in V_0. That is, if $\{v_i\}_{i=1}^{\infty}$ with $v_i \in V_0$ and $v_i \to v$ implies $v \in V_0$. If so, V_0 is also a Hilbert space under the same norm as V.

Let $(\cdot, \cdot)$ be the inner product on the Hilbert space V. Given a closed linear subspace V_0 of V, the orthogonal complement $V_0^{\perp} = \{v \in V : (v, v_0) = 0, \forall v_0 \in V_0\}$ is also a closed linear subspace of V. Moreover, we have the orthogonal decomposition $V = V_0 \oplus V_0^{\perp}$. As a consequence, for any $u \in V$, there is a unique function $u_0 \in V_0$ that is closest to u with respect to the induced norm $\| \cdot \|_V^2 = (\cdot, \cdot)$. Indeed, u_0 is characterized by the best approximation result

$$\|u - u_0\|_V \leq \|u - v_0\|_V, \quad \forall v_0 \in V_0 \tag{7.16}$$

7.1.3 L^p Spaces

In finite element analysis the most important Banach and Hilbert spaces are vector spaces of functions and originate from the following families of functions.

The $L^p(\Omega)$ function spaces are defined by

$$L^p(\Omega) = \{v : \Omega \to \mathbb{R} : \|v\|_{L^p(\Omega)} < \infty\} \tag{7.17}$$

where

$$\|v\|_{L^p(\Omega)} = \left(\int_{\Omega} |v|^p \, dx \right)^{1/p}, \quad 1 \leq p < \infty \tag{7.18}$$

and

$$\|v\|_{L^{\infty}(\Omega)} = \sup_{x \in \Omega} |v(x)|, \quad p = \infty \tag{7.19}$$

Here, the integral is a so called Lebesgue integral which allows integration of more general functions than the classical Riemann integral. In fact, these functions do not need to have well-defined point values. It is possible to show that $L^p(\Omega)$ are Banach spaces for all $1 \leq p \leq \infty$. We shall not go into the theory of Lebesgue integration here, since it is quite involved, and since the completeness of the $L^p(\Omega)$ spaces is the main property that will be needed later.

If $p = 2$, then $\|v\|_{L^2(\Omega)}^2 = (v, v)_{L^2(\Omega)}$, where $(u, v)_{L^2(\Omega)} = \int_{\Omega} uv \, dx$, is the $L^2(\Omega)$ inner product and, thus, $L^2(\Omega)$ is a Hilbert space. If $p \neq 2$, then $\| \cdot \|_{L^p(\Omega)}$ is not given by any inner product, and, thus, $L^p(\Omega)$ is only a Banach space.

7.1.4 Weak Derivatives

Because functions in Hilbert spaces are not in general regular enough for the standard definition of the derivative to make sense we shall introduce a concept called weak derivative where the derivative is defined in an average sense. To this end, let $C^k(\Omega)$ be the space of all $k < \infty$ times continuously differentiable functions in Ω. Also, let $\mathcal{D}(\Omega)$ be the space of all infinitely differentiable compactly supported functions in Ω. That is,

$$\mathcal{D}(\Omega) = \{\varphi \in C^\infty(\Omega) : \mathrm{supp}(\varphi) \subset\subset \Omega\} \tag{7.20}$$

where the support of φ is the closure of the open set $\{x \in \Omega : \varphi(x) \neq 0\}$. A function φ is said to have compact support with respect to Ω if $\mathrm{supp}(\varphi)$ is a compact set ω that is a subset of the interior of Ω. A subset $\omega \subset \Omega$ is compact if and only if it is closed and bounded. We write $\mathrm{supp}(\varphi) \subset\subset \Omega$ to signify that $\mathrm{supp}(\varphi) \subset \omega \subset \Omega$ with ω compact, and say that the support of φ is compactly contained in Ω. Thus, the functions in $\mathcal{D}(\Omega)$ can be differentiated an infinite number of times, and are non-zero strictly within Ω. In particular, they vanish sufficiently close to the boundary $\partial\Omega$.

To simplify the derivative notation let us introduce the concept of a multi-index. A multi-index $\alpha = (\alpha_1, \alpha_2, \dots, \alpha_d)$ is a d-tuple of non-negative integers α_i. The order $|\alpha|$ of α is defined by

$$|\alpha| = \sum_{i=1}^{d} \alpha_i \tag{7.21}$$

We let

$$D^\alpha \varphi = \prod_{i=1}^{d} \left(\frac{\partial}{\partial x_i} \right)^{\alpha_i} \varphi, \quad \varphi \in \mathcal{D}(\Omega) \tag{7.22}$$

denote the classical partial derivative. Using partial integration and the fact that boundary terms vanish, since the support of φ is compact, we derive the following identity

$$\int_\Omega \frac{\partial u}{\partial x_i} \varphi dx = - \int_\Omega u \frac{\partial \varphi}{\partial x_i} dx, \quad \forall \varphi \in \mathcal{D}(\Omega) \tag{7.23}$$

for any $u \in C^1(\Omega)$. Repeating this formula we obtain

$$\int_\Omega (D^\alpha u) \varphi dx = (-1)^{|\alpha|} \int_\Omega u D^\alpha \varphi dx, \quad \forall \varphi \in \mathcal{D}(\Omega) \tag{7.24}$$

for any $u \in C^{|\alpha|}(\Omega)$. Here, we note that only the right hand side requires the strong regularity $u \in C^{|\alpha|}(\Omega)$. We finally introduce the space of locally integrable functions

$$L^1_{loc}(\Omega) = \{v : v \in L^1(K), \forall K \subset\subset \Omega\} \tag{7.25}$$

Let $u \in L^1_{loc}(\Omega)$. If there is a function $g \in L^1_{loc}(\Omega)$ such that

$$\int_\Omega g \varphi dx = (-1)^{|\alpha|} \int_\Omega u D^\alpha \varphi dx, \quad \forall \varphi \in \mathcal{D}(\Omega) \tag{7.26}$$

then we say that g is the weak derivative $D^\alpha u$ of u. We note that both the left and right hand side integrals are well-defined. For instance, for the right hand side, we have

$$\left|\int_\Omega u D^\alpha \varphi dx\right| \le \|u\|_{L^1(K)} \|D^\alpha \varphi\|_{L^\infty(K)} < \infty \tag{7.27}$$

where $K = \text{supp}(\varphi)$.

As an example, let $\Omega = (-1, 1)$, and let $u = 3 - |x|$. The weak derivative $g = D^1 u$ is equal to

$$g = \begin{cases} 1, & -1 < x \le 0 \\ -1, & 0 < x < 1 \end{cases} \tag{7.28}$$

In order to verify this we need to prove that

$$\int_{-1}^{1} g\varphi\, dx = -\int_{-1}^{1} u D^1 \varphi\, dx, \quad \forall \varphi \in \mathcal{D}(\Omega) \tag{7.29}$$

Integrating by parts, we get

$$-\int_{-1}^{1} u D^1 \varphi dx = -\int_{-1}^{0} u D^1 \varphi dx - \int_{0}^{1} u D^1 \varphi dx \tag{7.30}$$

$$= \int_{-1}^{0} D^1 u \varphi dx - \Big[u\varphi\Big]_{-1}^{0} + \int_{0}^{1} D^1 u \varphi dx - \Big[u\varphi\Big]_{0}^{1} \tag{7.31}$$

$$= \int_{-1}^{0} \varphi dx - (3\varphi(0) - 2\varphi(-1)) - \int_{0}^{1} \varphi dx - (2\varphi(1) - 3\varphi(0)) \tag{7.32}$$

$$= \int_{-1}^{1} g\varphi dx \tag{7.33}$$

Here, we used that $\varphi(-1) = \varphi(1) = 0$, since φ has compact support.

Weak and classical derivatives share many properties, such as linearity, the chain rule, and differentiation of products, for instance.

7.1.5 Sobolev Spaces

Let $u \in L^1_{loc}(\Omega)$ and assume that all weak derivatives $D^\alpha u$ with $|\alpha| \le k$, where k is a non-negative integer, exist. We define the Sobolev norm of u by

$$\|u\|_{W_k^p(\Omega)} = \left(\sum_{|\alpha| \le k} \|D^\alpha u\|^p_{L^p(\Omega)}\right)^{1/p}, \quad 1 \le p < \infty \tag{7.34}$$

and

$$\|u\|_{W_k^p(\Omega)} = \max_{|\alpha|\leq k} \|D^\alpha u\|_{L^\infty(\Omega)}, \quad p = \infty \tag{7.35}$$

The Sobolev space $W_k^p(\Omega)$ is then defined by

$$W_k^p(\Omega) = \{u \in L^1_{loc}(\Omega) : \|u\|_{W_k^p(\Omega)} < \infty\} \tag{7.36}$$

Thus, $W_k^p(\Omega)$ is the space of $L^p(\Omega)$ functions u, whose weak derivatives $D^\alpha u$, $|\alpha| \leq k$, also lies in $L^p(\Omega)$.

In this context we also define the Sobolev semi-norms

$$|u|_{W_k^p(\Omega)} = \left(\sum_{|\alpha|=k} \|D^\alpha u\|^p_{L^p(\Omega)} \right)^{1/p}, \quad 1 \leq p < \infty \tag{7.37}$$

and

$$|u|_{W_k^p(\Omega)} = \max_{|\alpha|=k} \|D^\alpha u\|_{L^\infty(\Omega)}, \quad p = \infty \tag{7.38}$$

If $\partial\Omega$ is sufficiently regular, it can be shown that $C^\infty(\Omega)$ is dense in $W_k^p(\Omega)$ for $1 \leq p < \infty$. Loosely speaking this means that smooth functions lie arbitrarily close to Sobolev functions. In fact, any Sobolev function can be approximated by a sequence of smooth functions. Indeed, for any $u \in W_k^p(\Omega)$ there is $u_i \in C^\infty(\Omega)$ such that $u_i \rightarrow u$ as $i \rightarrow \infty$. This approximation holds within the interior of the domain Ω, and up to the boundary provided that $\partial\Omega$ has C^1 continuity. As a consequence, we may alternatively define $W_k^p(\Omega)$ as the completion of $C^\infty(\Omega)$ with respect to the Sobolev norm $\|\cdot\|_{W_k^p(\Omega)}$. In other words, $W_k^p(\Omega)$ contains all limits of all sequences of $C^\infty(\Omega)$ functions with norm less than infinity.

The fact that complicated Sobolev functions can be approximated by simple smooth functions is often used to avoid working with weak derivatives. In doing so, the basic idea is to first establish the desired results for smooth functions and then generalize these results to Sobolev functions by taking limits.

The Sobolev spaces are Banach spaces for all $1 \leq p \leq \infty$. However, for $p = 2$, $W_k^2(\Omega)$ is also a Hilbert space with inner product and norm

$$(u, v)_{W_k^2(\Omega)} = \sum_{|\alpha|\leq k} (D^\alpha u, D^\alpha v)_{L^2(\Omega)} \tag{7.39}$$

$$\|u\|^2_{W_k^2(\Omega)} = \sum_{|\alpha|\leq k} \|D^\alpha u\|^2_{L^2(\Omega)} \tag{7.40}$$

The case $p = 2$ is the most common in finite element analysis and we shall use the notation $H^k(\Omega) = W_k^2(\Omega)$ to emphasize the Hilbert space property. Thus, with $k = 1$, we have the familiar space

$$H^1(\Omega) = \{v \in L^2(\Omega) : D^1 u \in L^2(\Omega)\} \tag{7.41}$$

with inner product and norm

$$(u, v)_{H^1(\Omega)} = (u, v)_{L^2(\Omega)} + (\nabla u, \nabla v)_{L^2(\Omega)} \tag{7.42}$$

$$\|u\|^2_{H^1(\Omega)} = \|u\|^2_{L^2(\Omega)} + \|\nabla u\|^2_{L^2(\Omega)} \tag{7.43}$$

The regularity of functions in Hilbert spaces depends on the space dimension d of the domain $\Omega \subset \mathbb{R}^d$. Regarding $H^1(\Omega)$ it can be shown that its functions are continuous for $d = 1$, may lack values at certain isolated points for $d = 2$, and be discontinuous along a curve for $d = 3$.

7.1.6 Traces

The properties of Sobolev functions on the boundary $\partial\Omega$ is a delicate matter. This has to do with the fact that such functions are defined by the Lebesgue integral only up to a set of measure zero, and that $\partial\Omega$ has precisely zero measure. This means that it is meaningless to speak about the restriction of a Sobolev function to the boundary, since we may to a large extent alter its boundary values without changing its status as a member of the space. Thus, the question arise how to define the restriction of a Sobolev function on the domain boundary? The answer is of course trivial if the function is continuous. We simply evaluate the function on the boundary. However, this leads to the idea that for a general Sobolev function we can first make a smooth approximation from within the domain, and then evaluate this approximation on the boundary. This is called the trace of the function. Loosely speaking, any $w \in W_1^p(\Omega)$ can be restricted to $v \in L^p(\partial\Omega)$. Conversely, any $v \in L^p(\partial\Omega)$ can be extended to $w \in W_1^p(\Omega)$. Indeed, assuming that $\partial\Omega$ is sufficiently regular, say C^1, the so-called trace operator

$$\gamma : W_1^p(\Omega) \to L^p(\partial\Omega) \tag{7.44}$$

is well-defined and satisfies the following stability estimate

$$\|\gamma v\|_{L^p(\partial\Omega)} \leq C\,\|v\|^{1-1/p}_{L^p(\Omega)}\|v\|^{1/p}_{W_1^p(\Omega)} \tag{7.45}$$

The case $p = 2$ is of particular importance from which we infer the Trace inequality

$$\|\gamma v\|_{L^2(\partial\Omega)} \leq C\,\|v\|^{1/2}_{L^2(\Omega)}\|v\|^{1/2}_{H^1(\Omega)} \tag{7.46}$$

We sometimes need to use the Trace inequality on a domain K whose diameter scales with a parameter h_K. A typical example is a triangle element. In order to get the correct asymptotic behavior as $h_K \to 0$ we need to take the parameter h_K into account. This yields

$$\|\gamma v\|^2_{L^2(\partial K)} \leq C\left(h_K^{-1}\|v\|^2_{L^2(K)} + h_K\|\nabla v\|^2_{L^2(K)}\right) \tag{7.47}$$

Using the trace operator we can define the constrained Hilbert space

$$H^1_0(\Omega) = \{v \in H^1(\Omega) : (\gamma v)|_{\partial\Omega} = 0\} \tag{7.48}$$

which consists of all functions in $H^1(\Omega)$ that vanish in the sense of the trace operator on $\partial\Omega$. On this space the full norm $\|\cdot\|_{H^1(\Omega)}$ may be replaced by the semi-norm $|\cdot|_{H^1(\Omega)}$, since the only constant function in $H^1_0(\Omega)$ is the zero function.

Often, the trace operator is omitted and $(\gamma v)|_{\partial\Omega}$ is simply written $v|_{\partial\Omega}$.

7.2 Interpolation of Functions in Hilbert Spaces

To approximate functions in the infinite dimensional Hilbert spaces we often use interpolation onto finite dimensional polynomial spaces. However, the interpolation operator π we have considered so far depends on pointwise evaluation and requires the function that we interpolate to be continuous. Now, because Hilbert space functions are not continuous in general we need to construct interpolation operators for these functions as well. We give a brief description of two types of such interpolants. For simplicity, we restrict our attention to interpolation with piecewise linear continuous functions defined on a shape regular triangulation $\mathcal{K}$ of Ω. As usual, let V_h be the space of continuous piecewise linears on $\mathcal{K}$.

7.2.1 The Clement Interpolant

Let $v \in C^0(\Omega)$ be a continuous function and recall that the usual interpolant $\pi v \in V_h$ of v is defined by

$$\pi v = \sum_{i=1}^{n_p} v(N_i)\,\varphi_i \tag{7.49}$$

where $\{N_i\}_{i=1}^{n_p}$ is the set of nodes, $\{\varphi_i\}_{i=1}^{n_p}$ is the standard hat function basis in V_h, and n_p is the number of nodes. In order to extend this definition to more general functions we shall replace the nodal values $v(N_i)$ by local averages of v. We define for each node N_i a patch of elements $\mathcal{N}(N_i)$ consisting of all elements that share the node N_i. We let $P_1(\mathcal{N}(N_i))$ denote the linear polynomials on $\mathcal{N}(N_i)$ and define the average using L^2-projection as follows: find $P_i v \in P_1(\mathcal{N}(N_i))$ such that

$$\int_{\mathcal{N}(N_i)} P_i v w \, dx = \int_{\mathcal{N}(N_i)} v w \, dx, \quad \forall w \in P_1(\mathcal{N}(N_i)) \tag{7.50}$$

The so-called Clement interpolant $\pi_C : L^1(\Omega) \to V_h$ is then defined by

$$\pi_C v = \sum_{i=1}^{n_p} P_i v(N_i) \, \varphi_i \tag{7.51}$$

This interpolant is due to Clement [20]. It satisfies the interpolation error estimate

$$\|v - \pi_C v\|_{H^m(K)} \leq C h^{k-m} |v|_{H^k(\mathcal{N}(K))}, \quad m = 0, 1, \quad m \leq k \leq 2, \quad \forall K \in \mathcal{K} \tag{7.52}$$

where $\mathcal{N}(K)$ is the neighborhood of element K consisting of all elements that share a node with K.

7.2.2 The Scott-Zhang Interpolant

The Clement interpolant is not a projection and it does not satisfy boundary conditions. Scott and Zhang [61] proposed another approach that solved these problems. Essentially, the idea is to represent the evaluation of a polynomial in a node as an inner product with a certain weight function. Here, we first associate one triangle K_i, that has N_i as one of its nodes, to each node N_i. Let $P_1(K_i)$ be the space of linear polynomials on K_i. We note that the mapping

$$v \mapsto v(N_i) \tag{7.53}$$

is a linear functional on $P_1(K_i)$ and that there exists $\chi_i \in P_1(K_i)$ such that

$$v(N_i) = \int_{K_i} v \chi_i \, dx, \quad \forall v \in P_1(K_i) \tag{7.54}$$

In this identity only the left hand side requires v to be continuous to be well defined. The right hand side is well defined for $v \in L^1(K)$. We can thus define the Scott-Zhang interpolation operator $\pi_{SZ} : L^1(\Omega) \to V_h$ as follows

$$\pi_{SZ} v = \sum_{i=1}^{n_p} \int_{K_i} v \chi_i \, dx \, \varphi_i \tag{7.55}$$

By construction $\pi_{SZ} v = v$ for $v \in V_h$, and, thus, π_{SZ} is a projection. Furthermore, the same interpolation error estimates as for the Clement interpolant holds. That is,

$$\|v - \pi_{SZ} v\|_{H^m(K)} \leq C h^{k-m} |v|_{H^k(\mathcal{N}(K))}, \quad m = 0, 1, \quad m \leq k \leq 2, \quad \forall K \in \mathcal{K} \tag{7.56}$$

In the following, we shall always assume that $\pi = \pi_{SZ}$ is the Scott-Zhang interpolation operator.

7.3 The Abstract Setting

Let us now put the pieces together and formulate an abstract variational problem posed in a Hilbert space and prove that it has a unique solution. Our reason for doing so is that many important, so-called elliptic, partial differential equations lead to such variational problems, and that we can obtain existence and uniqueness results for this whole class of problems with a single unified mathematical theory.

7.3.1 An Abstract Variational Problem

Let V be a Hilbert space and let $a(\cdot,\cdot) : V \times V \to \mathbb{R}$ be a bilinear form such that

- $a(\cdot,\cdot)$ is coercive (or elliptic). That is, there is a constant m such that

$$m\|v\|_V^2 \le a(v,v), \quad \forall v \in V \tag{7.57}$$

- $a(\cdot,\cdot)$ is continuous. That is, there is a constant C_a such that

$$a(u,v) \le C_a \|u\|_V \|v\|_V, \quad \forall u,v \in V \tag{7.58}$$

Further, let $l(\cdot) : V \to \mathbb{R}$ be a continuous linear functional. That is, there is a constant C_l such that

$$l(v) \le C_l \|v\|_V, \quad \forall v \in V \tag{7.59}$$

Our abstract variational problem takes the form: find $u \in V$ such that

$$a(u,v) = l(v), \quad \forall v \in V \tag{7.60}$$

7.3.2 The Riesz Representation Theorem

In order to analyze our abstract variational problem we need the following famous and fundamental result.

Theorem 7.1 (Riesz Representation Theorem). *Let V be a Hilbert space with inner product $(\cdot,\cdot)$. Every continuous linear form $l(\cdot)$ on V can be uniquely represented as*

$$l(v) = (u,v) \tag{7.61}$$

for some $u \in V$.

Proof. If $l(\cdot) = 0$, then $u = 0$, so let us assume that $l(\cdot) \neq 0$. Let N be the null space $N = \{v \in V : l(v) = 0\}$. Then, V admits the decomposition $V = N + N^{\perp}$, with $N^{\perp}$ the orthogonal complement to N with respect to the inner product on V. Pick $w \in N^{\perp}$. Then, $l(w) \neq 0$, and, for any $v \in V$,

$$l\left(v - \frac{l(v)}{l(w)}w\right) = 0 \tag{7.62}$$

Thus, $v - \frac{l(v)}{l(w)}w \in N$, which implies

$$\left(v - \frac{l(v)}{l(w)}w, w\right) = 0 \tag{7.63}$$

From this equation we obtain

$$l(v) = \frac{l(w)}{\|w\|_V^2}(w, v) \tag{7.64}$$

Hence, $l(v) = (u, v)$ with

$$u = \frac{l(w)}{\|w\|_V^2}\, w \tag{7.65}$$

It remains to show that u is unique. Suppose that u_1 and u_2 both satisfy $l(v) = (u_i, v)$, $i = 1, 2$, for all $v \in V$. Then, by subtraction $(u_1 - u_2, v) = 0$. Choosing $v = u_1 - u_2$ yields $\|u_1 - u_2\|_V = 0$, so $u_1 = u_2$. □

In the special case that the bilinear form $a(\cdot,\cdot)$ is symmetric existence and uniqueness of a solution to the abstract variational problem (7.60) immediately follows from the Riesz representation theorem due to the fact that $a(\cdot,\cdot)$ defines an inner product on V. Hence, there is a unique $u \in V$ for each $l \in V^*$.

7.3.3 An Equivalent Minimization Problem

If the bilinear form $a(\cdot,\cdot)$ is symmetric, then the solution $u \in V$ to the abstract variational problem (7.60) is also the minimizer of a minimization problem. In fact, these two problems are equivalent.

Theorem 7.2. *Let V be a Hilbert space, and let $a(\cdot,\cdot)$ be a symmetric coercive continuous bilinear form on V, and let $l(\cdot)$ be a continuous linear form on V. Then, the abstract variational problem* (7.60) *is equivalent to the minimization problem: find $u \in V$ such that*

$$F(u) = \min_{v \in V} F(v) \tag{7.66}$$

where

$$F(v) = \frac{1}{2}a(v,v) - l(v) \tag{7.67}$$

Proof. Let u be the minimizer of (7.66) and consider the function $g : \mathbb{R} \to \mathbb{R}$ such that

$$g(\epsilon) = F(u + \epsilon v) \tag{7.68}$$

for a fixed, but arbitrary, $v \in V$. Expanding g we have

$$g(\epsilon) = \frac{1}{2}a(u+\epsilon v, u+\epsilon v) - l(u+\epsilon v) \tag{7.69}$$

$$= \frac{1}{2}(a(u,u) + 2\epsilon a(u,v) + \epsilon^2 a(v,v)) - l(u) - \epsilon l(v) \tag{7.70}$$

where we have used the symmetry $a(u,v) = a(v,u)$. By construction, $g(\epsilon)$ assumes its minimum for $\epsilon = 0$, which implies $g'(0) = 0$. Thus, differentiating g with respect to ϵ we have

$$g'(\epsilon) = a(u,v) - \epsilon a(v,v) - l(v) \tag{7.71}$$

implying

$$g'(0) = a(u,v) - l(v) = 0 \tag{7.72}$$

for each $v \in V$. This is exactly our abstract variational problem.

Now, let u instead be the solution to the abstract variational problem (7.60). Expanding $F(u+v)$, with $v \in V$ arbitrary, we have

$$F(u+v) = \frac{1}{2}a(u+v, u+v) - l(u+v) \tag{7.73}$$

$$= \frac{1}{2}(a(u,u) + 2a(u,v) + a(v,v)) - l(u) - l(v) \tag{7.74}$$

$$= F(u) + a(u,v) - l(v) + \frac{1}{2}a(v,v) \tag{7.75}$$

$$= F(u) + \frac{1}{2}a(v,v) \tag{7.76}$$

$$\geq F(u) \tag{7.77}$$

where we have used that $a(u,v) = l(v)$ for all $v \in V$. From this we conclude that $F(u+w)$ attains its minimum value for $v = 0$.

Hence, (7.60) and (7.66) are equivalent. □

In mechanics applications the minimization problem can be through of as a demand for minimal energy. Indeed, the functional $F(\cdot)$ represents the mechanical energy of the physical system under consideration, and the solution u a displacement from a state of rest into one of equilibrium of forces.

7.3.4 The Lax-Milgram Lemma

The extension of the Riesz representation theorem to non-symmetric bilinear forms $a(\cdot,\cdot)$ is known as the Lax-Milgram lemma.

Theorem 7.3 (Lax-Milgram Lemma). *Let V be a Hilbert space with inner product $(\cdot,\cdot)$, and let $a(\cdot,\cdot)$ be a coercive continuous bilinear form on V, and let $l(\cdot)$ be a continuous linear form on V. Then, there exist a unique solution $u \in V$ to the abstract variational problem: find $u \in V$ such that*

$$a(u, v) = l(v), \quad \forall v \in V \tag{7.78}$$

Proof. Since $l(\cdot)$ is a continuous linear functional on V, there is a $b \in V$ such that $l(v) = (b, v)$ for all $v \in V$ by the Riesz representation theorem. Also, since $a(u,\cdot)$ is a linear and continuous functional on V for each $u \in V$ on V, there is a $w \in V$ such that $a(u, v) = (w, v)$ for all $v \in V$, again, by the Riesz representation theorem. Thus, let $Au = w$. The operator $A : V \to V$ is linear and continuous. The linearity follows from

$$(A(\lambda u + \mu w), v) = a(\lambda u + \mu w, v) = \lambda(Au, v) + \mu(Aw, v) \tag{7.79}$$

for any $u, v \in V$ and $\lambda, \mu \in \mathbb{R}$. The continuity follows from

$$\|Au\|_V^2 = (Au, Au) = a(u, Au) \leq C\|u\|_V\|Au\|_V \tag{7.80}$$

or

$$\|Au\|_V \leq C\|u\|_V \tag{7.81}$$

Thus, instead of the variational equation $a(u, v) = l(v)$ on V, let us consider the operator equation $Au = b$ in V^*. For this equation to be solvable we require that $b \in V$ is in the range $R(A) = \{w \in V : w = Av \text{ for some } v\}$ of A for any b. To show this, we use the coercivity

$$m\|u\|_V^2 \leq a(u, u) = (Au, u) \leq C\|u\|_V\|Au\|_V \tag{7.82}$$

or

$$m\|u\|_V \leq C\|Au\|_V \tag{7.83}$$

We want $R(A) = V$ and, consequently, $R(A)^\perp = N(A) = \{0\}$, where $N(A) = \{w \in V : Aw = 0\}$ is the null space of A. Assume that $Av_i \to w$ as $i \to \infty$. Then, $m\|v_i - v_j\|_V \leq C\|Av_i - Av_j\|_V \to 0$ as $i, j \to \infty$, so $\{v_i - v_j\}$ is a Cauchy sequence. Set $v = \lim_{i\to\infty} v_i$. This limit exist, since V is complete. Also, since A is continuous, we have $Av = w$, which shows that $R(A)$ is a closed subspace of V. To show that $R(A)$ is the whole of V, assume $z \in N(A)$. Then, we have

$$m\|z\|_V \leq C\|Az\|_V = 0 \tag{7.84}$$

which shows that z is zero, and that $N(A)$ is the zero set $\{0\}$. Thus, since $V = R(A) \oplus N(A)$, we conclude that $R(A) = V$. Hence, there exist a solution $u \in V$ to $Au = b$ for any $b \in V$.

It remains to show that u is unique. Suppose that u_1 and u_2 both satisfy $Au_i = b$, $i = 1, 2$. Then $m\|u_1 - u_2\|_V \leq C\|A(u_1 - u_2)\|_V = 0$, so $u_1 = u_2$. □

There are several variants and extensions of the Lax-Milgram lemma. For example, to complex-valued problems.

7.3.5 Application to Elliptic Partial Differential Equations

Let us demonstrate the usability of the Lax-Milgram lemma by working through some examples. For ease of notation, let $(\cdot,\cdot)$ and $\|\cdot\|$ denote the $L^2(\Omega)$ inner product and norm on the domain Ω, respectively.

7.3.5.1 Poisson's Equation

Let us first revisit Poisson's equation

$$-\Delta u = f, \quad x \in \Omega, \qquad u = 0, \quad x \in \partial\Omega \tag{7.85}$$

As we have seen before, the bilinear and linear form for this equation are given by

$$a(u, v) = (\nabla u, \nabla v) \tag{7.86}$$

$$l(v) = (f, v) \tag{7.87}$$

and the appropriate Hilbert space is $V = H_0^1(\Omega)$ with norm $\|v\|_V = \|v\|_{H_0^1(\Omega)} = \|\nabla v\|$.

The coercivity of $a(\cdot,\cdot)$ follows trivially.

$$a(u, u) = (\nabla u, \nabla u) = \|\nabla u\|^2 \geq m\|u\|_V^2 \tag{7.88}$$

with $m = 1$.

The continuity of $a(\cdot,\cdot)$ follows from the Cauchy-Schwarz inequality.

$$a(u,v) = (\nabla u, \nabla v) \leq \|\nabla u\|\,\|\nabla v\| \leq \|u\|_V\,\|v\|_V \tag{7.89}$$

The continuity of $l(\cdot)$ follows from the Cauchy-Schwarz inequality, again, and the Poincaré inequality.

$$l(v) = (f,v) \leq \|f\|\,\|v\| \leq C\|f\|\,\|\nabla v\| \leq C\|f\|\,\|v\|_V = C\|v\|_V \tag{7.90}$$

Here, we have absorbed the norm of f into the constant C at the end. This shows that it is natural to demand $f \in L^2(\Omega)$, since $\|f\|$ might not be well-defined otherwise.

Hence, the requirements of the Lax-Milgram lemma are fulfilled.

7.3.5.2 The Diffusion-Reaction Equation

As a second example we consider the Diffusion-Reaction equation

$$-\Delta u + cu = f, \quad x \in \Omega, \qquad n \cdot \nabla u = 0, \quad x \in \partial\Omega \tag{7.91}$$

where $c \in L^2(\Omega)$ is a given positive function with minimum value $c_0 > 0$, and $f \in L^2(\Omega)$ is a given function. The bilinear and linear forms for this equation are given by

$$a(u,v) = (\nabla u, \nabla v) + (cu, v) \tag{7.92}$$

$$l(v) = (f,v) \tag{7.93}$$

and the appropriate Hilbert space is $V = H^1(\Omega)$ with norm $\|v\|_V^2 = \|v\|_{H^1(\Omega)}^2 = \|\nabla v\|^2 + \|v\|^2$.

The coercivity of $a(\cdot,\cdot)$ follows from

$$a(u,u) = (\nabla u, \nabla u) + (cu, u) \tag{7.94}$$

$$\geq \|\nabla u\|^2 + c_0\|u\|^2 \tag{7.95}$$

$$\geq \min(1, c_0)\,(\|\nabla u\|^2 + \|u\|^2) \tag{7.96}$$

$$\geq m\|u\|_V^2 \tag{7.97}$$

with $m = \min(1, c_0)$. Also, the continuity of $a(\cdot,\cdot)$ follows from

$$a(u,v) = (\nabla u, \nabla v) + (cu, v) \tag{7.98}$$

$$\leq \|\nabla u\|\,\|\nabla v\| + \|c\|_{L^\infty(\Omega)}\|u\|\,\|v\| \tag{7.99}$$

$$\leq C(\|\nabla u\|\,\|\nabla v\| + \|u\|\,\|v\|) \tag{7.100}$$

$$\leq C(\|\nabla u\|^2 + \|u\|^2)^{1/2}(\|\nabla v\|^2 + \|v\|^2)^{1/2} \tag{7.101}$$

$$\leq C\|u\|_V \|v\|_V \tag{7.102}$$

The continuity of $l(\cdot)$ follows similarly from

$$l(v) = (f, v) \leq \|f\| \, \|v\| \leq C\|f\| \, (\|\nabla v\| + \|v\|) \leq C\|v\|_V \tag{7.103}$$

7.3.5.3 Laplace Equation with Mixed Boundary Conditions

As a final example we consider Laplace equation with mixed boundary conditions

$$-\Delta u = 0, \quad x \in \Omega, \quad u = 0, \quad x \in \Gamma_D, \quad n \cdot \nabla u = g_N, \quad x \in \Gamma_N \tag{7.104}$$

where $g_N \in L^2(\Gamma_N)$ is a given function on Γ_N. The bilinear and linear forms for this equation are given by

$$a(u, v) = (\nabla u, \nabla v) \tag{7.105}$$

$$l(v) = (g_N, v)_{L^2(\Gamma_N)} \tag{7.106}$$

and the appropriate Hilbert space is $V = \{v \in H^1(\Omega) : v|_{\Gamma_D} = 0\}$ with norm $\|v\|_V = \|\nabla v\|$. The coercivity and continuity of $a(\cdot, \cdot)$ on V is obvious. However, the continuity of $l(\cdot)$ requires us to estimate the test function $v \in V$ on the boundary segment Γ_N. To this end, we first use the Cauchy-Schwarz inequality, then the Trace inequality, and finally the Poincaré inequality.

$$l(v) = (g_N, v)_{L^2(\Gamma_N)} \tag{7.107}$$

$$\leq \|g_N\|_{L^2(\Gamma_N)} \|v\|_{L^2(\Gamma_N)} \tag{7.108}$$

$$\leq C\|g_N\|_{L^2(\Gamma_N)} (\|v\| + \|\nabla v\|) \tag{7.109}$$

$$\leq C\|g_N\|_{L^2(\Gamma_N)} \|\nabla v\| \tag{7.110}$$

$$\leq C\|g_N\|_{L^2(\Gamma_N)} \|v\|_V \tag{7.111}$$

$$= C\|v\|_V \tag{7.112}$$

7.3.6 *A General Linear Second Order Elliptic Problem*

Let us consider the general problem

$$Lu = f, \quad \text{in } \Omega \tag{7.113a}$$

$$u = 0, \quad \text{on } \partial\Omega \tag{7.113b}$$

where L is the linear second order differential operator

$$Lu = \sum_{i,j=1}^{d} -\frac{\partial}{\partial x_i}\left(a_{ij}\frac{\partial u}{\partial x_j}\right) + \sum_{i=1}^{d} b_i \frac{\partial u}{\partial x_i} + cu \tag{7.114}$$

where a_{ij}, $i, j = 1, 2, \ldots, d$, b_i, $i = 1, 2, \ldots, d$, and c are given coefficients depending only on the space coordinate x.

We shall assume that there is a constant $a_0 > 0$ such that

$$a_0 \sum_{i=1}^{d} x_i^2 \leq \sum_{i,j=1}^{d} a_{ij} x_i x_j \tag{7.115}$$

for any $x \in \Omega$, and that these coefficients are symmetric in the sense that $a_{ij} = a_{ji}$.

We shall also assume that there is a constant $c_0 > 0$ such that

$$\frac{1}{2}c - \sum_{i=1}^{d} \frac{\partial b_i}{\partial x_i} > c_0 \tag{7.116}$$

for any $x \in \Omega$.

In compact form, we may write

$$Lu = -\nabla \cdot (a\nabla u) + b \cdot \nabla u + cu \tag{7.117}$$

where a is the $d \times d$ matrix with entries a_{ij}, and b is the $d \times 1$ vector with entries b_i. This kind of partial differential equation is a called elliptic.

The assumption (7.115) is the same as assuming that the $d \times d$ matrix a is symmetric and positive definite.

Multiplying $f = Lu = -\nabla \cdot (a\nabla u) + b \cdot \nabla u + cu$ with a function $v \in V = H_0^1(\Omega)$, and integrating by parts using Green's formula, we obtain the weak form: find $u \in V$ such that

$$a(u, v) = l(v), \quad \forall v \in V \tag{7.118}$$

where the bilinear form and linear form are defined by

$$a(u, v) = (a\nabla u, \nabla v) + (b \cdot \nabla u, v) + (cu, v) \tag{7.119}$$

$$l(v) = (f, v) \tag{7.120}$$

Let us verify that the assumptions on the coefficients a_{ij}, b_i, and c are necessary to fulfill the requirements of the Lax-Milgram lemma. It is easy to show that $a(\cdot, \cdot)$ is continuous on V.

$$a(u, v) = (a\nabla u, \nabla v) + (b \cdot \nabla u, v) + (cu, v) \tag{7.121}$$

$$\leq C(\|a\|_{L^\infty(\Omega)}\|\nabla u\|\,\|\nabla v\| + \|b\|_{L^\infty(\Omega)}\|\nabla u\|\,\|v\| + \|c\|_{L^\infty(\Omega)}\|u\|\,\|v\|) \tag{7.122}$$

$$\leq C\,\|u\|_V\,\|v\|_V \tag{7.123}$$

where $\|a\|_{L^\infty(\Omega)} = \max_{1\leq i,j\leq d} \|a_{ij}\|_{L^\infty(\Omega)}$ and $\|b\|_{L^\infty(\Omega)} = \max_{1\leq i\leq d} \|b_i\|_{L^\infty(\Omega)}$. It is harder to show that $a(\cdot,\cdot)$ is coercive on V. To this end, we first use the chain rule to get

$$(\nabla\cdot(bu^2), 1) = (\nabla\cdot bu, u) + 2(b\cdot\nabla u, u) \tag{7.124}$$

The divergence theorem and the boundary condition $u = 0$ on $\partial\Omega$ then gives us

$$0 = (n\cdot bu^2, 1)_{L^2(\partial\Omega)} = (\nabla\cdot(bu^2), 1) \tag{7.125}$$

Finally, as a consequence,

$$a(u,u) = (a\nabla u, \nabla u) - \frac{1}{2}((\nabla\cdot b)u, u) + (cu, u) \tag{7.126}$$

$$\geq a_0\|\nabla u\|^2 + c_0\|u\|^2 \tag{7.127}$$

$$\geq m\|u\|_V^2 \tag{7.128}$$

with $m = \min(a_0, c_0)$.

It is obvious that, $l(\cdot)$ is continuous on V.

7.4 Abstract Finite Element Approximation

7.4.1 Abstract Finite Element Method

From the Lax-Milgram lemma we know that the solution u to the weak form (7.60) exist and is unique. Thus, let us seek to approximate it using finite elements. To this end, let $V_h \subset V$ be a finite dimensional subspace of V typically consisting of continuous piecewise linear polynomials on a mesh $\mathcal{K}$ of Ω with mesh size h.

The finite element approximation of (7.60) takes the form: find $u_h \in V_h$ such that

$$a(u_h, v) = l(v), \quad \forall v \in V_h \tag{7.129}$$

7.4.2 The Stiffness Matrix and Load Vector

Let $\{\varphi_i\}_{i=1}^n$ be a basis for V_h. The finite element method is equivalent to

$$a(u_h, \varphi_i) = l(\varphi_i), \quad i = 1, 2, \ldots, n \tag{7.130}$$

Indeed, if (7.129) holds for v anyone of the basis functions φ_i, $i = 1, 2, \ldots, n$, then it also holds for v a linear combination of these basis functions. Conversely, since any $v \in V_h$ is precisely such a linear combination, (7.130) implies (7.129).

Since the finite element solution $u_h \in V_h$ it can written as the linear combination

$$u_h = \sum_{j=1}^{n} \xi_j \varphi_j \tag{7.131}$$

where ξ_j, $j = 1, 2, \ldots, n$, are n unknown coefficients to be determined.

Inserting (7.131) into (7.130) we get

$$b_i = l(\varphi_i) = \sum_{j=1}^{n} \xi_j a(\varphi_j, \varphi_i) = \sum_{j=1}^{n} A_{ij} \xi_j, \quad i = 1, 2, \ldots, n \tag{7.132}$$

which is n linear algebraic equations for the unknowns ξ_j, $j = 1, 2, \ldots, n$. In matrix form, we write this

$$A\xi = b \tag{7.133}$$

where the entries of the $n \times n$ matrix A are given by $A_{ij} = a(\varphi_j, \varphi_i)$, and the entries of the $n \times 1$ vector b are given by $b_i = l(\varphi_i)$. For historical reasons, A and b are referred to as stiffness matrix and load vector. Solving the linear system $A\xi = b$ for the unknown $n \times 1$ vector ξ containing the unknowns ξ_j yields the finite element solution $u_h = \sum_{j=1}^{n} \xi_j \varphi_j$.

7.4.3 Galerkin Orthogonality

To extract information about the error $e = u - u_h$ we can subtract the finite element method (7.129) from the weak form (7.60). In doing so, since $V_h \subset V$, we obtain the Galerkin orthogonality

$$a(e, v) = 0, \quad \forall v \in V_h \tag{7.134}$$

Thus, the error e is orthogonal to V_h with respect to the inner product $a(\cdot, \cdot)$.

7.4.4 A Priori Error Estimates

The Galerkin orthogonality yields the following abstract best approximation result known as Cea's lemma.

Theorem 7.4 (Cea's Lemma). *The error e satisfies the best approximation result*

$$\|e\|_V \le \frac{C_a}{m}\|u - v\|_V, \quad \forall v \in V_h \tag{7.135}$$

Proof. Starting from the coercivity of $a(\cdot,\cdot)$ we have, for any $v \in V_h$,

$$m\|e\|_V^2 \le a(e,e) \tag{7.136}$$
$$= a(e, u - u_h) \tag{7.137}$$
$$= a(e, u - v + v - u_h) \tag{7.138}$$
$$= a(e, u - v) + a(e, v - u_h) \tag{7.139}$$
$$= a(e, u - v) \tag{7.140}$$
$$\le C_a\|e\|_V\|u - v\|_V \tag{7.141}$$

where we have used that $a(e, v - u_h) = 0$ due to Galerkin orthogonality, and the continuity of $a(\cdot,\cdot)$ in the last line. Dividing by $\|e\|_V$ concludes the proof. □

We can quantify Cea's lemma by first choosing v as the interpolant $\pi u \in V_h$ of u, and then use interpolation theory to estimate the difference $u - \pi u$. Assuming that $\|u - \pi u\|_V \le Ch|u|_{H^2(\Omega)}$, we immediately have the following abstract a priori error estimate.

Theorem 7.5. *The error e satisfies the a priori estimate*

$$\|e\|_V \le Ch|u|_{H^2(\Omega)} \tag{7.142}$$

Thus, the error measured in the energy norm on V tends to zero as the mesh size h tends to zero. Hence, the convergence of the finite element method is asserted.

To obtain a priori error estimates in the L^2-norm, instead of the V-norm, it is generally necessary to use a duality argument.

7.4.5 A Posteriori Error Estimates

A priori estimates are not computable as they involve the unknown solution u. By contrast, a posteriori estimates can be used to actually compute bounds on the error e.

To derive abstract posteriori estimates we note that, for any $v \in V_h$,

$$m\|e\|_V^2 \le a(e,e) \tag{7.143}$$
$$= a(u,e) - a(u_h,e) \tag{7.144}$$
$$= l(e) - a(u_h,e) \tag{7.145}$$

Introducing the weak residual $R(u_h) \in V^*$, defined by

$$(R(u_h), v) = l(v) - a(u_h, v), \quad \forall v \in V \tag{7.146}$$

we obtain the following error representation

$$m\|e\|_V^2 \leq (R(u_h), e) \tag{7.147}$$

and thus

$$m\|e\|_V^2 \leq \sup_{v \in V} \frac{(R(u_h), v)}{\|v\|_V} \|e\|_V = \|R(u_h)\|_{V^*}\|e\|_V \tag{7.148}$$

Dividing by $\|e\|_V$ we have the a posteriori estimate

$$\|e\|_V \leq \frac{1}{m}\|R(u_h)\|_{V^*} \tag{7.149}$$

Unfortunately the dual norm $\|R(u_h)\|_{V^*}$ is complicated to compute due to the supremum, but, of course, we still wish to derive a computable estimate. To this end, consider for simplicity Poisson's equation $-\Delta u = f$ in Ω with $u = 0$ on $\partial\Omega$, and assume that we are using a standard finite element method based on continuous piecewise polynomial approximation. In this case, we have $V = H_0^1(\Omega)$, and proceed as follows. First, we have from Galerkin orthogonality, and elementwise integration by parts

$$(R(u_h), v) = l(v) - a(u_h, v) \tag{7.150}$$

$$= l(v - \pi v) - a(u_h, v - \pi v) \tag{7.151}$$

$$= \sum_{K \in \mathcal{K}} (f + \Delta u, v - \pi v)_K + \frac{1}{2}([n \cdot \nabla u_h], v - \pi v)_{\partial K} \tag{7.152}$$

$$\leq \sum_{K \in \mathcal{K}} \|f + \Delta u\|_K \|v - \pi v\|_K + \frac{1}{2}\|[n \cdot \nabla u_h]\|_{\partial K \setminus \partial \Omega} \|v - \pi v\|_{\partial K} \tag{7.153}$$

Next, using the Scott-Zhang interpolant we have

$$\|v - \pi v\|_K = Ch\|\nabla v\|_{\mathcal{N}(K)} \tag{7.154}$$

which together with the Trace inequality yields

$$\|v - \pi v\|_{\partial K}^2 \leq C\left(h_K^{-1}\|v - \pi v\|_K^2 + h_K\|\nabla(v - \pi v)\|_K^2\right) \tag{7.155}$$

$$\leq C h_K \|\nabla v\|_{\mathcal{N}(K)}^2 \tag{7.156}$$

Now, combining the above results we arrive at

$$
(R(u_h), v) \leq C \left(\sum_{K \in \mathcal{K}} h_K^2 \|f + \Delta u\|_K^2 + \frac{1}{4} h_K \|[n \cdot \nabla u_h]\|_{\partial K \setminus \partial \Omega}^2 \right)^{1/2}
$$

$$
\left(\sum_{K \in \mathcal{K}} \|\nabla v\|_{\mathcal{N}(K)}^2 \right)^{1/2} \tag{7.157}
$$

$$
= C \left(\sum_{K \in \mathcal{K}} h_K^2 \|f + \Delta u\|_K^2 + \frac{1}{4} h_K \|[n \cdot \nabla u_h]\|_{\partial K \setminus \partial \Omega}^2 \right)^{1/2} \|v\|_V \tag{7.158}
$$

Finally, dividing by $\|v\|_V$ and taking the supremum we get the following estimate

$$
\|e\|_V \leq C \|R(u_h)\|_{V^*} \leq C \left(\sum_{K \in \mathcal{K}} h_K^2 \|f + \Delta u\|_K^2 + \frac{1}{4} h_K \|[n \cdot \nabla u_h]\|_{\partial K \setminus \partial \Omega}^2 \right)^{1/2} \tag{7.159}
$$

We finally remark that to obtain a posteriori estimates for any quantities other than the V-norm it is generally necessary to solve dual problems.

7.5 Further Reading

There are many references on functional analysis. We mention Kreyszig [48] and Rudin [56]. Texts on partial differential equations include Evans [29] and Folland [30]. Sobolev spaces are treated by Adams and Fournier [1]. The functional analysis and Sobolev space theory necessary for finite elements can also be found in Brenner and Scott [60].

7.6 Problems

Exercise 7.1. Write out all terms in the sum $\sum_{|\alpha|=2} D^\alpha u$.

Exercise 7.2. Calculate the weak derivative of the function

$$
g = \begin{cases} x, & 0 < x < 1 \\ 1, & 1 < x < 2 \end{cases}
$$

Exercise 7.3. Write down the inner product and norm for the Hilbert space $H^2(\Omega)$.

Exercise 7.4. Is $L^2(\Omega)$ a subset of $H_0^1(\Omega)$? What about $H^2(\Omega)$?

Exercise 7.5. Let $v = \log(\log(1/|x|))$ on the disc $\Omega = \{(r,\theta) : 0 \leq r \leq R,\ 0 \leq \theta < 2\pi\}$ with $R \leq 1/e$. Verify that $v \in H^1(\Omega)$. Is $v \in C^0(\Omega)$?

Exercise 7.6. Consider the normed space $C^0(I)$ of all continuous functions on the interval $I = [0,1]$ with norm $\|v\|_{L^\infty(I)} = \sup_{x \in I} |v(x)|$. Let $f(x) = 1$ and $g(x) = x$. Does the Parallelogram law hold for f and g? Is $C^0(I)$ an inner product space?

Exercise 7.7. Show that the solution $u \in V$ to the weak form (7.60) satisfies the stability estimate $\|u\|_V \leq C_l/m$.

Exercise 7.8. Use the Poincaré inequality to show that $\|v\|_{H^1(\Omega)}$ and $|v|_{H_0^1(\Omega)}$ are equivalent norms on $H_0^1(\Omega)$. In particular, verify that $\|\nabla v\| = 0$ implies $v = 0$.

Exercise 7.9. What numerical values do the constants m, C_a, and C_l have for the problem $-\Delta u = xy^2$ on the square $\Omega = [-1,2] \times [0,3]$ assuming $u = 0$ on $\partial\Omega$?

Exercise 7.10. Consider

$$a(u,v) = v^T Au, \qquad l(v) = v^T b, \qquad V = \mathbb{R}^n$$

where A is a real $n \times n$ matrix, b is a real $n \times 1$ vector, and $\|\cdot\|_V = \|\cdot\|_2$ the Euclidean 2-norm.

(a) Assuming that $a(\cdot,\cdot)$ is coercive on V, what can be said about the eigenvalues of A?
(b) Show that there exist a unique solution $u \in V$ to the linear system $Au = b$.

Exercise 7.11. Verify the Trace inequality (7.46) for the particular choice $v = x$ on the square $\Omega = [0,L]^2$ with side length L. How does the constant C in the inequality depend on L?

Exercise 7.12. Decide if the requirements for the Lax-Milgram lemma are fulfilled for the problem $-\nabla \cdot (a\nabla u) = 1$ in $\Omega = [0,1]^2$, with $u = 0$ on $\partial\Omega$, assuming that a is the 2×2 matrix

$$a = \begin{bmatrix} 4 & 1 \\ 1 & 2 \end{bmatrix}$$

Chapter 8
The Finite Element

Abstract In this chapter we study the concept of a finite element in some more detail. We begin with the classical definition of a finite element as the triplet of a polygon, a polynomial space, and a set of functionals. We then show how to derive shape functions for the most common Lagrange elements. The isoparametric mapping is introduced as a tool to allow for elements with curved boundaries, and to simplify the computation of the element stiffness matrix and load vector. We finish by presenting some more exotic elements.

8.1 Different Types of Finite Elements

8.1.1 Formal Definition of a Finite Element

Formally, a finite element consists of the triplet:

- A polygon $K \subset \mathbb{R}^d$.
- A polynomial function space P on K.
- A set of $n = \dim(P)$ linear functionals $L_i(\cdot)$, $i = 1, 2, \dots, n$, defining the so-called degrees of freedom.

The polygon K is of different type depending on if the space dimension d is $1, 2$, or 3. The most common types of polygons in use are lines, triangles, quadrilaterals, tetrahedrons, and bricks. Occasionally, prisms are used. Each polygon stems from a mesh $\mathcal{K} = \{K\}$ of the computational domain Ω. Triangle and tetrahedron meshes are able to easily represent domains with curved boundaries, while quadrilateral and brick meshes are easy to implement in a computer. Prisms are primarily used for domains with cylindrical symmetries, such as pipes, for instance.

Let us equip P with a basis $\{S_j\}_{j=1}^n$. The basis functions S_j are generally called shape functions.

M.G. Larson and F. Bengzon, *The Finite Element Method: Theory, Implementation, and Applications*, Texts in Computational Science and Engineering 10,
DOI 10.1007/978-3-642-33287-6_8, © Springer-Verlag Berlin Heidelberg 2013

A finite element is said to be unisolvent if the functionals can uniquely determine the shape functions. Unisolvency can be thought of as a necessary compatibility condition for $L_i(\cdot)$, P, and K. By definition, it is equivalent to $L_i(v) = 0 \equiv v = 0$, for all $v \in P$ and all i. Even though it can be a bit hard to establish unisolvency, the actual calculation of the shape functions is easy as it simply amounts to solving a linear system. Indeed, the shape functions S_j are determined from the n linear algebraic equations

$$L_i(S_j) = \delta_{ij}, \quad i, j = 1, 2, \ldots, n \tag{8.1}$$

By taking a linear combination of shape functions and coefficients we get a polynomial or finite element function in P on each polygon K.

Besides specifying the shape functions on each polygon K, the functionals also specify the behavior of the these functions between adjacent polygons. For instance, if we want finite element functions that are continuous on the whole mesh $\mathcal{K}$, then we must take care in choosing functionals, so that the resulting shape functions become continuous, especially across the polygon boundary ∂K. Indeed, the functionals determine both the local and the global properties of the finite element space V_h.

The particular choice of functionals $L_i(\cdot)$ give rise to groups or families of finite elements sharing similar properties (e.g., continuity) although they might be different in other aspects (e.g., polynomial order). The Lagrange family is the most popular and widely used. The defining functionals are

$$L_i(v) = v(N_i), \quad i = 1, 2, \ldots, n \tag{8.2}$$

where $N_i, i = 1, 2, \ldots, n$ are n carefully selected node points. The functionals are the simplest possible in the sense that each of them only consist of a point evaluation of v. Because each shape function is associated with a particular node, the resulting set of shape functions is called a nodal basis. For $d = 2$ with $P = P_1(K)$ the space of linear polynomials on a triangle K, these node points are the triangle vertices, and the shape functions S_j, $j = 1, 2, 3$, are the familiar hat functions. However, the above functionals can be used to define shape functions in any dimension and for any polynomial order provided that the node points are chosen appropriately. We remark that for linear tetrahedrons the nodes are the tetrahedron vertices.

The Lagrange elements are continuous, but have discontinuous derivatives across element boundaries. Thus, they are C^0 continuous elements, which suffice to approximate functions in H^1. However, in some applications it is necessary to use more regular elements. An example of such an element is the so-called Argyris element, which has C^1 continuity. Because all its first derivatives are continuous, this element can be used to approximate functions in H^2, which is the appropriate space for some fourth order problems, such as $\Delta^2 u = 0$, for instance. Not surprisingly, construction of the Argyris element is more elaborate than for the Lagrange element. Indeed, on a triangle there are 21 defining functionals involving point evaluation of the first, normal, and second order derivatives for the Argyris

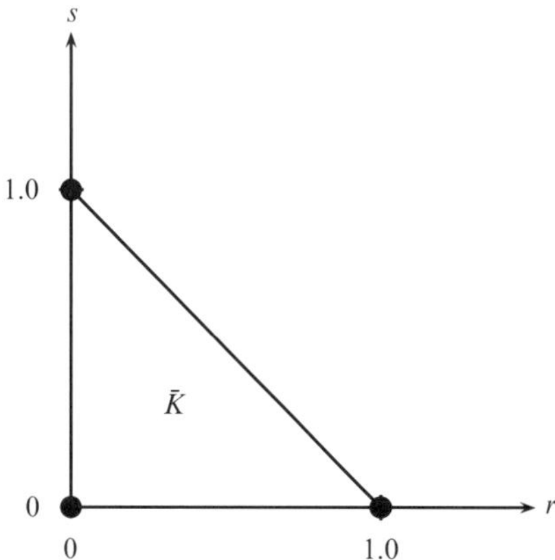

Fig. 8.1 Node points for the linear Lagrange element on the reference *triangle* $\bar{K}$

element. Other element types with higher continuity properties include the Hermite elements, and the Morley element, which we shall return to shortly.

8.1.2 Shape Functions for the Linear Lagrange Triangle

Let us derive the shape functions for the triangular linear Lagrange finite element. To this end, let $\bar{K}$ be the domain $\bar{K} = \{(r, s) : 0 < r, s < 1, r + s < 1\}$, that is, the triangle with vertices at origo, $(1, 0)$, and $(0, 1)$, see Fig. 8.1. We shall refer to this as the reference triangle.

By definition, the appropriate polynomial space P is the space of linear polynomials $P_1(\bar{K})$ on $\bar{K}$, and the defining functionals are given by

$$L_1(v) = v(0, 0), \quad L_2(v) = v(1, 0), \quad L_3(v) = v(0, 1) \tag{8.3}$$

That is, the nodes are the three vertices of $\bar{K}$.

Perhaps the simplest basis for $P_1(K)$ is the canonical basis $\{1, r, s\}$, so anyone of the three shape functions S_j, $j = 1, 2, 3$, can be expressed as a linear combination of 1, r, and s. For example, S_1 can be written $S_1 = c_1 + c_2 r + c_3 s$, where c_i, $i = 1, 2, 3$ are coefficients to be determined. In doing so, we demand that $L_i(S_1) = \delta_{i1}$, which yields the 3×3 linear system

$$e_1 = \begin{bmatrix} 1 \\ 0 \\ 0 \end{bmatrix} = \begin{bmatrix} L_1(1) & L_1(r) & L_1(s) \\ L_2(1) & L_2(r) & L_2(s) \\ L_3(1) & L_3(r) & L_3(s) \end{bmatrix} \begin{bmatrix} c_1 \\ c_2 \\ c_3 \end{bmatrix} = \begin{bmatrix} 1 & 0 & 0 \\ 1 & 1 & 0 \\ 1 & 0 & 1 \end{bmatrix} \begin{bmatrix} c_1 \\ c_2 \\ c_3 \end{bmatrix} = Vc \tag{8.4}$$

for the unknown coefficients c_i. Note that the entries of V are very simple to evaluate. For example, the first row is point evaluation of the functions 1, r, and s at origo. This immediately gives us $V_{11} = L_1(1) = 1$, $V_{12} = L_1(r) = 0$, $V_{13} = L_1(s) = 0$, and so on. The matrix V is generally called a Vandermonde matrix. Computing $V^{-1}e_1$ we readily obtain $c = [1,\ -1,\ -1]^T$, from which we deduce that $S_1 = c_1 + c_2 r + c_3 s = 1 - r - s$. Proceeding similarly for the shape functions S_2 and S_3 we eventually find that

$$S_1 = 1 - r - s \tag{8.5}$$

$$S_2 = r \tag{8.6}$$

$$S_3 = s \tag{8.7}$$

which we recognize as the usual hat functions on $\bar{K}$.

Let us list a routine for evaluating the linear shape functions and their partial derivatives at a point (r, s) in $\bar{K}$.

```
function [S,dSdr,dSds] = P1shapes(r,s)
S=[1-r-s; r; s];
dSdr=[-1; 1; 0];
dSds=[-1; 0; 1];
```

8.1.3 Shape Functions for the Quadratic Lagrange Triangle

For the triangular quadratic Lagrange finite element, the polynomial space P is the space of quadratic polynomials $P_2(\bar{K})$ on the reference triangle K, and the defining functionals are given by

$$L_1(v) = v(0,0), \quad L_2(v) = v(1,0), \quad L_3(v) = v(0,1) \tag{8.8}$$

$$L_4(v) = v(0.5,0.5), \quad L_5(v) = v(0,0.5), \quad L_6(v) = v(0.5,0) \tag{8.9}$$

In other words, the nodes are the three vertices, and the mid-points of the three edges of K, see Fig. 8.2.

Since a general polynomial of two variables has six coefficients, there must be six shape functions S_j, $j = 1, 2, \dots, 6$. To see this, note that the canonical basis for $P_2(\bar{K})$ is $\{1, r, s, r^2, rs, s^2\}$, and that N_j is a linear combination of these monomials. Thus, we have $S_1 = c_1 + c_2 r + c_3 s + c_4 r^2 + c_5 rs + c_6 s^2$ for example. To determine the coefficients c_i, $i = 1, 2, \dots, 6$, we again demand that $L_i(S_1) = \delta_{i1}$, which yields the 6×6 linear system

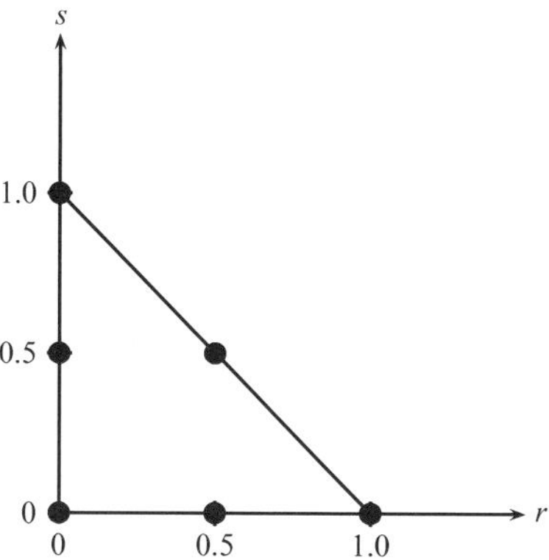

Fig. 8.2 Node points for the quadratic Lagrange element on the reference *triangle* $\bar{K}$

$$e_1 = \begin{bmatrix} 1 \\ 0 \\ 0 \\ 0 \\ 0 \\ 0 \end{bmatrix} = \begin{bmatrix} L_1(1) & L_1(r) & L_1(s) & L_1(r^2) & L_1(rs) & L_1(s^2) \\ L_2(1) & L_2(r) & L_2(s) & L_2(r^2) & L_2(rs) & L_2(s^2) \\ L_3(1) & L_3(r) & L_3(s) & L_3(r^2) & L_3(rs) & L_3(s^2) \\ L_4(1) & L_4(r) & L_4(s) & L_4(r^2) & L_4(rs) & L_4(s^2) \\ L_5(1) & L_5(r) & L_5(s) & L_5(r^2) & L_5(rs) & L_5(s^2) \\ L_6(1) & L_6(r) & L_6(s) & L_6(r^2) & L_6(rs) & L_6(s^2) \end{bmatrix} \begin{bmatrix} c_1 \\ c_2 \\ c_3 \\ c_4 \\ c_5 \\ c_6 \end{bmatrix} \tag{8.10}$$

$$= \begin{bmatrix} 1 & 0 & 0 & 0 & 0 & 0 \\ 1 & 1 & 0 & 1 & 0 & 0 \\ 1 & 0 & 1 & 0 & 0 & 1 \\ 1 & 0.5 & 0.5 & 0.25 & 0.25 & 0.25 \\ 1 & 0 & 0.5 & 0 & 0 & 0.25 \\ 1 & 0.5 & 0 & 0.25 & 0 & 0 \end{bmatrix} \begin{bmatrix} c_1 \\ c_2 \\ c_3 \\ c_4 \\ c_5 \\ c_6 \end{bmatrix} = Vc \tag{8.11}$$

from which it follows that $c = [1, -3, -4, 2, 4, 2]^T$.

The other shape functions can be found in a similar fashion. Their explicit formulas are

$$S_1 = 1 - 3r - 3s + 2r^2 + 4rs + 2s^2 \tag{8.12}$$

$$S_2 = 2r^2 - r \tag{8.13}$$

$$S_3 = 2s^2 - s \tag{8.14}$$

$$S_4 = 4rs \tag{8.15}$$

$$S_5 = 4s - 4rs - 4s^2 \tag{8.16}$$

$$S_6 = 4r - 4r^2 - 4rs \tag{8.17}$$

Let us list a routine for evaluating the quadratic shape functions and their partial derivatives at a point (r, s) in $\bar{K}$.

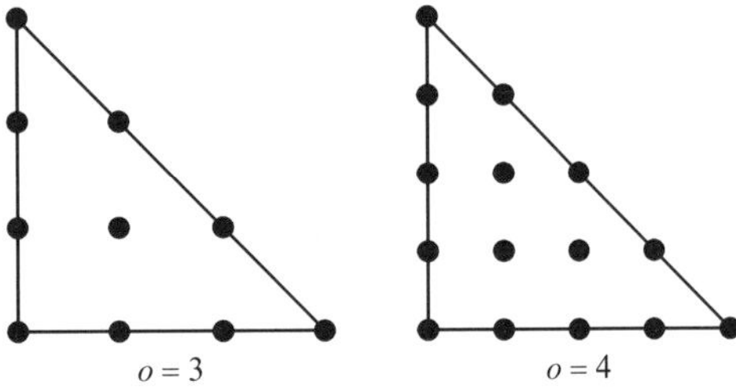

Fig. 8.3 Node points for the cubic and quartic Lagrange elements

```
function [S,dSdr,dSds] = P2shapes(r,s)
S=[1-3*r-3*s+2*r^2+4*r*s+2*s^2;
   2*r^2-r;
   2*s^2-s;
   4*r*s;
   4*s-4*r*s-4*s^2;
   4*r-4*r^2-4*r*s];
dSdr=[-3+4*r+4*s; 4*r-1; 0; 4*s; -4*s; 4-8*r-4*s];
dSds=[-3+4*r+4*s; 0; 4*s-1; 4*r; 4-4*r-8*s; -4*r];
```

8.1.4 Higher Order Triangular Lagrange Elements

The procedure for computing Lagrange shape functions on the reference triangle $\bar{K}$ generalizes to higher order. If the order of the polynomial space P is o, then there are $n = (o+1)(o+2)/2$ nodes and shape functions. Moreover, the nodes are positioned in a lattice called the principal lattice on the reference triangle $\bar{K}$. We have already seen this lattice for $o = 1$ and 2. Figure 8.3 shows it also for $o = 3$ and 4. The generalization to any higher order should be obvious.

8.1.5 Shape Functions for the Bilinear Quadrilateral Element

As already mentioned, isoparametric finite elements can also be constructed on quadrilaterals. To do so, let $\bar{Q}$ be the reference square $\bar{Q} = \{(r,s) : -1 < r,s < 1\}$, as shown in Fig. 8.4, and let $P(\bar{Q})$ be the space of so-called bilinear functions spanned by the canonical basis $\{1, r, s, rs\}$. Defining, again, the functionals $L_i(v) = v(N_i)$, where the nodes N_i, $i = 1,2,3,4$ are the four corners of the square $\bar{Q}$, it is a simple task to verify that the shape functions take the form

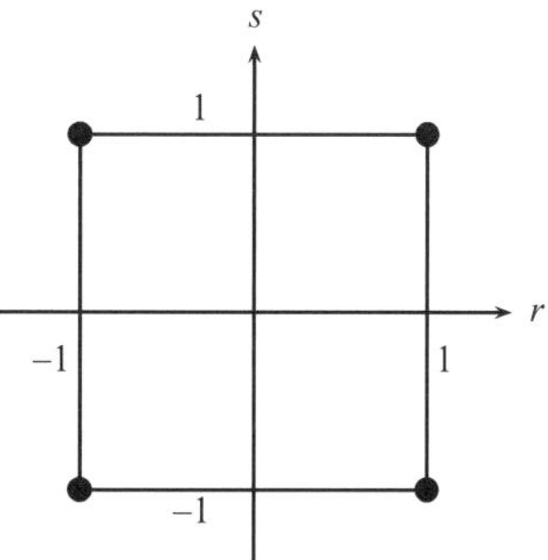

Fig. 8.4 Node points for the bilinear element on the reference *square* $\bar{Q}$

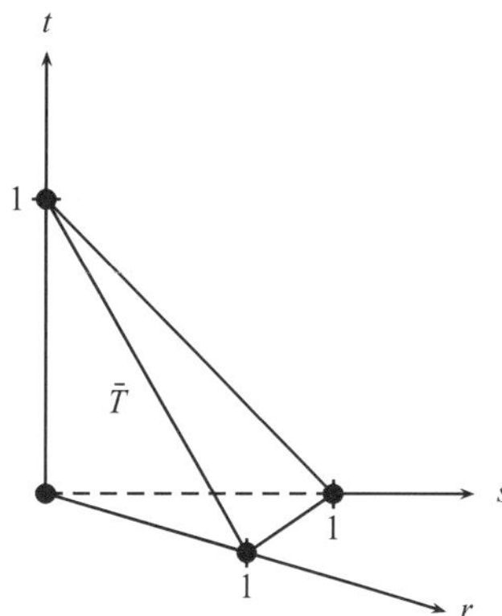

Fig. 8.5 Node points for the linear Lagrange element on the reference tetrahedron $\bar{T}$

$$S_1 = \tfrac{1}{4}(1-r)(1-s) \tag{8.18}$$

$$S_2 = \tfrac{1}{4}(1+r)(1-s) \tag{8.19}$$

$$S_3 = \tfrac{1}{4}(1+r)(1+s) \tag{8.20}$$

$$S_4 = \tfrac{1}{4}(1-r)(1+s) \tag{8.21}$$

8.1.6 Shape Functions for the Linear Lagrange Tetrahedron

To construct isoparametric Lagrange finite elements on tetrahedrons, let $\bar{T} = \{(r,s,t) : 0 < r,s,t < 1, r+s+t < 1\}$ be the reference tetrahedron with vertices at origo, $(1,0,0)$, $(0,1,0)$, and $(0,0,1)$, see Fig. 8.5.

In the linear case $P = P_1(\bar{T})$, the nodes N_i, $i = 1,2,3,4$, are the vertices of $\bar{T}$, and the functionals are given by $L_i(v) = v(N_i)$, where $v = c_0 + c_1x_1 + c_2x_2 + c_3x_3$. This yields the shape functions

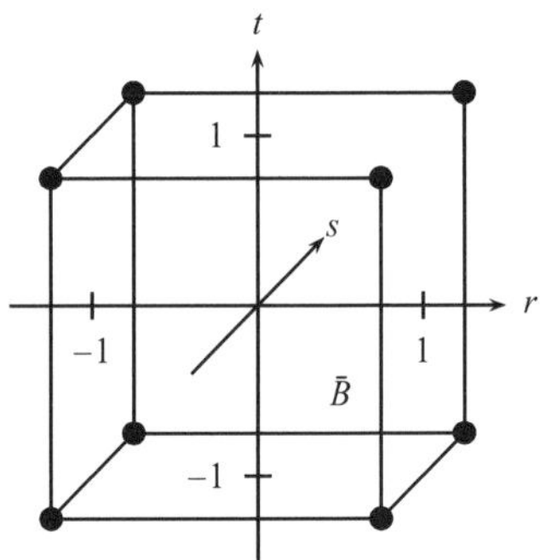

Fig. 8.6 Node points for the trilinear element on the reference brick $\bar{B}$

$$S_1 = 1 - r - s - t \tag{8.22}$$

$$S_2 = r \tag{8.23}$$

$$S_3 = s \tag{8.24}$$

$$S_4 = t \tag{8.25}$$

Higher order elements on tetrahedra can be constructed in the same way as shown above for triangles.

8.1.7 Shape Functions for the Trilinear Brick Element

There are also isoparametric finite elements on bricks. Let $\bar{B} = \{(r, s, t) : -1 < r, s, t < 1\}$, as shown in Fig. 8.6, and let $P(\bar{B})$ be the space of so-called trilinear functions spanned by the canonical basis $\{1, r, s, t, rs, rt, st, rst\}$. Letting also the functionals be $L_i(v) = v(N_i)$, with the nodes N_i, $i = 1, 2, \ldots, 8$, the corners of the brick, we obtain the shape functions

$$S_1 = \tfrac{1}{8}(1 - r)(1 - s)(1 - t) \tag{8.26}$$

$$S_2 = \tfrac{1}{8}(1 + r)(1 - s)(1 - t) \tag{8.27}$$

$$S_3 = \tfrac{1}{8}(1 + r)(1 + s)(1 - t) \tag{8.28}$$

$$S_4 = \tfrac{1}{8}(1 - r)(1 + s)(1 - t) \tag{8.29}$$

$$S_5 = \tfrac{1}{8}(1 - r)(1 - s)(1 + t) \tag{8.30}$$

$$S_6 = \tfrac{1}{8}(1 + r)(1 - s)(1 + t) \tag{8.31}$$

$$S_7 = \tfrac{1}{8}(1 + r)(1 + s)(1 + t) \tag{8.32}$$

$$S_8 = \tfrac{1}{8}(1 - r)(1 + s)(1 + t) \tag{8.33}$$

8.2 The Isoparametric Mapping

Up to now we have used various tricks to integrate the entries of the element stiffness matrix and load vector. However, this approach quickly gets cumbersome for higher order elements. Also, to improve the geometry representation of the computational domain we would like to be able to work with elements with curved boundaries. Fortunately, it turns out that these two obstacles can be overcome by combining the idea of numerical quadrature with so-called isoparametric elements. As we shall see, the combination of these two ideas allows for a simple and uniform treatment of the elemental assembly procedure. In the following, we shall present the isoparametric mapping for triangle elements, although all concepts directly carry over to many other element types, such as quadrilaterals, tetrahedrons, and bricks, for instance.

The setting up of the isoparametric map is easily described. Suppose we have a triangle K with nodes at $N_i = (x_1^{(i)}, x_2^{(i)})$, $i = 1, 2, \dots, n$. We will refer to K as to the physical element, as opposed to the reference element $\bar{K}$. Now, the basic idea is to use the shape functions S_j on $\bar{K}$ to define the geometry of K through the formulas

$$x_1(r,s) = \sum_{i=1}^{n} x_1^{(i)} S_i(r,s) \tag{8.34}$$

$$x_2(r,s) = \sum_{i=1}^{n} x_2^{(i)} S_i(r,s) \tag{8.35}$$

In other words, given a point (r,s) in $\bar{K}$ the above formulas maps it to a corresponding physical point (x_1, x_2) in K. We say that the coordinates x_1 and x_2 are parametrized by r and s. This is the two dimensional isoparametric mapping. It is shown by Fig. 8.7 for a quadratic Lagrange triangle. In three dimensions the isoparametric map is similar, but uses three parameters r, s, t to map a point (x_1, x_2, x_3) from the reference to the physical element.

The isoparametric map implies that the boundary ∂K is curved whenever the nodes N_i on an edge of K do not lie on a straight line. In fact, the boundary of K is defined by interpolating the nodes on the edges by a polynomial of order o, where o is the polynomial order of the shape functions S_j.

As usual, any finite element function v on K is also expressed using the shape functions, viz.,

$$v(r,s) = \sum_{i=1}^{n} v_i S_i(r,s) \tag{8.36}$$

where $v_i = v(N_i)$ are the n nodal values of v.

Since the stiffness matrix involves partial derivatives of v we use the chain rule to differentiate v with respect to r and s, yielding

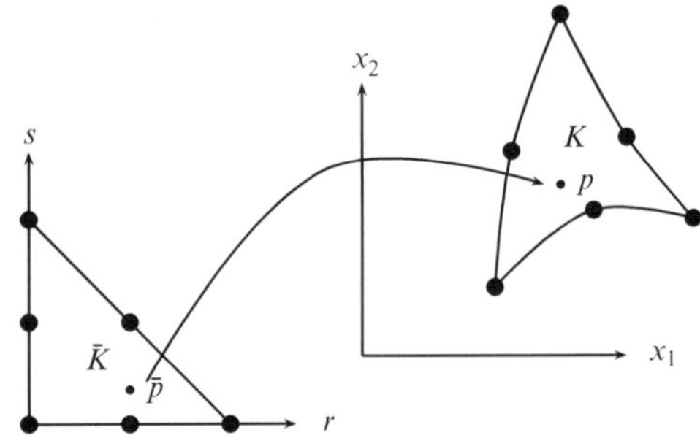

Fig. 8.7 Isoparametric map of a quadratic Lagrange *triangle*. Point $\bar{p}$ is mapped to p

$$\frac{\partial v}{\partial x_1} = \frac{\partial v}{\partial r}\frac{\partial r}{\partial x_1} + \frac{\partial v}{\partial s}\frac{\partial s}{\partial x_1} \tag{8.37}$$

$$\frac{\partial v}{\partial x_2} = \frac{\partial v}{\partial r}\frac{\partial r}{\partial x_2} + \frac{\partial v}{\partial s}\frac{\partial s}{\partial x_2} \tag{8.38}$$

In matrix form we can write this as

$$\begin{bmatrix} \frac{\partial v}{\partial x_1} \\ \frac{\partial v}{\partial x_2} \end{bmatrix} = \begin{bmatrix} \frac{\partial r}{\partial x_1} & \frac{\partial s}{\partial x_1} \\ \frac{\partial r}{\partial x_2} & \frac{\partial s}{\partial x_2} \end{bmatrix} \begin{bmatrix} \frac{\partial v}{\partial r} \\ \frac{\partial v}{\partial s} \end{bmatrix} = J^{-1} \begin{bmatrix} \frac{\partial v}{\partial r} \\ \frac{\partial v}{\partial s} \end{bmatrix} \tag{8.39}$$

where we have introduced the Jacobian matrix J, defined by

$$J = \begin{bmatrix} \frac{\partial x_1}{\partial r} & \frac{\partial x_2}{\partial r} \\ \frac{\partial x_1}{\partial s} & \frac{\partial x_2}{\partial s} \end{bmatrix} \tag{8.40}$$

Here, the explicit expressions for the entries of J are given by

$$J_{11} = \frac{\partial x_1}{\partial r} = \sum_{i=1}^{n} \frac{\partial S_i}{\partial r} x_1^{(i)} \tag{8.41}$$

$$J_{12} = \frac{\partial x_2}{\partial r} = \sum_{i=1}^{n} \frac{\partial S_i}{\partial r} x_2^{(i)} \tag{8.42}$$

$$J_{21} = \frac{\partial x_1}{\partial s} = \sum_{i=1}^{n} \frac{\partial S_i}{\partial s} x_1^{(i)} \tag{8.43}$$

$$J_{22} = \frac{\partial x_2}{\partial s} = \sum_{i=1}^{n} \frac{\partial S_i}{\partial s} x_2^{(i)} \tag{8.44}$$

Thus, we can compute the partial derivatives of v at any physical point (x_1, x_2) in K, or equivalently, at the corresponding reference point (r, s) in $\bar{K}$, by solving the 2×2 linear system (8.39).

We remark that the invertability of J depends on the quality of K. If J^{-1} exist, then the isoparametric map is one-to-one.

For the triangular linear Lagrange finite element the Jacobian matrix is given by

$$J = \begin{bmatrix} x_1^{(2)} - x_1^{(1)} & x_2^{(2)} - x_2^{(1)} \\ x_2^{(3)} - x_2^{(1)} & x_2^{(3)} - x_2^{(1)} \end{bmatrix} \tag{8.45}$$

Further, the determinant of J is given by

$$\det(J) = 2|K| \tag{8.46}$$

This is to be expected, since we might recall from calculus that the determinant of a mapping is the area scale between the image and range of the mapping (i.e., two domains K and $\bar{K}$). Now, the area of $\bar{K}$ is $1/2$. Hence, the factor 2 in front of $|K|$. Note that $\det(J)$ is constant for this particular element.

A routine for computing the Jacobian J at a point (r, s), given the n node coordinates $N_i = (x_1^{(i)}, x_2^{(i)})$, is given below.

```
function [S,dSdx,dSdy,detJ] = Isopmap(x,y,r,s,shapefcn)
[S,dSdr,dSds]=shapefcn(r,s);
j11=dot(dSdr,x); j12=dot(dNdr,y);
j21=dot(dSds,x); j22=dot(dNds,y);
detJ=j11*j22-j12*j21;
dSdx=( j22*dSdr-j12*dSds)/detJ;
dSdy=(-j21*dSdr+j11*dSds)/detJ;
```

Here, `shapefun` is assumed to be a function handle, which can be either of the subroutines `P1shapes` and `P2shapes`, depending on if we want to evaluate linear or quadratic shape functions.

8.2.1 Numerical Quadrature Revisited

The entries of the stiffness matrix and load vector involves integrals over the physical element K. However, since we want to compute on the reference element $\bar{K}$ we have to study how the isoparametric map $(x_1, x_2) \mapsto (r, s)$ affects integrals. To do so, we recall the change of variables formula

$$\int_K f(x_1, x_2)\, dx = \int_{\bar{K}} f(r, s) \det(J(r, s))\, drds \tag{8.47}$$

which allows us to integrate over $\bar{K}$ instead of K.

Now, approximating the integral over $\bar{K}$ by a quadrature formula we have

$$\int_{\hat{K}} f(r,s)\det(J(r,s))\,drds \approx \sum_{q=1}^{n_q} w_q f(r_q,s_q)\det(J(r_q,s_q)) \tag{8.48}$$

where N_q is the number of quadrature points, w_q the quadrature weights, and (r_q, s_q) the quadrature points.

The construction of efficient quadrature rules on triangles is difficult and still to some extent unexplored territory. Below, we present a routine, which tabulates so-called Gauss quadrature weights and points on the reference triangle $\bar{K}$ up to precision four (i.e., polynomials of maximal degree four can be integrated exactly). The weights are scaled so that they sum to one. As a consequence, the determinant $\det(J)$ needs to be divided by two to integrate correctly.

```
function [rspts,qwgts] = Gausspoints(precision)
switch precision
    case 1
     qwgts=[1];
     rspts=[1/3 1/3];
    case 2
     qwgts=[1/3 1/3 1/3];
     rspts=[1/6 1/6;
    2/3 1/6;
    1/6 2/3];
    case 3
     qwgts=[-27/48 25/48 25/48 25/48];
     rspts=[1/3 1/3;
    0.2 0.2;
    0.6 0.2;
    0.2 0.6];
    case 4
     qwgts=[0.223381589678011
    0.223381589678011
    0.223381589678011
    0.109951743655322
    0.109951743655322
    0.109951743655322];
     rspts=[0.445948490915965 0.445948490915965;
    0.445948490915965 0.108103018168070;
    0.108103018168070 0.445948490915965;
    0.091576213509771 0.091576213509771;
    0.091576213509771 0.816847572980459;
    0.816847572980459 0.091576213509771];
    otherwise
     error('Quadrature precision too high on triangle')
end
```

As a small example of use, let us integrate the mass matrix $M^K = (S_i, S_j)_K$ on a triangle K with vertices at $(0,0)$, $(3,0)$ and $(-2,4)$, using linear Lagrange shape functions.

```
[rspts,qwgts]=Gausspoints(2) % quadrature rule
x=[0 3 -2]; % node x-coordinates
y=[0 0  4]; %      y-
MK=zeros(3,3); % allocate element mass matrix
for q=1:length(qwgts) % quadrature loop
  r=rspts(q,1); % r coordinate
  s=rspts(q,2); % s
  [S,dSdx,dSdy,detJ]=Isopmap(x,y,r,s,@P1shapes); % map
  wxarea=qwgts(q)*detJ/2; % weight times det(J)
  MK=MK+(S*S')*wxarea; % compute and add integrand to MK
end
```

8.2.2 Renumbering the Mesh for Quadratic Nodes

As we have seen triangular Lagrange finite elements have $n = (o + 1)(o + 2)/2$ nodes per element. To correctly assemble the stiffness matrix and load vector it is therefore necessary to modify the original mesh to include all nodes. In this section we show how this can be done efficiently for the special case $o = 2$. Recall that quadratic Lagrange elements have nodes at the vertices and the mid-points of the edges. As the vertex nodes are already numbered by the mesh generator `initmesh` the problem boils down to numbering the edge nodes. To do so, we first record the node to edge incidence by using a sparse matrix A. More precisely, if there is an edge between vertex i and j, then we set $A_{ij} = -1$. Using the standard point and triangle matrices `p` and `t` this can efficiently be done with the code snippet

```
np=size(t,2); % number of vertices
nt=size(t,2); % number of triangles
i=t(1,:); % i=1st vertex within all elements
j=t(2,:); % j=2nd
k=t(3,:); % k=3rd
A=sparse(j,k,-1,np,np);   % 1st edge is between (j,k)
A=A+sparse(i,k,-1,np,np); % 2nd                (i,k)
A=A+sparse(i,j,-1,np,np); % 3rd                (i,j)
```

Since the edge between vertex i and j trivially also lies between vertex j and i we should have $A_{ij} = A_{ji} = -1$. To ensure this we add the transpose A^T to A and look for negative matrix entries, that is,

```
A=-((A+A.')<0);
```

We can look at the stored matrix entries (i.e., created edges) by typing

```
A=triu(A); % extract upper triangle of A
[r,c,v]=find(A) % rows, columns, and values(=-1)
```

Now, to number the edges we simply take the matrix values, which are all -1, and renumber them consecutively, staring from 1. Then, we reassemble the upper triangle part of A. Finally, we expand A to symmetric form by, again, adding A^T to A.

```
v=[1:length(v)]; % renumber values (ie. edges)
A=sparse(rows,cols,entries,np,np); % reassemble A
A=A+A'; % expand A to a symmetric matrix
```

The edge numbers for the three edges of each element can now be read form A, viz.,

```
edges=zeros(nt,3);
for k=1:nt
  edges(k,:)=[A(t(2,k),t(3,k))
              A(t(1,k),t(3,k))
              A(t(1,k),t(2,k))]';
end
```

In the Appendix we list a routine called `Tri2Edge` containing the above code.

Using the edge numbering routine it is straight forward to insert the new nodes into the point and triangle matrices `p` and `t`.

```
function [p,t] = ChangeP1toP2Mesh(p,t)
np=size(p,2); % number of nodes
edges=Tri2Edge(p,t); % get element edge numbers
edges=edges+np; % change edges to new nodes
i=t(1,:); j=t(2,:); k=t(3,:);
e=edges(:,1);
p(1,e)=0.5*(p(1,j)+p(1,k)); % edge node coordinates
p(2,e)=0.5*(p(2,j)+p(2,k));
e=edges(:,2);
p(1,e)=0.5*(p(1,i)+p(1,k));
p(2,e)=0.5*(p(2,i)+p(2,k));
e=edges(:,3);
p(1,e)=0.5*(p(1,i)+p(1,j));
p(2,e)=0.5*(p(2,i)+p(2,j));
t(7,:)=t(4,:); % move subdomain info, resize t
t(4:6,:)=edges'; % insert edge nodes into t
```

Upon return the quadratic mesh overwrites the linear one in `p` and `t`.

For higher order Lagrange elements it is necessary to insert more nodes on the edges, but this is fairly simple once these have been properly numbered. Further, higher order elements also contains interior nodes, but these are trivial to number uniquely.

8.2.3 Assembly of the Isoparametric Quadratic Stiffness Matrix

We next show how to assemble the usual stiffness matrix on a mesh renumbered for isoparametric Lagrange finite elements of order 2.

```
function [A,M,F] = IsoP2StiffnessAssembler(p,t,force)
[rspts,qwgts]=Gausspoints(4); % quadrature rule
np=size(p,2); % number of nodes
nt=size(t,2); % number of elements
A=sparse(np,np); % allocate stiffness matrix
for i=1:nt % loop over elements
    nodes=t(1:6,i); % node numbers
    x=p(1,nodes); % node x-coordinates
    y=p(2,nodes); %      y-
    AK=zeros(6,6); % elements stiffness
    for q=1:length(qwgts) % quadrature loop
        r=rspts(q,1); % quadrature r-coordinate
        s=rspts(q,2); %            s-
        [S,dSdx,dSdy,detJ]=Isopmap(x,y,r,s,@P2shapes);
        wxarea=qwgts(q)*detJ/2; % weight times area
        AK=AK+(dSdx*dSdx'...
               +dSdy*dSdy')*wxarea; % element stiffness
    end
    A(nodes,nodes)=A(nodes,nodes)+AK;
end
```

To call this routine one can type, for example,

```
[p,e,t]=initmesh('squareg');
[p,t]=ChangeP1toP2Mesh(p,t);
A=IsoP2StiffnessAssembler(p,t);
```

8.3 Some Exotic Finite Elements

Finite elements are often invented for a particular purpose. For example, they might be designed for a specific application area, or to mimic certain properties of a particular function space. In this section we shall briefly look at a few exotic elements, which are tailor made to approximate a certain Hilbert space, or that are otherwise somewhat peculiar. For simplicity, we shall restrict attention mainly to triangular elements.

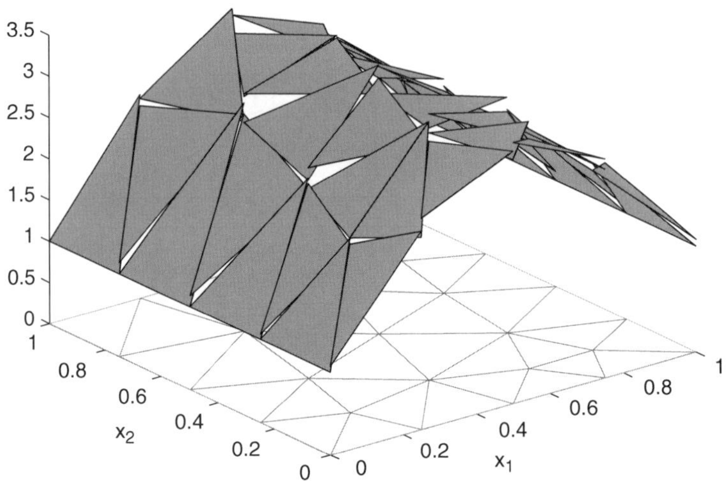

Fig. 8.8 Crouzeix-Raviart interpolant of $1 + 2\sin(3x_1)$ on a mesh of the unit *square*

8.3.1 The Morley Element

The Morley element is perhaps the simplest example of a finite element with higher continuity. However, it is neither C^1 nor even C^0. On a triangle K with vertices v_i, $i = 1, 2, 3$, and edge mid-points m_i, the polynomial space is $P_2(K)$, and the defining functionals are given by

$$L_i(v) = v(v_i) \tag{8.49}$$

$$L_{i+3}(v) = n \cdot \nabla v(m_i) \tag{8.50}$$

That is, the values of v at the vertices v_i, and the normal derivative of v at the edge mid-points m_i.

The Morley element finds application in solid mechanics.

8.3.2 The Crouzeix-Raviart Element

The Crouzeix-Raviart element is a linear element, which is only continuous at the mid-points of the triangle edges m_i. Figure 8.8 shows a mesh of the unit square and the Crouzeix-Raviart interpolant of $1 + 2\sin(3x_1)$.

On a triangle K, the polynomial space for the Crouzeix-Raviart element is $P_1(K)$, and the defining functionals are given by

$$L_i(v) = (v, 1)_{L^2(E_i)}/|E_i|, \quad i = 1, 2, 3 \tag{8.51}$$

where E_i is triangle edge i. These degrees of freedom are the mean values of v over the edges.

Now, since the mean of a linear function v over any edge E_i is the value of v at the edge mid-point m_i, we can alternatively define the functionals by

$$L_i(v) = v(m_i), \quad i = 1, 2, 3 \tag{8.52}$$

From this we see that the mid-points m_i acts as nodes for the Crouzeix-Raviart shape functions S_i^{CR} in the sense that S_i^{CR} has a unit value at m_j if $i = j$, while being zero if $i \neq j$.

The explicit expressions for the Crouzeix-Raviart shape functions are given by

$$S_i^{CR} = -\varphi_i + \varphi_j + \varphi_k \tag{8.53}$$

where φ_i are the usual hat functions, and with cyclic permutation of the indices $\{i, j, k\}$ over $\{1, 2, 3\}$. We observe that since the gradient of a hat function is the constant vector $\nabla\hat{\varphi}_i = [b_i, c_i]^T$, the gradients of the Crouzeix-Raviart shape functions take the form

$$\nabla S_i^{CR} = \begin{bmatrix} -b_i + b_j + b_k \\ -c_i + c_j + c_k \end{bmatrix} \tag{8.54}$$

For later use, let us write a routine to compute these derivatives.

```
function [area,Sx,Sy] = CRGradients(x,y)
[area,b,c]=HatGradients(x,y);
Sx=[-b(1)+b(2)+b(3); b(1)-b(2)+b(3); b(1)+b(2)-b(3)];
Sy=[-c(1)+c(2)+c(3); c(1)-c(2)+c(3); c(1)+c(2)-c(3)];
```

Here, we have used the subroutine `HatGradients` to compute b_i and c_i. Input is the triangle vertex coordinates `x` and `y`, and output the Crouzeix-Raviart shape function derivatives `Sx` and `Sy`. The triangle area `area` is also computed.

We remark that the Crouzeix-Raviart space is not a subspace of H^1 due its discontinuous nature. This is actually somewhat surprising, since Crouzeix-Raviart functions are often used to approximate precisely functions in H^1. A discrete space that is not a subspace of the continuous space on which the variational equation is posed is called non-conforming. Thus the Crouzeix-Raviart element provide a non-conforming approximation of H^1. The use of a non-conforming element is called a variational crime.

The Crouzeix-Raviart element finds application in fluid mechanics.

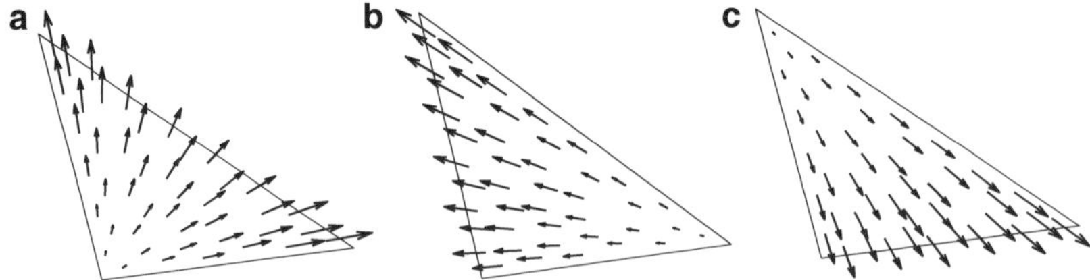

Fig. 8.9 RT_0 shape functions on a *triangle*. (**a**) $S_1^{RT_0}$, (**b**) $S_2^{RT_0}$, and (**c**) $S_3^{RT_0}$

8.3.3 The Lowest Order Raviart-Thomas Element

Not all finite elements are scalar and there are also vector valued elements. As the name suggests, vector valued elements are used to approximate vector valued functions. One such element is the Raviart-Thomas element, which is designed to mimic the Hilbert space

$$H(\text{div};\Omega) = \{v \in [L^2(\Omega)]^2 : \nabla \cdot v \in L^2(\Omega)\} \tag{8.55}$$

That is, the space of all vectors with bounded divergence. A simple application of Green's formula shows that all such functions must have continuous normal components, which is the basic design feature of the Raviart-Thomas element.

Actually there is a whole family of Raviart Thomas elements, but we shall only study the simplest of them called the RT_0 element. On a triangle K the polynomial space for RT_0 is given by $P = [P_0(K)]^2 + [x_1,\ x_2]^T P_0(K)$, that is, all vectors of the form

$$v = \begin{bmatrix} a_1 \\ a_2 \end{bmatrix} + b \begin{bmatrix} x_1 \\ x_2 \end{bmatrix} \tag{8.56}$$

for some coefficients a_1, a_2, and b. Further, the functionals are given by

$$L_i(v) = (n^{E_i}, v)_{L^2(E_i)}/|E_i|, \quad i = 1,2,3 \tag{8.57}$$

where n^{E_i} is the unit normal on edge E_i on K.

Closed form formulas for the RT_0 shape functions can be derived and are given by

$$S_i^{RT_0} = \frac{|E_i|}{2|K|} \begin{bmatrix} x_1 - x_1^{(i)} \\ x_2 - x_2^{(i)} \end{bmatrix}, \quad i = 1,2,3 \tag{8.58}$$

where $(x_1^{(i)}, x_2^{(i)})$ are the coordinates of the vertex opposite edge E_i. See Fig. 8.9.

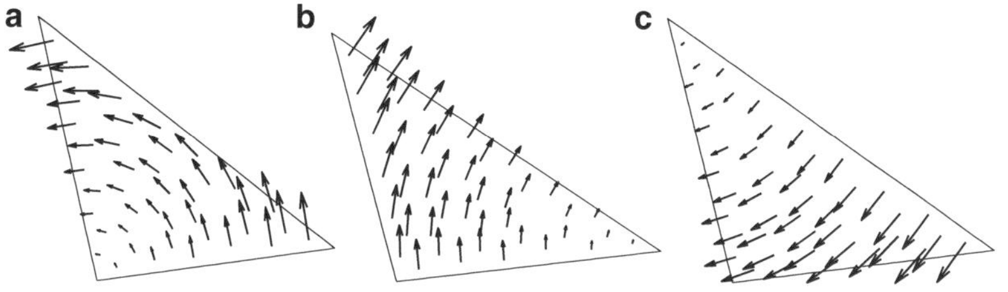

Fig. 8.10 Lowest order Nédélec shape functions on a *triangle*. (**a**) S_1^{ND}, (**b**) S_2^{ND}, and (**c**) S_3^{ND}

If the normal is chosen consistently on each edge in the mesh, then by construction is the RT_0 shape functions are continuous in the normal direction across the edge of any two adjacent elements. This ensures that the RT_0 functions belong to $H(\text{div}; \Omega)$.

The isoparametric map can not be used for RT_0 elements, since the divergence is not preserved by this mapping.

The Raviart-Thomas elements typically find application in acoustics and elasticity.

8.3.4 The Lowest Order Nédélec Element

The Nédélec, or edge, elements is another example of a family of vector valued finite elements. The Nédélec element is designed to mimic the Hilbert space

$$H(\text{curl}; \Omega) = \{v \in [L^2(\Omega)]^2 : \nabla \times v \in L^2(\Omega)\} \tag{8.59}$$

That is, the space of all vectors with bounded curl.

On a triangle K the polynomial space is $P = [P_0(K)]^2 + [x_2, \ -x_1]^T P_0(K)$, and the defining functionals are given by

$$L_i(v) = (t^{E_i}, v)_{L^2(E_i)}/|E_i|, \quad i = 1, 2, 3 \tag{8.60}$$

where t^{E_i} is a unit tangent vector on edge E_i. We shall assume that these tangent vectors are directed so that the triangle perimeter is traversed counter clockwise when moving along them.

Closed form formulas for the Nédélec shape functions can be derived and are given by

$$S_i^{ND} = |E_i|(\varphi_j \nabla \varphi_k - \varphi_k \nabla \varphi_j) \tag{8.61}$$

where φ_i are the usual hat functions and with cyclic permutation of the indices $\{i, j, k\}$ over $\{1, 2, 3\}$. See Fig. 8.10.

The curl of S_i^{ND} is given by

$$\nabla \times S_i^{ND} = \frac{|E_i|}{|K|} \tag{8.62}$$

which follows from the definition of the Nédélec shape functions, Stokes theorem, and the fact that $\nabla \times S_i^{ND}$ is a constant. To see this, note that

$$|E_i| = (t^{E_i}, S_i^{ND})_{L^2(E_i)} = (t, S_i^{ND})_{L^2(\partial K)} = (\nabla \times S_i^{ND}, 1)_{L^2(K)} = \nabla \times S_i^{ND}|K| \tag{8.63}$$

The Nédélec shape functions are continuous in the tangent direction on the element edges provided that the tangent vector is chosen consistently on each edge in the mesh.

The isoparametric map can not be used for Nédélec elements, since the curl is not preserved by this mapping.

The Nédélec elements finds application in electromagnetics.

We remark that it is straight forward to extend the Nédélec element to three dimensions on tetrahedra. Indeed, on a tetrahedron K the shape functions are still given by (8.61), with edge i lying between tetrahedron vertex j and k. Of course, the involved hat functions are then three- and not two-dimensional.

8.4 Further Reading

The definition of a finite element as the triplet of a polygon, a polynomial space and a set of functionals is due to Ciarlet, and is used in his book [19] to define the most common types of finite elements. The same in done by Brenner and Scott in [60]. The definition of more exotic finite elements can be found in Brezzi and Fortin [31]. Advanced topics, such as the construction of hierarchical finite elements, are presented in Solin [62].

8.5 Problems

Exercise 8.1. Work out formulas for the cubic Lagrange shape functions on the reference triangle $\bar{K}$.

Exercise 8.2. Calculate the entries of the 4×4 element stiffness matrix A^K using bilinear shape functions on the square $S = [0, h]^2$. What is the dependence of the side length h?

Exercise 8.3. Show that the bilinear element is not unisolvent if the four nodes are placed at $(-1, 0)$, $(0, -1)$, $(1, 0)$, and $(0, 1)$ on the reference square $\bar{Q}$.

Exercise 8.4. Write a routine for assembling the mass matrix using Lagrange shape functions of order 2.

Exercise 8.5. Draw the Crouzeix-Raviart shape functions on the reference triangle $\bar{K}$.

Exercise 8.6. Calculate the Crouzeix-Raviart interpolant of $f = 2x_1x_2 + 4$ on the reference triangle $\bar{K}$.

Exercise 8.7. Show that the curl of the Nédelec shape function S_i^{ND} is given by $\nabla \times S_i^{ND} = 2|E_i|\nabla\varphi_j \times \nabla\varphi_k$.

Exercise 8.8. How does the isoparametric map look in three dimensions?

Chapter 9
Non-linear Problems

Abstract Many real-world problems are governed by non-linear mathematical models. The drying of paint, the weather, or the mixing of fluids are just some examples of non-linear phenomenons. In fact, most of the physical, biological, and chemical processes going on around us everyday are described by more or less non-linear laws of nature. Thus, extensions of the finite element method to non-linear equations are of special interest. In this chapter we study the standard methods for tackling non-linear partial differential equations discretized with finite elements, namely, Newton's method and its simplified variant Piccard, or, fixed-point iteration.

9.1 Piccard Iteration

Piccard, or fixed-point, iteration is perhaps the most primitive technique for solving non-linear equations. It is applicable to equations of the form

$$x = g(x) \tag{9.1}$$

where we for simplicity assume that g is a scalar non-linear function of a single variable x. The basic idea is to take a first rough guess $x^{(0)}$ at the solution, say, $\bar{x}$, and then to compute successively until convergence, for $k = 0, 1, 2, \ldots$,

$$x^{(k+1)} = g(x^{(k)}) \tag{9.2}$$

M.G. Larson and F. Bengzon, *The Finite Element Method: Theory, Implementation, and Applications*, Texts in Computational Science and Engineering 10,
DOI 10.1007/978-3-642-33287-6_9,

This leads to the following algorithm, which is Piccard's iteration scheme:

Algorithm 25 Piccard iteration for a scalar non-linear equation

Choose a staring guess $x^{(0)}$, and a desired accuracy ϵ.
for $k = 0, 1, 2, \ldots$ **do**
  Compute the next solution approximation from $x^{(k+1)} = g(x^{(k)})$.
  if $|x^{(k+1)} - x^{(k)}| < \epsilon$ **then**
    Stop.
  end if
end for

It turns out that this algorithm will converge if the operator g is a contraction mapping, that is, if there exist a constant $L < 1$ such that $\|g(x) - g(y)\| \leq L\|x - y\|$ for all x and y. To see this, first note that the exact solution satisfies $\bar{x} = g(\bar{x})$, and is therefore a so-called fixed point. By subtracting this from (9.2) we then have $\|x^{(k+1)} - \bar{x}\| = \|g(x^{(k)}) - g(\bar{x})\| \leq L\|x^{(k)} - \bar{x}\| \leq L^{k+1}\|x^{(0)} - \bar{x}\|$, from which convergence indeed follows if $L < 1$.

Piccard iteration is simple to implement, but its rate of convergence is often slow.

9.2 Newton's Method

Besides Piccard iteration there is also Newton's method for solving non-linear equations. Newton's method is more complicated than the Piccard iteration technique, but it usually converges much faster. It is applicable to equations of the form

$$g(x) = 0 \tag{9.3}$$

where we again assume that g is a scalar non-linear function of the single variable x.

In deriving Newton's method the first step is to assume that the solution $\bar{x}$ can be written as the sum

$$\bar{x} = x^0 + \delta x \tag{9.4}$$

where x^0 is some known guess of $\bar{x}$ and δx a correction. Assuming also that x^0 is close to $\bar{x}$ so that δx is small and using Taylor expansion of $g(x)$ around $\bar{x}$, we have

$$g(\bar{x}) = g(x^0 + \delta x) = g(x^0) + g'(x^0)\,\delta x + O(\delta x^2) \tag{9.5}$$

Neglecting second order terms, and using that $g(\bar{x}) = 0$, we further have

$$0 \approx g(x^0) + g'(x^0)\,\delta x \tag{9.6}$$

Now, the key observation is that this is a linear equation for δx, as opposed to the non-linear equation for x. Thus, by evaluating

$$\delta x = -g(x^0)/g'(x^0) \tag{9.7}$$

and adding δx to x^0 we expect to get a better approximation of $\bar{x}$ than just x^0 alone, at least if x^0 is close to $\bar{x}$. Iterating this line of reasoning leads to the following algorithm, which is Newton's method:

Algorithm 26 Newton's method for a scalar non-linear equation

Choose a staring guess $x^{(0)}$, and a desired accuracy ϵ.
for $k = 0, 1, 2, \ldots$ **do**
    Compute the correction $\delta x^{(k)} = -g(x^{(k)})/g'(x^{(k)})$.
    Update the solution guess $x^{(k+1)} = x^{(k)} + \delta x^{(k)}$.
    if $|\delta x^{(k)}| < \epsilon$ **then**
        Stop.
    end if
end for

In difficult cases it is sometimes good to use a damped update of the form $x^{(k+1)} = x^{(k)} + \alpha\delta x^{(k)}$ with $0 < \alpha < 1$ a damping parameter.

Newton's method is popular because it usually converges rapidly. It can be shown that

$$\|x^{(k+1)} - \bar{x}\| \leq C\|x^{(k)} - \bar{x}\|^2 \tag{9.8}$$

when $x^{(k)}$ is sufficiently close to $\bar{x}$. From this relation we see that the asymptotic rate of convergence is quadratic, which is very fast for any numerical method. However, quadratic convergence can only be obtained if g is differentiable and g' is non-zero near $\bar{x}$. These are sometimes unrealistic expectations on g, since many real-world problems are not smooth.

The primary drawback of Newton's method is that it requires information about the derivative g', which can be costly to compute.

9.3 The Non-linear Poisson Equation

Having derived Newton's method for a non-linear scalar equation we shall now do the same for a non-linear partial differential equation. To this end, we first linearize the continuous problem, and then apply finite element discretization. As model problem we use the non-linear Poisson equation

$$-\nabla \cdot (a(u)\nabla u) = f, \quad \text{in } \Omega \tag{9.9a}$$

$$u = 0, \quad \text{on } \partial\Omega \tag{9.9b}$$

where a and f are given coefficients. The non-linearity is due to the coefficient $a = a(u)$, which depends on the unknown solution u. In order to fulfill the Lax-Milgram lemma, we assume that $a(u)$ is a well-behaved positive function. Typically, $a(u)$ is, or can be approximated, by a polynomial in u.

9.3.1 The Newton-Galerkin Method

Multiplying $f = -\nabla\cdot(a(u)\nabla u)$ by a test function $v \in V = H_0^1(\Omega)$, and integrating by parts using Green's formula, we obtain the weak form of (9.1): find $u \in V$ such that

$$(a(u)\nabla u, \nabla v) = (f, v), \quad \forall v \in V \tag{9.10}$$

Newton's method is in the context of non-linear partial differential equations also known as the Newton-Galerkin method, and to derive it we first write u as the sum

$$u = u^0 + \delta u \tag{9.11}$$

where u^0 is a some known approximation of u, and δu is a correction. Inserting this into (9.10) then gives us

$$(a(u^0 + \delta u)\nabla(u^0 + \delta u), \nabla v) = (f, v), \quad \forall v \in V \tag{9.12}$$

Making a Taylor expansion of $a(u) = a(u^0 + \delta u)$ around u^0 we have

$$a(u^0 + \delta u) = a(u^0) + a_u'(u^0)\,\delta u + O(\delta u^2) \tag{9.13}$$

Substituting this into (9.12) we further have

$$((a(u^0) + a_u'(u^0)\,\delta u + O(\delta u^2))\nabla(u^0 + \delta u), \nabla v) = (f, v), \quad \forall v \in V \tag{9.14}$$

Neglecting all terms quadratic in δu, we end up with a linear equation for the correction δu: find $\delta u \in V$ such that

$$(a(u^0)\nabla\delta u + a_u'(u^0)\,\delta u\nabla u^0, \nabla v) = (f, v) - (a(u^0)\nabla u^0, \nabla v), \quad \forall v \in V \tag{9.15}$$

Once we have solved (9.15) for δu, the Newton-Galerkin method is then to iterate with $u^0 + \delta u$ as new solution guess.

9.3.2 Finite Element Approximation

Let $\mathcal{K} = \{K\}$ be a mesh of Ω, and let $V_h \subset V$ be the usual space of continuous piecewise linears on $\mathcal{K}$. Replacing V with V_h we obtain the finite element approximation of the weak form (9.15): find $\delta u_h \in V_h$ such that

$$(a(u^0)\nabla\delta u_h + a'_u(u_h^0)\,\delta u_h \nabla u_h^0, \nabla v) = (f, v) - (a(u^0)\nabla u^0, \nabla v), \quad \forall v \in V_h \tag{9.16}$$

Here, we have tacitly assumed that u^0 can be expressed as a function u_h^0 in the finite element space V_h.

A basis for V_h is given by the set of hat functions $\{\varphi_i\}_{i=1}^{n_i}$ associated with the n_i interior nodes.

The finite element method (9.16) is equivalent to

$$(a(u^0)\nabla\delta u_h + a'_u(u_h^0)\delta u_h \nabla u_h^0, \nabla\varphi_i) = (f, \varphi_i) - (a(u^0)\nabla u^0, \nabla\varphi_i), \quad i = 1, \dots, n_i \tag{9.17}$$

Also, we can write δu_h as the sum

$$\delta u_h = \sum_{j=1}^{n_i} d_j \varphi_j \tag{9.18}$$

Inserting (9.18) into (9.17) we get

$$\sum_{j=1}^{n_i} d_j (a(u_h^0)\nabla\varphi_j + a'_u(u_h^0)\varphi_j \nabla u_h^0, \nabla\varphi_i) = (f, \varphi_i) - (a(u_h^0)\nabla u^0, \nabla\varphi_i), \quad i = 1, \dots, n_i \tag{9.19}$$

which is just a set of n_i linear algebraic equations for the unknowns d_j. In matrix form, we write this

$$Jd = r \tag{9.20}$$

where J is the $n_i \times n_i$ so-called Jacobian matrix with entries

$$J_{ij} = (a(u_h^0)\nabla\varphi_j, \nabla\varphi_i) + (a'_u(u_h^0)\varphi_j \nabla u_h^0, \nabla\varphi_i), \quad i, j = 1, \dots, n_i \tag{9.21}$$

and r is the $n_i \times 1$ so-called residual vector with entries

$$r_i = (f, \varphi_i) - (a(u_h^0)\nabla u_h^0, \nabla\varphi_i), \quad i = 1, \dots, n_i \tag{9.22}$$

We can now formulate a discrete Newton-Galerkin method with the following algorithm:

Algorithm 27 Newton-Galerkin method for the non-linear Poisson equation

1: Choose a starting guess $u_h^{(0)} \in V_h$, and a desired tolerance ϵ.
2: **for** $k = 0, 1, 2, \dots$ **do**
3: Assemble the Jacobian matrix $J^{(k)}$ and the residual vector $r^{(k)}$ with entries

$$J_{ij}^{(k)} = (a(u_h^{(k)})\nabla\varphi_j, \nabla\varphi_i) + (a_u'(u_h^{(k)})\varphi_j \nabla u_h^{(k)}, \nabla\varphi_i) \tag{9.23}$$

$$r_i^{(k)} = (f, \varphi_i) - (a(u_h^{(k)})\nabla u_h^{(k)}, \nabla\varphi_i) \tag{9.24}$$

4: Solve the linear system

$$J^{(k)} d^{(k)} = r^{(k)} \tag{9.25}$$

5: Set $u_h^{(k+1)} = u_h^{(k)} + \delta u_h^{(k)}$, where $\delta u_h^{(k)} = \sum_{j=1}^{n_i} d_j^{(k)} \varphi_j$.
6: **if** $\|\delta u_h^{(k)}\| < \epsilon$ **then**
7: Stop.
8: **end if**
9: **end for**

Here, we terminate the iteration process when the correction $\delta u_h^{(k)}$ is small, which indicates that the iteration error $u_h^{(k+1)} - u_h^{(k)}$ is small, but we could equally well stop iterating when the residual $r^{(k)}$ is small, which would indicate that the equation is well satisfied by $u_h^{(k+1)}$. Both these termination criteria are natural and it usually does not matter which one is used.

9.3.3 Computer Implementation

Below we list a MATLAB code for assembling the Jacobian matrix (9.24) and the residual vector (9.25). The computation of the derivative a_u' is done using numeric differentiation.

```
function [J,r] = JacResAssembler2D(p,e,t,u,Afcn,Ffcn)
i=t(1,:); j=t(2,:); k=t(3,:); % triangle vertices
xc=(p(1,i)+p(1,j)+p(1,k))/3; % triangle centroids
yc=(p(2,i)+p(2,j)+p(2,k))/3;
% Evaluate u, a, a', and f.
tiny=1.e-8;
uc=(u(i)+u(j)+u(k))/3;
a=Afcn(uc); % a(u)
da=Afcn(uc+tiny); % a(u+tiny)
da=(da-a)/tiny; % da(u)/du
```

```
f=Ffcn(xc,yc); % f
[ux,uy]=pdegrad(p,t,u); % grad u
np=size(p,2); nt=size(t,2);
% Assemble Jacobian and residual
J=sparse(np,np); r=zeros(np,1);
for i=1:nt
    nodes=t(1:3,i);
    x=p(1,nodes); y=p(2,nodes);
    [area,b,c]=HatGradients(x,y);
    rK=(f(i)*ones(3,1)/3-a(i)*(ux(i)*b+uy(i)*c))*area;
    JK=(a(i)*(b*b'+c*c')+da(i)*(ux(i)*b+uy(i)*c)*ones(1,3)/3)*area;
    J(nodes,nodes)=J(nodes,nodes)+JK;
    r(nodes)=r(nodes)+rK;
end
% Enforce zero Dirichlet BC.
fixed=unique([e(1,:) e(2,:)]); % boundary nodes
for i=1:length(fixed)
    n=fixed(i); % a boundary node
    J(n,:)=0; % zero out row n of the Jacobian, J
    J(n,n)=1; % set diagonal entry J(n,n) to 1
    r(n)=0;   % set residual entry r(n) to 0
end
```

Input to this routine is the usual point, edge, and connectivity matrices `p`, `e`, and `t`, and a vector `u` containing the nodal values of the current approximation $u^{(k)}$. The coefficients a and f are assumed to be defined by two separate subroutines `Afcn` and `Ffcn` defined elsewhere and passed via function handles. Output is the assembled Jacobian matrix `J` and the residual vector `r`.

As a numerical experiment, let us compute the finite element solution to the non-linear Poisson equation (9.1) on the unit square $\Omega = [0, 1]^2$ with $a(u) = 1/8 + u^2$, and $f = 1$. The necessary code is listed below.

```
function NewtonPoissonSolver2D()
g=Rectg(0,0,1,1);
[p,e,t]=initmesh(g,'hmax',0.05); % create mesh
xi=zeros(size(p,2),1); % initial zero guess
for k=1:5 % non-linear loop
  [J,r]=JacResAssembler2D(p,e,t,xi,@Afcn,@Ffcn);
  d=J\r; % solve for correction
  xi=xi+d; % update solution
  sprintf('|d|=%f, |r|=%f', norm(d), norm(r))
end
pdesurf(p,t,xi)

function z = Afcn(u)
z=0.125+u.^2;

function z = Ffcn(x,y)
z=x.^0; % =1
```

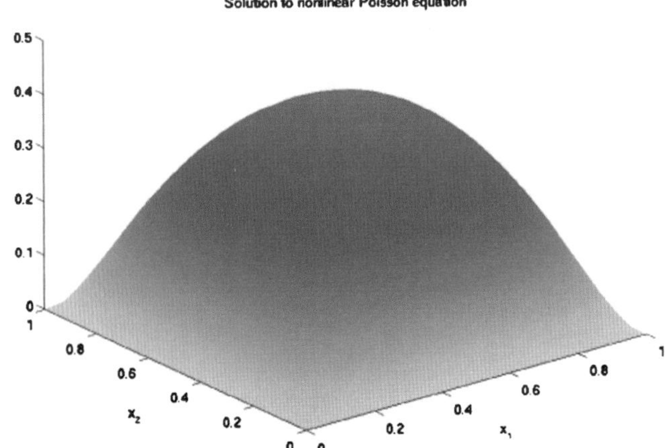

Fig. 9.1 Computed solution to the non-linear Poisson equation $-\nabla \cdot ((1/8 + u^2)\nabla u) = 1$ on the unit square $\Omega = [0, 1]^2$, with zero boundary conditions

Table 9.1 Norm of correction and residual in each Newton step

k	$\|d^{(k)}\|$	$\|r^{(k)}\|$
1	8.767519	0.038330
2	1.769684	0.038821
3	0.326485	0.007199
4	0.012397	0.000393
5	0.000022	0.000001

In Fig. 9.1 we show the computed solution u_h. Due to the non-linearity $a = 1/8 + u^2$, u_h is flatter near its maximum and has steeper gradients near the boundary, than in the linear case $a = 1/8$.

At each Newton step k we monitor the two-norm of the vectors $d^{(k)}$ and $r^{(k)}$ holding the nodal values of the correction and the residual. Table 9.1 shows these numbers. Clearly, the convergence is very rapid.

9.3.4 Piccard Iteration as a Simplified Newton Method

The reassembly of the Jacobian $J^{(k)}$ at each stage k of Newton's method is expensive, and we would like to avoid it, if possible. To this end we make the brutal approximation of replacing $J^{(k)}$ with the stiffness matrix $A^{(k)}$. The rationale for doing so is that $J_{ij}^{(k)}$ contains the term $(a(u^{(k)})\nabla\varphi_j, \nabla\varphi_i)$, which, by definition, is the stiffness $A_{ij}^{(k)}$. This gives the simplified Newton method

$$\xi^{(k+1)} = \xi^{(k)} + d^{(k)} \tag{9.26}$$

$$= \xi^{(k)} + J^{(k)^{-1}} r^{(k)} \tag{9.27}$$

$$= \xi^{(k)} + A^{(k)^{-1}}(b - A^{(k)}\xi^{(k)}) \tag{9.28}$$

$$= A^{(k)^{-1}}b \tag{9.29}$$

We recognize this as Piccard iteration on the non-linear system $A(\xi)\xi = b$. But this system is precisely what we get when applying finite elements to the non-linear Poisson equation $-\nabla \cdot (a(u)\nabla u) = f$. Hence, Piccard iteration can be seen as a simplified Newton method. Of course, due to all the cheats, this method will work only for mild non-linearities.

9.4 The Bistable Equation

Instead of deriving a Newton method by first linearizing the continuous problem and then discretizing with finite elements, there is, of course, also the possibility of doing these things in reverse order. That is, applying Newton's method after finite element discretization. Let us do this on the following non-linear time-dependent equation called the Bistable equation.

$$\dot{u} - \epsilon \Delta u = u - u^3, \quad \text{in } \Omega \times J \tag{9.30a}$$

$$n \cdot \nabla u = 0, \quad \text{in } \partial\Omega \times J \tag{9.30b}$$

$$u = u_0, \quad \text{in } \Omega, \text{ for } t = 0 \tag{9.30c}$$

Here, $\epsilon > 0$ is a small number, $J = (0, T]$ is the time interval, and u_0 a given initial condition. Obviously, this is a non-linear equation due to the cubic term u^3.

9.4.1 Weak Form

The weak form of (9.30) reads: find u such that for every fixed time t, $u \in H^1(\Omega)$ and

$$(\dot{u}, v) + \epsilon(\nabla u, \nabla v) = (f(u), v), \quad \forall v \in H^1(\Omega) \tag{9.31}$$

where $f(u) = u - u^3$.

9.4.2 Space Discretization

As always for transient problems we make the space discrete ansatz

$$u_h = \sum_{j=1}^{n_p} \xi_j(t)\varphi_j \tag{9.32}$$

where φ_j, $j = 1, \dots, n_p$, are the usual hat basis functions of V_h and n_p the number of nodes.

Substituting (9.32) into (9.31) and choosing $v = \varphi_i$, $i = 1, \dots, n_p$, we get a system of n_p ODE s

$$\sum_{j=1}^{n_p} \dot{\xi}_j(\varphi_j, \varphi_i) + \epsilon \sum_{j=1}^{n_p} \xi_j(\nabla\varphi_j, \nabla\varphi_i) = (f(u_h)), \varphi_i) \tag{9.33}$$

In matrix notation we write this

$$M\dot{\xi} + A\xi = b(\xi) \tag{9.34}$$

where M is the mass matrix, A is the stiffness matrix, and b a non-linear load vector, with entries

$$M_{ij} = (\varphi_j, \varphi_i) \tag{9.35}$$

$$A_{ij} = \epsilon(\nabla\varphi_j, \nabla\varphi_i) \tag{9.36}$$

$$b_i(\xi) = (f(u_h), \varphi_i) \tag{9.37}$$

9.4.3 *Time Discretization*

Applying backward Euler on the ODE system (9.34) we get the time stepping scheme

$$M\frac{\xi_l - \xi_{l-1}}{k_l} + A\xi_l = b(\xi_l) \tag{9.38}$$

or

$$(M + k_l A)\xi_l = M\xi_{l-1} + k_l b(\xi_l) \tag{9.39}$$

We shall now solve this non-linear system of equations using both Piccard iteration and Newton's method.

9.4.4 *Piccard Iteration*

Applying Piccard iteration to (9.39) yields the iteration scheme

$$\xi_l^{(k)} = (M + k_l A)^{-1}(M\xi_{l-1} + k_l b(\xi_l^{(k-1)})) \tag{9.40}$$

This iteration scheme has the structure of a double for loop over the indices l and k. The outer loop keeps track of the time level l, whereas the inner loop counts the non-linear iterate k. Indeed, for each time level l we solve the non-linear problem (9.39) by iterating over k. In doing so, the natural choice for the starting guess $\xi_l^{(0)}$ is the solution ξ_{l-1} from the previous time level. Once the inner loop has converged to a new solution $\xi_l^{(k)}$, the old solution ξ_{l-1} is overwritten and the outer loop is incremented. The double for loop is clearly seen in the code below, which solves the Bistable equation (9.30) on the unit square $\Omega = [0, 1]^2$ with the parameter $\epsilon = 0.01$ and the initial condition $u_0 = \cos(2\pi x_1^2)\cos(2\pi x_2^2)$.

```
function PiccardBiStableSolver2D()
g=Rectg(0,0,1,1);
[p,e,t]=initmesh(g,'hmax',0.025);
x=p(1,:)'; y=p(2,:)';
xi_old=cos(2*pi*x.^2).*cos(2*pi*y.^2); % IC
xi_new=xi_old;
dt=0.1; % time step
epsilon=0.01;
[A,M]=assema(p,t,1,1,0);
for l=1:100 % time loop
  for k=1:3 % non-linear loop
    xi_tmp=xi_new;
    b=M*(xi_tmp-xi_tmp.^3);
    xi_new=(M+dt*epsilon*A)\(M*xi_old+dt*b);
       fixpterror=norm(xi_tmp-xi_new)
     end
     xi_old=xi_new;
     pdesurf(p,t,xi_new)
     axis([0,1,0,1,-1,1]), caxis([-1,1]), pause(.1);
   end
```

In Fig. 9.2 we show a few snapshots of the finite element solution u_h at various times. The Bistable equation is a little peculiar because it has three steady states, $u = \pm 1$, and $u = 0$. The first two of these are stable, while the third is unstable. As a consequence, there is always a struggle between regions where the solution is 1 and regions where it is -1. In the end, however, one of these will win and the solution will always end up being constant and either 1 or -1. However, which of these states it will be is somewhat random and depends on the parameter ϵ, and in the discrete setting also the mesh size h, and the time step k_l. Indeed, from the figure we see that the final solution at $t = 25$ is constant -1.

9.4.5 Newton's Method

From the algebraic point of view Newton's method for (9.39) corresponds to solving the non-linear equations

$$r(\xi_l) = 0 \tag{9.41}$$

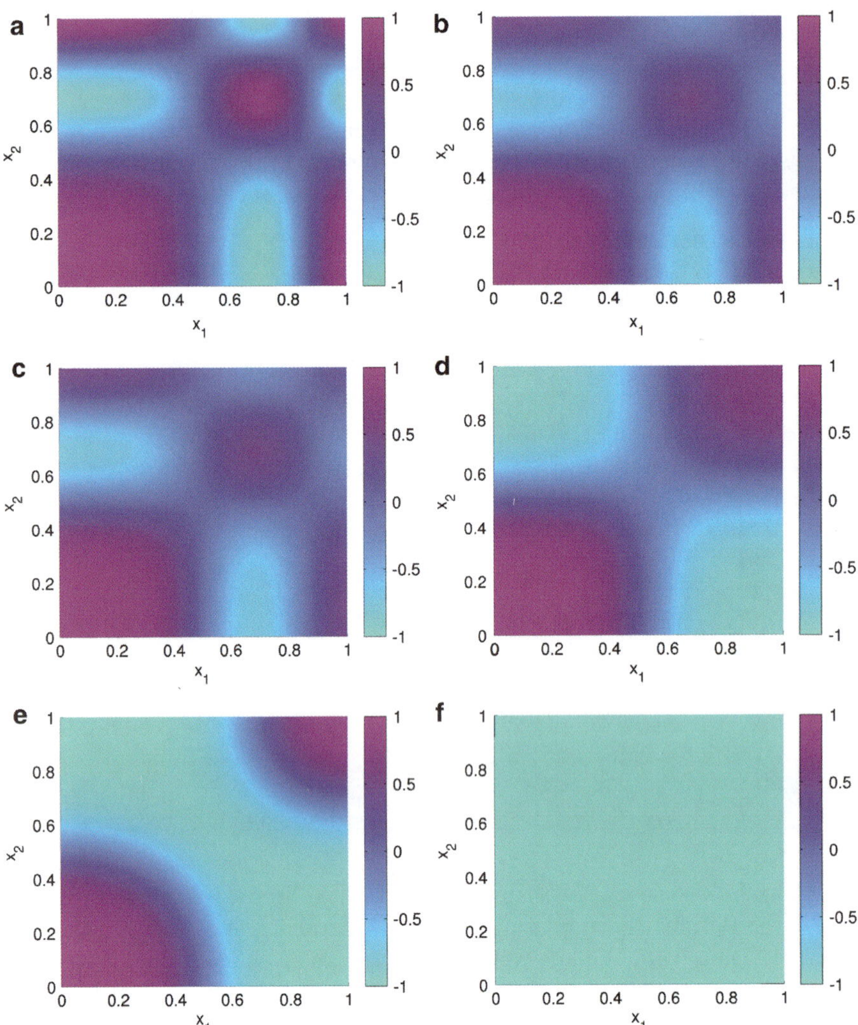

Fig. 9.2 Snapshots of the computed solution to the Bistable equation. (**a**) $t = 0.1$. (**b**) $t = 1$. (**c**) $t = 2.5$. (**d**) $t = 5$. (**e**) $t = 20$. (**f**) $t = 25$

where $r(\xi_l)$ is the residual vector

$$r(\xi_l) = (M + k_l A)\xi_l - M\xi_{l-1} - k_l b(\xi_l) \tag{9.42}$$

The entries of the Jacobian J are the partial derivatives $J_{ij} = \partial r_i/\partial(\xi_l)_j$, $i, j = 1, \ldots, n_p$. To compute J, we note that the first term $(M + k_l A)\xi_l$ in the left hand side of (9.42) is easy to differentiate with respect to ξ_l.

$$\frac{\partial((M + k_l A)\xi_l)_i}{\partial(\xi_l)_j} = M_{ij} + k_l A_{ij} \tag{9.43}$$

Continuing, the second term $M\xi_{l-1}$ does not depend on ξ_l, so its derivative with respect to this variable is 0. The third term $b(\xi_l)$, though, is a bit complicated. To differentiate it we use the chain rule.

$$\frac{\partial b(\xi_l)_i}{\partial \xi_j} = \frac{\partial}{\partial \xi_j}\int_\Omega f(\xi_l)\varphi_i\,dx = \int_\Omega \frac{\partial f}{\partial u_h}\frac{\partial u_h}{\partial \xi_j}\varphi_i\,dx = \int_\Omega \frac{\partial f}{\partial u}\varphi_j\varphi_i\,dx \tag{9.44}$$

Here, we have used that $\partial f/\partial u_h = \partial f/\partial u$. This is just a mass matrix $M_{f'}$ with weight $\partial f/\partial u$. Hence, the Jacobian J is given by

$$J = (M + k_l A) - k_l M_{f'} \tag{9.45}$$

A MATLAB implementation of Newton's method is shown below.

```
function NewtonBiStableSolver2D()
g=Rectg(0,0,1,1);
[p,e,t]=initmesh(g,'hmax',0.02);
x=p(1,:); y=p(2,:);
xi_old=(cos(2*pi*x.^2).*cos(2*pi*y.^2))';
xi_new=xi_old;
dt=0.1; % time step
epsilon=0.01;
[A,M]=assema(p,t,1,1,0);
for l=1:100 % time loop
    for k=1:3 % non-linear loop
        ii=t(1,:); jj=t(2,:); kk=t(3,:);
        xi_tmp=xi_new; % copy temporary solution to new
        xi_tmp_mid=(xi_tmp(ii)+xi_tmp(jj)+xi_tmp(kk))/3;
        f =(xi_tmp_mid-xi_tmp_mid.^3); % evaluate f
        df=1-3*xi_tmp_mid.^2; % evaluate derivative df of f
        [crap,Mdf,b]=assema(p,t,0,df',f');
        J=(M+dt*epsilon*A)-dt*Mdf; % Jacobian
        rho=(M+dt*epsilon*A)*xi_new ...
            -M*xi_old-dt*b; % residual
        xi_new=xi_tmp-J\rho; % Newton update
        error=norm(xi_tmp-xi_new)
    end
    xi_old=xi_new; % copy old solution to new
    pdesurf(p,t,xi_new)
    axis([0 1 0 1 -1 1]), caxis([-1,1]), pause(.25)
end
```

9.5 Numerical Approximations of the Jacobian

We finish this chapter by presenting two ways of approximating the Jacobian using only knowledge about the residual vector.

9.5.1 Difference Quotient Approximation

Replacing the derivative $J_{ij} = \partial r_i / \partial \xi_j$, $j, i = 1, \ldots, n_p$ by the simplest difference quotient we can compute column i of J from the formula

$$J_{:i} = \frac{r(\xi + \epsilon e_i) - r(\xi)}{\epsilon} \tag{9.46}$$

where $\epsilon > 0$ is a small parameter and e_i column i of the $n_p \times n_p$ identity matrix I. This is a simple, but very expensive, way of computing J, as it requites n_p assemblies of r.

9.5.2 Broyden's Method

A more elaborate way of numerically constructing a Jacobian is Broyden's method, which builds on the idea that $J^{(k)}$ approximately satisfies the so-called secant condition

$$J^{(k)}(\xi^{(k)} - \xi^{(k-1)}) \approx r^{(k)} - r^{(k-1)} \tag{9.47}$$

However, this is only n_p equations for the n_p^2 matrix entries $J_{ij}^{(k)}$. Thus, to uniquely determine $J^{(k)}$ we need to impose additional constraints. To this end, we require that $J^{(k)}$ should be a minimal modification of $J^{(k-1)}$ with respect to the so-called Frobenius matrix norm, defined by $\|A\|_F = \sqrt{\sum_{i,j=1}^{n_p} |A_{ij}|^2}$ for any $n_p \times n_p$ matrix A. This yields the following formula for $J^{(k)}$.

$$J^{(k)} = J^{(k-1)} + \frac{y^{(k)} - J^{(k-1)} s^{(k)}}{\|s^{(k)}\|^2} s^{(k)T} \tag{9.48}$$

where $s^{(k)} = \xi^{(k)} - \xi^{(k-1)}$ and $y^{(k)} = r^{(k)} - r^{(k-1)}$. Thus, the modification of $J^{(k-1)}$ is just the scaled outer product of the two vectors $y^{(k)} - J^{(k-1)} s^{(k)}$ and $s^{(k)}$.

Using (9.48) we can cheaply and automatically update the Jacobian at each stage k of Newton's method without assembling anything else but the residual vector. The convergence of the resulting method is somewhere between linear and quadratic.

Due to its construction $J^{(k)}$ can be efficiently and explicitly inverted using the Sherman Morrison formula.

A method closely related to Broyden's is the BFGS method. Unlike Broyden's, the BFGS method is symmetry preserving. Both these methods belong to the class of co-called quasi Newton methods.

9.6 Further Reading

Non-linear problems are inherently difficult to analyze and therefore most books on this topic are quite advanced. Books on finite elements for general non-linear continua include those by Reddy [52], Belytschko, Liu and Moran [9], and Bonet and Wood [14]. The simulation of non-linear solids and structures are treated, among others, by Crisfield and co-authors [22].

9.7 Problems

Exercise 9.1. Show that Newton's method applied to a square linear system $Ax = b$ converges in a single iteration.

Exercise 9.2. Derive Newton's method for the following non-linear problems:

(a) $-\Delta u = u + u^3$.
(b) $-\Delta u + \sin(u) = 1$.
(c) $-\nabla \cdot ((1 + u^2)\nabla u) = 1$.
(d) $-\Delta u = f(u)$, with $f(u) \in L^2(\Omega)$ a differentiable function of u.

For simplicity, assume zero Dirichlet boundary conditions.

Exercise 9.3. Use `NewtonPoissonSolver2D` to solve (9.9) on the unit square with $a = 0.1 + u^n$ and $f = 1$. Study the influence of the non-linear term u^n for the cases $n = 2$ and 6. Compare with the linear case $n = 0$.

Exercise 9.4. Derive a Newton method for the so-called predator-pray system

$$\begin{cases} -\Delta u_1 = u_1(1 - u_2) \\ -\Delta u_2 = -u_2(1 - u_1) \end{cases}$$

with $u_1 = u_2 = 0$ on the boundary.

Hint: Set $u_i = u_i^0 + \delta u_i$, $i = 1, 2$.

Chapter 10
Transport Problems

Abstract In this chapter we study the important transport equation that models transport of various physical quantities, such as density, momentum, and energy, for instance. In particular, the transportation of heat through convection is modeled by this equation. That is, the transfer of heat by some external physical process, such as air blown by a fan, or a moving fluid, for instance. Often, high convection takes place alongside low diffusion (i.e., uniform spreading) of heat, leading to large temperature gradients. As we shall see, this may cause numerical instabilities unless special care is taken. To do so, we introduce the Galerkin Least Squares (GLS) method, which is more robust than the standard Galerkin method. We illustrate with numerical examples.

10.1 The Transport Equation

The transport equation is given by

$$-\epsilon \Delta u + b \cdot \nabla u = f, \quad \text{in } \Omega \tag{10.1a}$$

$$u = 0, \quad \text{on } \partial\Omega \tag{10.1b}$$

where $\epsilon > 0$ is a (small) parameter, b a given vector field, and f is a given function. For this problem to be well-posed we must assume that $\nabla \cdot b = 0$. For simplicity, we also assume homogeneous Dirichlet boundary conditions. However, other types of boundary conditions are, of course, possible. Indeed, for the numerical experiments we shall use both Neumann and Robin boundary conditions.

In (10.1) each of the two operators $-\epsilon\Delta$, and $b \cdot \nabla$ play a specific role in determining what the solution u will look like, and can be given simple interpretations. Loosely speaking, the first one smears u proportionally to ϵ, while the second one transports u in the direction of the vector b. Therefore, we say that these operators

M.G. Larson and F. Bengzon, *The Finite Element Method: Theory, Implementation, and Applications*, Texts in Computational Science and Engineering 10,
DOI 10.1007/978-3-642-33287-6_10,

model the physical processes of diffusion, and convection, respectively. In fact, the transport equation is sometimes referred to as the Convection-Diffusion equation.

10.1.1 Weak Form

The weak form of (10.1) reads: find $u \in V = H_0^1(\Omega)$ such that

$$a(u, v) = l(v), \quad \forall v \in V \tag{10.2}$$

where the bilinear and linear forms $a(\cdot, \cdot)$ and $l(\cdot)$ are given by

$$a(u, v) = \epsilon(\nabla u, \nabla v) + (b \cdot \nabla u, v) \tag{10.3}$$

$$l(v) = (f, v) \tag{10.4}$$

The existence and uniqueness of $u \in V$ follows from the Lax-Milgram lemma, since $a(\cdot, \cdot)$ is continuous and coercive on V, and $l(\cdot)$ is continuous on V.

10.1.2 Standard Galerkin Finite Element Approximation

Let $V_h \subset V$ be the usual space of continuous piecewise linears. The standard so-called Galerkin finite element approximation of (10.2) reads: find $u_h \in V_h$ such that

$$a(u_h, v) = l(v), \quad \forall v \in V_h \tag{10.5}$$

Now, let $\{\varphi_i\}_{i=1}^{n_i}$ be the usual hat function basis for V_h, with n_i the number of interior nodes. Expanding the finite element ansatz $u_h = \sum_{j=1}^{n_i} \xi_j \varphi_j$, and choosing $v = \varphi_i$, $i = 1, 2, \dots, n_i$, in (10.5) we obtain the linear system for the unknown nodal values ξ_j of u_h

$$(A + C)\xi = b \tag{10.6}$$

where the matrix and vector entries are given by

$$A_{ij} = \epsilon(\nabla \varphi_j, \nabla \varphi_i) \tag{10.7}$$

$$C_{ij} = (b \cdot \nabla \varphi_j, \varphi_i) \tag{10.8}$$

$$b_i = (f, \varphi_i) \tag{10.9}$$

with $i, j = 1, 2, \dots, n_i$.

10.1.3 Computer Implementation

In the linear system (10.6) we note that the diffusion (i.e., stiffness) matrix A, and load vector b can be assembled using the built-in routine `assema`. However, we have no routine to assemble the convection matrix C. In order to write such a routine we note that the entries of the 3×3 element convection matrix C^K is approximately given by

$$\begin{aligned} C_{ij}^K &= (b \cdot \nabla \varphi_j, \varphi_i)_K \approx b(x_c) \cdot [b_j, c_j]^T (\varphi_i, 1)_K \\ &= b(x_c) \cdot [b_j, c_j]^T |K|/3, \quad i, j = 1, 2, 3 \end{aligned} \tag{10.10}$$

where $[b_j, c_j]^T$ is the gradient of hat function φ_j, and x_c the centroid of element K. This immediately translates into an assembly routine for C.

```
function C = ConvectionAssembler2D(p,t,bx,by)
np=size(p,2);
nt=size(t,2);
C=sparse(np,np);
for i=1:nt
    loc2glb=t(1:3,i);
    x=p(1,loc2glb);
    y=p(2,loc2glb);
    [area,b,c]=HatGradients(x,y);
    bxmid=mean(bx(loc2glb));
    bymid=mean(by(loc2glb));
    CK=ones(3,1)*(bxmid*b+bymid*c)'*area/3;
    C(loc2glb,loc2glb)=C(loc2glb,loc2glb)+CK;
end
```

Here, input is the usual point and connectivity matrix `p` and `t` and the components `bx` and `by` of the convection field. These components are and given as two $n_p \times 1$ vectors of nodal values with n_p the number of nodes. Output is the assembled global convection matrix `C`.

A main routine for solving the transport equation $-\epsilon \Delta u + [1, 1]^T \cdot \nabla u = 1$ on the square $\Omega = [-1, 1]^2$ with $u = 0$ on $\partial\Omega$ is listed below.

```
function TransportSolver2D()
epsilon=0.1; % diffusion parameter
[p,e,t]=initmesh('squareg','hmax',0.05); % mesh
np=size(p,2); % number of nodes
[A,unused,b]=assema(p,t,1,0,1); % diffusion matrix A
                                 % load vector b
bx=ones(np,1); by=ones(np,1); % convection field
C=ConvectionAssembler2D(p,t,bx,by); % convection matrix C
fixed=unique([e(1,:) e(2,:)]); % boundary nodes
```

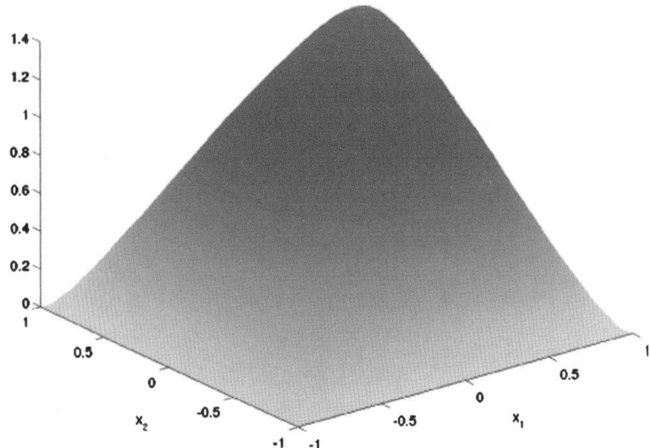

Fig. 10.1 Surface plot of u_h

```
free=setdiff([1:np],fixed); % interior nodes
b=b(free); % modify b for BC
A=A(free,free); C=C(free,free); % modify A and C for BC
xi=zeros(np,1); % solution vector
xi(free)=(epsilon*A+C)\b; % solve for free node values
pdesurf(p,t,U) % plot u
```

Running this code with $\epsilon = 0.1$ we get the results of Fig. 10.1. Notice how u_h is, so to speak, offset in the direction of b. This is even more clearly seen in Fig. 10.2, which shows isocontours of u_h. The compression of the isocontours seen in the upper right corner, where u_h must bend downwards to satisfy the boundary condition, is called a boundary layer.

10.1.4 The Need for Stabilization

Using the coercivity of $a(\cdot, \cdot)$ and the continuity of $l(\cdot)$ we have

$$\epsilon \|\nabla u\|^2 \leq a(u, u) = l(u) \leq C \|f\| \, \|\nabla u\| \tag{10.11}$$

That is, the stability estimate

$$\epsilon \|\nabla u\| \leq C \|f\| \tag{10.12}$$

From this we see that as ϵ decreases we loose control of the gradients of u. In other words, small perturbations of f can lead to a large local values of ∇u. Indeed, it is common for u to have thin regions called layers where it changes rapidly. As we

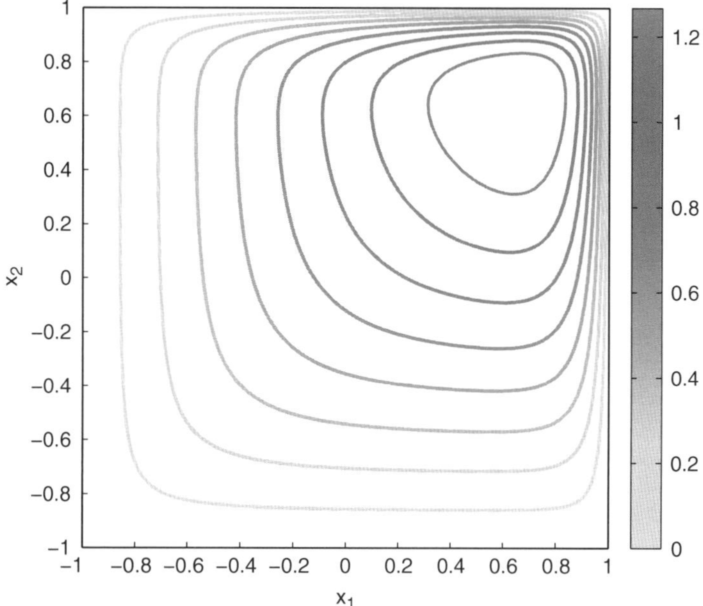

Fig. 10.2 Isocountours of u_h

have already seen, layers typically arise near a boundary, where u must adhere to a Dirichlet boundary condition. However, layers may also occur in the interior of the domain due to discontinuities in the coefficients ϵ, b, and f, for example.

Standard finite element methods have great difficulties in handling layers. In fact, layers may trigger oscillations throughout the whole computational domain that renders the finite element approximation useless. For example, consider the transport in one dimension, say,

$$-\epsilon u_{xx} + u_x = 1, \quad 0 < x < 1, \quad u(0) = u(1) = 0 \tag{10.13}$$

For small ϵ the analytical solution u to this equation looks like $u = x$, except near $x = 1$, where it abruptly changes from 1 to 0 in order to satisfy the boundary condition $u(1) = 0$. This change takes place over a small distance of length proportional to ϵ and is therefore a boundary layer.

Application of standard finite element discretization to (10.13) using a continuous piecewise linear approximation for u_h on a uniform mesh with $n + 1$ nodes and mesh size h leads to the system of equations

$$-\epsilon \frac{\xi_{i+1} - 2\xi_i + \xi_{i-1}}{h^2} + \frac{\xi_{i+1} - \xi_{i-1}}{2h} = 1, \quad i = 1, 2, \dots, n-1 \tag{10.14}$$

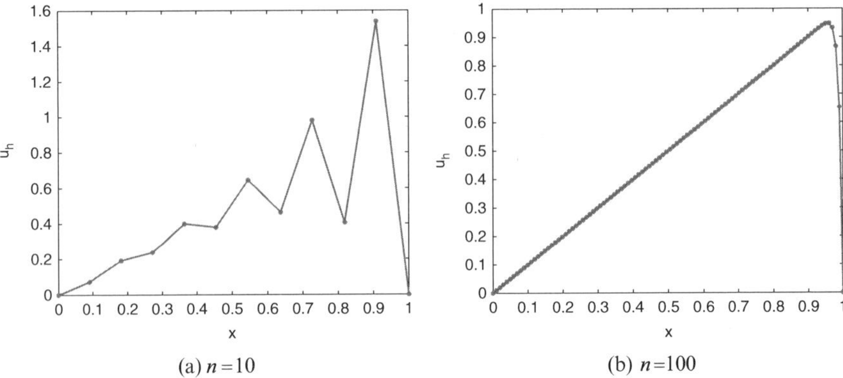

Fig. 10.3 Illustration of oscillations due to under resolution of the mesh (**a**). Increasing the number of elements resolves the issue and yields a good finite element solution (**b**). *Red asterisks* denote node values

where ξ_i are the nodal values of u_h with $\xi_0 = \xi_n = 0$. From this we see that if ϵ is small then information is only shared between every other node through the convective term. This opens up for the possibility of oscillations, since node $i + 1$ and $i - 1$ talk with each other, but not with node i. Furthermore, in a layer we know that there are naturally large variations between the node values ξ_i. Now, suppose that node $i - 1$ has value $\xi_{i-1} = -1$, whereas node i has value $\xi_i = 1$. Then, due to the finite element method (10.14), and neglecting the unit load which has a small influence on a fine mesh, we will get $\xi_{i+1} = -1$, $\xi_{i+2} = 1$, $\xi_{i+3} = -1$, and so on. That is, a highly oscillatory u_h. Figure 10.3 shows the finite element solution u_h for $\epsilon = 0.01$ on two meshes with $n = 10$ and $n = 100$ elements, respectively.

The onset of oscillations is a mesh resolution problem. It occurs only if the diffusion parameter ϵ is smaller than the mesh size h. In this case the diffusion acts on a length scale below the mesh size. As a consequence, small features in the solution cannot be accurately represented on the mesh, and this triggers the onset of oscillations. If the mesh size h can be decreased below ϵ, then no oscillations occur. The same oscillatory behavior is present also in higher dimension.

10.1.5 Least-Squares Stabilization

The forming of layers and the inability of the standard finite element method to deal with these calls for modification of the numerical method. Since the oscillations are due to the small diffusion parameter ϵ a simple way of stabilizing is to add more diffusion. In doing so, the general idea is to add as little as possible not to sacrifice accuracy, but as much as needed to obtain stability. A natural choice is to limit the smallest value of ϵ to h. In doing so, the stabilization will automatically decrease when using a finer mesh. This is known as isotropic stabilization or artificial diffusion. However, due to the perturbation of the equation, such a method can never be more than first order accurate in h. It turns out that a more accurate way to stabilize the equation is to use least squares stabilization.

To explain the least squares stabilization technique let us consider the abstract equation

$$Lu = f \tag{10.15}$$

where L is a differential operator, u the sought solution, and f a given function. We do not worry about boundary conditions for the moment.

The standard Galerkin method, is obtained by multiplying the differential equation by a test function v from a suitable function space V and integrate. This leads to the weak form: find $u \in V$ such that

$$(Lu, v) = (f, v), \quad \forall v \in V \tag{10.16}$$

We can interpret this as a demand for residual orthogonality $(r, v) = 0$ with $r = f - Lu$ is the residual. A potential problem with this is if the product (Lu, v) does not define a norm on V, in which case it is not associated with a minimization principle of some sort. This can happen if (Lu, v) is not coercive or symmetric on V. In this case, the numerical method resulting from finite element discretization might be unstable. Indeed, as we have seen, this is so for the transport equation.

Using instead the idea of least squares minimization we seek a solution $u \in V$, which is the minimizer of the problem

$$F(u) = \min_{v \in V} F(v) \tag{10.17}$$

where the functional $F(\cdot)$ is defined by

$$F(v) = \|Lv - f\|^2 \tag{10.18}$$

The first order optimality condition for this optimization problem takes the form: find $u \in V$ such that

$$(Lu, Lv) = (f, Lv), \quad \forall v \in V \tag{10.19}$$

From linear algebra we recognize this as the normal equations of the Least Squares method.

Here, the bilinear form (Lu, Lv) is always symmetric, trivially coercive with respect to the least squares norm $\|Lu\|$, and by setting $v = u$ we have the stability estimate $\|Lu\| \leq \|f\|$, which is a stronger stability compared to the Galerkin method.

The Galerkin Least Squares (GLS) method is obtained by combining the standard Galerkin and the Least Squares method. In effect, this amounts to replacing the test function v by $v+\delta Lv$, where δ is a parameter to be chosen suitably (i.e., for maximal accuracy). In doing so, we obtain the variational equation: find $u \in V$ such that

$$(Lu, v + \delta Lv) = (f, v + \delta Lv), \quad \forall v \in V \tag{10.20}$$

or

$$(Lu, v) + \delta(Lu, Lv) = (f, v) + \delta(f, Lv), \quad \forall v \in V \tag{10.21}$$

The hope is to combine the accuracy of the Galerkin method with the stability of the Least Squares method.

10.1.6 GLS Finite Element Approximation

In operator form the transport equation (10.1) can be written $Lu = f$ with $L = -\epsilon\Delta + b \cdot \nabla$. The Galerkin Least Squares finite element method therefore takes the form: find $u_h \in V_h$ such that

$$a_h(u_h, v) = l_h(v), \quad \forall v \in V_h \tag{10.22}$$

where the bilinear and linear form $a_h(\cdot,\cdot)$ and $l_h(\cdot)$ are defined by

$$a_h(u, v) = a(u, v) + \delta \sum_{K \in \mathcal{K}} (-\epsilon\Delta u + b \cdot \nabla u, -\epsilon\Delta v + b \cdot \nabla v)_{L^2(K)} \tag{10.23}$$

$$l_h(v) = l(v) + \delta \sum_{K \in \mathcal{K}} (f, -\epsilon\Delta v + b \cdot \nabla v)_{L^2(K)} \tag{10.24}$$

Here, we have written the GLS stabilization terms (Lu, Lv) and (f, Lv) as a sum over the elements K in the mesh $\mathcal{K}$. This is due to the fact that the term $-\epsilon\Delta v$ does not lie in $L^2(\Omega)$, since the second order derivatives of the test function v are unbounded across element boundaries. However, this term does lie in $L^2(K)$ on any $K \in \mathcal{K}$. In fact, to obtain well-defined integrals we only apply GLS stabilization only to the interior of the elements. Now, since we are using piecewise linear approximation we have $\Delta v = 0$, which implies that (10.23) and (10.24) reduces to simply

$$a_h(u, v) = \epsilon(\nabla u, \nabla v) + (b \cdot \nabla u, v) + \delta(b \cdot \nabla u, b \cdot \nabla v) \tag{10.25}$$

$$l_h(v) = (f, v) + \delta(f, b \cdot \nabla v) \tag{10.26}$$

Because $b \cdot \nabla u$ is the derivative of u in the flow direction (i.e., b) we see that $\delta(b \cdot \nabla u, b \cdot \nabla v)$ stabilizes the numerical method by adding diffusion proportional to

δ along the streamlines. Therefore, this low order GLS method is sometimes referred to as the Streamline-Diffusion (SD) method.

Next we note that $a_h(\cdot,\cdot)$ is coercive on V_h with respect to the norm

$$|||v|||^2 = \epsilon\|\nabla v\|^2 + \delta\|b\cdot\nabla v\|^2 \tag{10.27}$$

This norm is more natural to use than $\|\cdot\|_{H_0^1(\Omega)}$, since it is closely related to the bilinear form $a_h(\cdot,\cdot)$ and gives additional control over the streamline derivative.

Setting $v = u$ and observing that $(b\cdot\nabla v, v) = 0$, which follows from the fact that $0 = (b\cdot nv^2, 1)_{L^2(\partial\Omega)} = (\nabla\cdot(bv^2), 1) = ((\nabla\cdot b)v, v) + 2(b\cdot\nabla v, v)$ for any $v\in H_0^1(\Omega)$, and that $\nabla\cdot b = 0$, we obtain

$$a_h(v,v) = \epsilon\|\nabla v\|^2 + \delta\|b\cdot\nabla v\|^2 = |||v|||^2 \tag{10.28}$$

which shows that $a_h(\cdot,\cdot)$ is coercive on V_h. Hence, u_h exist and is unique.

From (10.28) we observe that the coercivity depends on the stabilization parameter δ, which, consequently, must not be too small for problems with small ϵ. However, for problems with large ϵ, δ may practically vanish as no stabilization is needed. A good choice of δ turns out to be

$$\delta = \begin{cases} Ch^2, & \text{if } \epsilon > h \\ Ch/\|b\|_{L^\infty(\Omega)}, & \text{if } \epsilon < h \end{cases} \tag{10.29}$$

where h is the mesh size. As we shall see, this choice is optimal and follows from the error analysis.

10.1.7 A Priori Error Estimate

The GLS method (10.22) is consistent in the sense that it is satisfied by the exact solution u, that is,

$$a_h(u,v) = l_h(v), \quad \forall v\in V_h \tag{10.30}$$

Subtracting (10.30) from (10.22) we immediately obtain the Galerkin orthogonality

$$a_h(u-u_h, v) = 0, \quad \forall v\in V_h \tag{10.31}$$

A difficulty with the error analysis of our GLS method stems from the fact that the continuous and discrete bilinear forms $a(\cdot,\cdot)$ and $a_h(\cdot,\cdot)$ are not the same. For example, we do not have coercivity of $a(\cdot,\cdot)$ in the norm $|||\cdot|||$. Therefore, the derivation of error estimates differs somewhat from the ordinary.

Let us first write the error

$$e = u - u_h = (u - \pi u) + (\pi u - u_h) \tag{10.32}$$

where $\pi u \in V_h$ is the interpolant of u. Application of the Triangle inequality then yields

$$|||e||| \leq |||u - \pi u||| + |||\pi u - u_h||| \tag{10.33}$$

Here, we shall estimate $|||u - \pi u|||$ using interpolation error estimates, and $|||\pi u - u_h|||$ using the approximation properties of the GLS method. Indeed, using the coercivity of $a_h(\cdot, \cdot)$ on V_h, and the Galerkin orthogonality (10.31) with $v = u_h - \pi u$, we find

$$|||u_h - \pi u|||^2 = a_h(u_h - \pi u, u_h - \pi u) \tag{10.34}$$

$$= a_h(u_h - \pi u, u_h - \pi u) + a_h(u - u_h, u_h - \pi u) \tag{10.35}$$

$$= a_h(u - \pi u, u_h - \pi u) \tag{10.36}$$

Let us estimate each of the three terms in $a_h(u - \pi u, u_h - \pi u)$ separately using the trivial estimates $\sqrt{\epsilon}\|\nabla v\| \leq |||v|||$ and $\sqrt{\delta}\|b \cdot \nabla v\| \leq |||v|||$. First, we have

$$\epsilon(\nabla(u - \pi u), \nabla(u_h - \pi u)) \leq \sqrt{\epsilon}\|\nabla(u - \pi u)\| \sqrt{\epsilon}\|\nabla(u_h - \pi u)\| \tag{10.37}$$

$$\leq C\sqrt{\epsilon}h|u|_{H^2(\Omega)}|||u_h - \pi u||| \tag{10.38}$$

Then, integrating by parts, using that $u = \pi u = 0$ on $\partial\Omega$, and that $\nabla \cdot b = 0$, we further have

$$(b \cdot \nabla(u - \pi u), u_h - \pi u) = (n(u - \pi u), b(u_h - \pi u))_{L^2(\partial\Omega)}$$

$$- (u - \pi u, \nabla \cdot (b(u_h - \pi u))) \tag{10.39}$$

$$\leq \|u - \pi u\| \, \|b \cdot \nabla(u_h - \pi u)\| \tag{10.40}$$

$$\leq Ch^2|u|_{H^2(\Omega)}|||u_h - \pi u|||/\sqrt{\delta} \tag{10.41}$$

Finally, we have

$$\delta(b \cdot \nabla(u - \pi u), b \cdot \nabla(u_h - \pi u)) \leq \delta\|b \cdot \nabla(u - \pi u)\| \, \|b \cdot \nabla(u_h - \pi u)\| \tag{10.42}$$

$$\leq \sqrt{\delta}\|b\|_{L^\infty(\Omega)}\|\nabla(u - \pi u)\| \, |||u_h - \pi u||| \tag{10.43}$$

$$\leq C\sqrt{\delta}h|u|_{H^2(\Omega)}|||u_h - \pi u||| \tag{10.44}$$

Now, in the case of high convection $\epsilon \leq Ch$, and, consequently, $\delta = Ch$. Thus, the three right hand sides of (10.38), (10.41), and (10.44) are all of order $h^{3/2}$, and we see that

$$|||u_h - \pi u||| \leq Ch^{3/2}|u|_{H^2(\Omega)} \tag{10.45}$$

It remains to estimate $|||u - \pi u|||$. Repeating the above estimates with $u_h - \pi u$ replaced by $u - \pi u$ it is easily seen that this term is also of order $h^{3/2}$. Hence, we have shown the following a priori estimate.

Theorem 10.1. *The finite element solution* u_h*, defined by* (10.22)*, satisfies the estimate*

$$|||u - u_h||| \leq Ch^{3/2}|u|_{H^2(\Omega)} \tag{10.46}$$

10.1.8 Computer Implementation

10.1.8.1 Heat Transfer in a Fluid Flow

Assembly of the SD matrix S_d stemming from the GLS term $(b \cdot \nabla u, b \cdot \nabla v)$ is easy. We list a routine for doing so below.

```
function Sd = SDAssembler2D(p,t,bx,by)
np=size(p,2);
nt=size(t,2);
Sd=sparse(np,np);
for i=1:nt
  loc2glb=t(1:3,i);
  x=p(1,loc2glb);
  y=p(2,loc2glb);
  [area,b,c]=HatGradients(x,y);
  bxmid=mean(bx(loc2glb));
  bymid=mean(by(loc2glb));
  SdK=(bxmid*b+bymid*c)*(bxmid*b+bymid*c)'*area;
  Sd(loc2glb,loc2glb)=Sd(loc2glb,loc2glb)+SdK;
end
```

Input is the same as for the routine `ConvectionAssembler2D`. Output is the global SD matrix Sd.

We now study a real-world application with more general boundary conditions, namely, heat transfer in a fluid flow. This kind of physical problem is of interest when designing heat exchangers or electronic devices, for instance. More precisely,

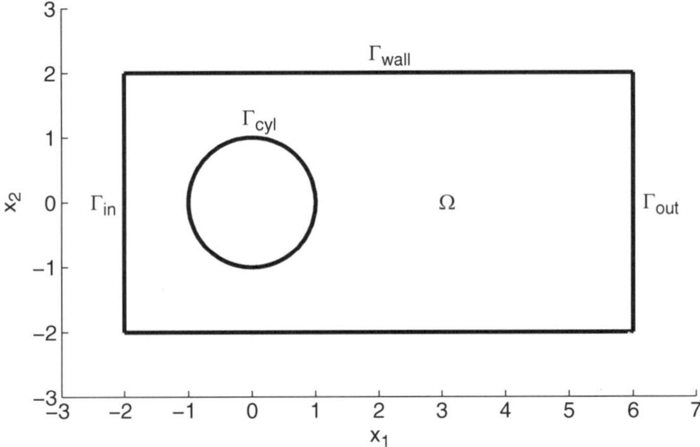

Fig. 10.4 Geometry of the channel domain and boundaries

we consider a heated object submerged into a channel with a flowing fluid. See Fig. 10.4. The channel is rectangular and fluid is flowing from left to right round a heated circle.

The fluid flow is unaffected by the temperature and given by the velocity field

$$b_1 = U_\infty \left(1 - \frac{x_1^2 - x_2^2}{(x_1^2 + x_2^2)^2}\right) \tag{10.47}$$

$$b_2 = -2U_\infty \frac{x_1 x_2}{(x_1^2 + x_2^2)^2} \tag{10.48}$$

where $U_\infty = 1$ is the free stream velocity of the fluid. Figure 10.5 shows a glyph plot of b.

Let us write a routine to evaluate the vector field b.

```
function [bx,by] = FlowField(x,y)
a=1; % cylinder radius
Uinf=1; % free stream velocity
r2=x.^2+y.^2; % radius vector squared
bx=Uinf*(1-a^2*(x.^2-y.^2)./r2.^2); % x-component of b
by=-2*a^2*Uinf*x.*y./r2.^2;          % y-
```

We assume that the cylinder is kept at constant temperature 1. Further, the walls of the channel are insulated so that no heat can flow across them. In other words, the normal heat flux $n \cdot q$ is zero on the walls, where q is given by Fourier's law

$$q = -\epsilon \nabla u + bu \tag{10.49}$$

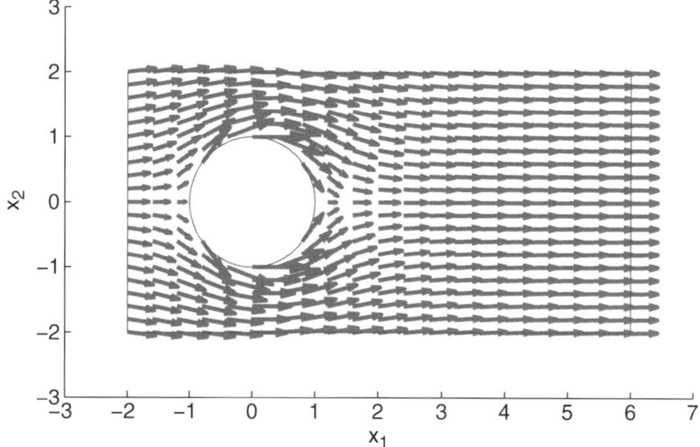

Fig. 10.5 Glyphs showing the fluid velocity field b

At the outflow we ignore the diffusion, so that $\epsilon n \cdot \nabla u = 0$. Finally, at the inflow the fluid has zero temperature.

All in all, we have the following transport equation and boundary conditions for the fluid temperature u.

$$-\epsilon \Delta u + b \cdot \nabla u = 0, \quad \text{in } \Omega \tag{10.50a}$$

$$u = 0, \quad \text{on } \Gamma_{\text{in}} \tag{10.50b}$$

$$u = 1, \quad \text{on } \Gamma_{\text{cyl}} \tag{10.50c}$$

$$-\epsilon n \cdot \nabla u = 0, \quad \text{on } \Gamma_{\text{out}} \tag{10.50d}$$

$$n \cdot (-\epsilon \nabla u + bu) = 0, \quad \text{on } \Gamma_{\text{wall}} \tag{10.50e}$$

In order to simplify the computer implementation we first approximate the Dirichlet conditions (10.50b) and (10.50c) using the Robin conditions $-\epsilon n \cdot \nabla u = 10^6 u$ on Γ_{in} and $-\epsilon n \cdot \nabla u = 10^6(u - 1)$ on Γ_{cyl}, respectively. Multiplying $0 = -\epsilon \Delta u + b \cdot \nabla u$ by $v \in V = H^1(\Omega)$ and integrating both the diffusive and convective term by parts, we then have

$$0 = \epsilon(\nabla u, \nabla v) - \epsilon(n \cdot \nabla u, v)_{L^2(\partial\Omega)} - (u, b \cdot \nabla v) + (n \cdot bu, v)_{L^2(\partial\Omega)} \tag{10.51}$$

$$= \epsilon(\nabla u, \nabla v) + 10^6(u, v)_{L^2(\Gamma_{\text{in}})} + 10^6(u - 1, v)_{L^2(\Gamma_{\text{cyl}})}$$
$$- (u, b \cdot \nabla v) + (n \cdot bu, v)_{L^2(\Gamma_{out})} \tag{10.52}$$

As a consequence, the weak form of (10.50) reads: find $u \in V$ such that

$$\begin{aligned} &\epsilon(\nabla u, \nabla v) + 10^6(u,v)_{L^2(\Gamma_{\text{in}})} + 10^6(u,v)_{L^2(\Gamma_{\text{cyl}})} \\ -&(u, b\cdot\nabla v) + (n\cdot bu, v)_{L^2(\Gamma_{out})} = 10^6(1,v)_{L^2(\Gamma_{\text{cyl}})}, \quad \forall v \in V \end{aligned} \tag{10.53}$$

To approximate V let $V_h \subset V$ be the usual space of all continuous piecewise linears. Adding the least squares term $\delta(b\cdot\nabla u, b\cdot\nabla v)$ to the weak form we obtain the GLS finite element approximation: find $u_h \in V_h$ such that

$$\begin{aligned} &\epsilon(\nabla u, \nabla v) + 10^6(u,v)_{L^2(\Gamma_{\text{in}})} + 10^6(u,v)_{L^2(\Gamma_{\text{cyl}})} \\ -&(u, b\cdot\nabla v) + (n\cdot bu, v)_{L^2(\Gamma_{out})} + \delta(b\cdot\nabla u, b\cdot\nabla v) = 10^6(1,v)_{L^2(\Gamma_{\text{cyl}})}, \quad \forall v \in V_h \end{aligned} \tag{10.54}$$

We observe that the left hand side boundary terms can be written $(\kappa u, v)_{L^2(\partial\Omega)}$ with κ defined by

$$\kappa = \begin{cases} 10^6, & \text{on } \Gamma_{\text{in}} \cup \Gamma_{\text{cyl}} \\ b\cdot n, & \text{on } \Gamma_{\text{out}} \\ 0, & \text{elsewhere} \end{cases} \tag{10.55}$$

or, in the MATLAB language,

```
function k = Kappa2(x,y)
k=0;
if x<-1.99 % inflow
  k=1e6;
end
if sqrt(x^2+y^2)<1.01 % cylinder
  k=1e6;
end
if x>5.99 % outflow
  [bx,by]=FlowField(x,y);
  nx=1; ny=0; % normal components
  k=bx*nx+by*ny; % kappa = dot(b,n)
end
```

Similarly, the right hand side boundary term may be written $(\kappa g_D + g_N, v)_{L^2(\partial\Omega)}$ with κ as above, and g_D and g_N defined by

```
function g = gD2(x,y)
g=0;
if sqrt(x^2+y^2)<1.01, g=1; end

function g = gN2(x,y)
g=0;
```

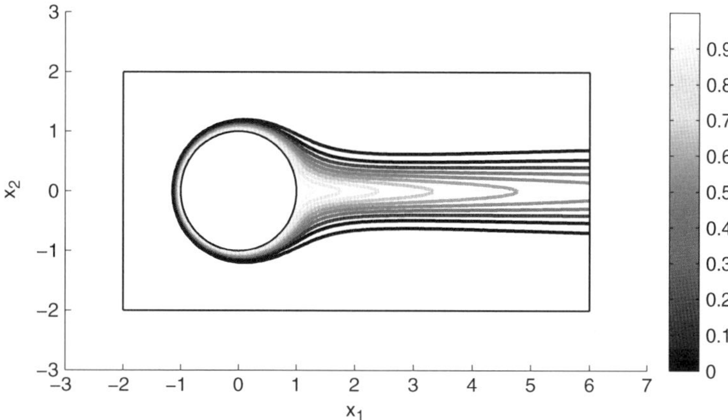

Fig. 10.6 Isocontours of the temperature u_h in the fluid

We can now compute all boundary terms with a call to `RobinAssembler2D` with function handles to `Kappa2`, `gD2`, and `gN2`, as arguments.

Finally, we notice that the convection matrix stemming from the term $-(u, b \cdot \nabla v)$ is just the negative transpose of the matrix assembled by `ConvectionAssembler2D`.

Putting all the pieces together we obtain the following main routine.

```
function HeatFlowSolver2D()
channel=RectCircg(); % channel geometry
epsilon=0.01; % diffusion parameter
h=0.1; % mesh size
[p,e,t]=initmesh(channel,'hmax',h); % create mesh
A=assema(p,t,1,0,0); % stiffness matrix
x=p(1,:); y=p(2,:); % node coordinates
[bx,by]=FlowField(x,y); % evaluate vector field b
C=ConvectionAssembler2D(p,t,bx,by); % convection matrix
Sd=SDAssembler2D(p,t,bx,by); % GLS stabilization matrix
[R,r]=RobinAssembler2D(p,e,@Kappa2,@gD2,@gN2); % Robin BC
delta=h; % stabilization parameter
U=(epsilon*A-C'+R+delta*Sd)\r; % solve linear system
pdecont(p,t,U), axis equal % plot solution
```

Note that the mesh size $h = 0.1$, while the diffusion parameter $\epsilon = 0.01$, which can lead to potential problems with oscillations. However, by choosing the stabilization parameter δ proportional to h we get additional diffusion along the streamlines of b that prevents the forming of oscillations. Running this code we get the result of Fig. 10.6.

From this figure it is clearly seen how the temperature behind the cylinder is transported downstream by the fluid flow, whereas a boundary layer is formed in front of the cylinder. As expected the temperature is decreasing downstream due to the artificial diffusion and no oscillations are visible. A nice detail visible is that the outflow appears transparent in the sense that the temperature isocontours seem unaffected by the domain boundary and the truncation of the computational domain.

10.2 Further Reading

A basic analysis of the SD method can be found in the book by Johnson [45]. A more recent and deeper analysis of GLS methods can be found in the book by Roos, Stynes, and Tobiska [55].

10.3 Problems

Exercise 10.1. Compute the least squares solution to the linear system $Ax = b$ with

$$A = \begin{bmatrix} 1 & 0 \\ 0 & 1 \\ 1 & 1 \end{bmatrix}, \quad x = \begin{bmatrix} x_1 \\ x_2 \end{bmatrix}, \quad b = \begin{bmatrix} 1 \\ 2 \\ 6 \end{bmatrix}$$

Which norm is minimized by this solution?

Exercise 10.2. Verify that a standard one-dimensional finite element method for the transport equation on a uniform mesh yields the linear system (10.14).

Exercise 10.3. Derive a GLS method for the problem

$$-\epsilon \Delta u + b \cdot \nabla u + cu = f, \quad x \in \Omega, \qquad u = 0, \quad x \in \partial\Omega$$

Use a standard continuous piecewise linear approximation of the solution. What does the linear system resulting from finite element discretization look like?

Exercise 10.4. Fill in the details of the a priori error estimate (10.45) for the GLS method.

Exercise 10.5. Use `TransportSolver2D` to verify that standard Galerkin is unstable also in higher dimensions. Choose the diffusion parameter $\epsilon = 0.01$ and the mesh size $h = 0.05$, for example.

Exercise 10.6. Show that $|||\cdot|||$ is a norm on $H_0^1(\Omega)$.

Exercise 10.7. Show that the bilinear form $a_h(\cdot,\cdot)$, defined by (10.25), is continuous in the norm $|||\cdot|||$ on the space of continuous piecewise linears V_h. Do the same for the linear form $l_h(\cdot)$, defined by (10.26).

Chapter 11
Solid Mechanics

Abstract Solid mechanics is arguably one of the most important areas of application for finite elements. Indeed, finite element analysis is used together with computer aided design (CAD) to optimize and speed up the design and manufacturing process of practically all mechanical structures, ranging from bearings to airplanes. In this chapter we derive the equations of linear elasticity and formulate finite element approximations of them. We do this in the abstract setting of elliptic partial differential equations introduced before and prove existence and uniqueness of the solution using the Lax-Milgram lemma. A priori and a posteriori error estimates are also proved. Some effort is laid on explaining the implementation of the finite element method. We also touch upon thermal stress – and modal analysis.

11.1 Governing Equations

11.1.1 Cauchy's Equilibrium Equation

Consider a volume Ω occupied by an elastic material and let ω denote an arbitrary subdomain of Ω with boundary $\partial\omega$ and exterior normal n. Two types of forces can act on ω. First, there are forces acting on the whole volume, so called body forces. These are described by a force density f, which expresses force per unit volume. The most common body force is gravity with $f = [0, 0, -9.82]^T$. Second, there are forces acting on the boundary $\partial\omega$. These are assumed to have the form $\sigma \cdot n$, where σ is the so called stress tensor $\sigma \cdot n$ denotes the vector with components $(\sigma \cdot n)_i = \sum_{j=1}^{3} \sigma_{ij} n_j$. The stress tensor is a second order tensor with components σ_{ij}, $i, j = 1, 2, 3$, where σ_{ij} expresses the force per unit area in direction x_i on a surface with unit normal in direction x_j. It follows from conservation of angular momentum that the stress tensor is symmetric with six independent components.

M.G. Larson and F. Bengzon, *The Finite Element Method: Theory, Implementation, and Applications*, Texts in Computational Science and Engineering 10,
DOI 10.1007/978-3-642-33287-6_11, © Springer-Verlag Berlin Heidelberg 2013

Summing body forces and contact forces we obtain the total net force F on Ω

$$F = \int_{\omega} f dx + \int_{\partial\omega} \sigma \cdot n ds \tag{11.1}$$

Using the divergence theorem on the surface integral we obtain

$$F = \int_{\omega} (f + \nabla \cdot \sigma)\, dx \tag{11.2}$$

In equilibrium $F = 0$, and since ω is arbitrary, we conclude that

$$f + \nabla \cdot \sigma = 0 \tag{11.3}$$

which is Cauchy's equilibrium equation. It is a system of equations, given in component form by

$$f_1 + \frac{\sigma_{11}}{\partial x_1} + \frac{\sigma_{12}}{\partial x_2} + \frac{\sigma_{13}}{\partial x_3} = 0 \tag{11.4a}$$

$$f_2 + \frac{\sigma_{21}}{\partial x_1} + \frac{\sigma_{22}}{\partial x_2} + \frac{\sigma_{23}}{\partial x_3} = 0 \tag{11.4b}$$

$$f_3 + \frac{\sigma_{31}}{\partial x_1} + \frac{\sigma_{32}}{\partial x_2} + \frac{\sigma_{33}}{\partial x_3} = 0 \tag{11.4c}$$

As there are six independent stress components, the above system with three equations needs to be closed by a material specific so-called constitutive equation, which relates the stress in the material to its deformation.

11.1.2 Constitutive Equations and Hooke's Law

The displacement of a material particle is defined as the vector $u = x - x_0$, where x is the current and x_0 the initial position of the particle. Under assumption of small displacement gradients the measure of deformation, the so-called strain, is given by the strain tensor

$$\varepsilon = \frac{1}{2}(\nabla u + \nabla u^T) \tag{11.5}$$

which in component form reads

$$\varepsilon_{ij} = \frac{1}{2}\left(\frac{\partial u_i}{\partial x_j} + \frac{\partial u_j}{\partial x_i}\right), \quad i, j = 1, 2, 3 \tag{11.6}$$

A desirable property for the strain tensor is that it is invariant under rigid body motion, that is, translation and rotation. For the linear strain tensor above this is, however, only partially true since it vanishes for translations, but not for proper rotations – only linearized rotations. Hence, in the linearized theory of elasticity, a critical assumption is that any rotations are small. The set of rigid body translations and linearized rotations are commonly referred to as the rigid body modes.

The diagonal components ε_{ii} is the relative change in length along the x_i-axis, and the off diagonal components ε_{ij} are to the first order proportional to the change in angle between the coordinate axes x_i and x_j, which initially are orthogonal.

The constitutive equation relating stress to deformation in linear elastic materials is called Hooke's law. It is a linear relationship, which in its most general form reads $\sigma_{ij} = \sum_{kl} C_{ijkl}\varepsilon_{kl}$, where C_{ijkl} is a fourth order tensor with up to 36 independent components, or elastic moduli, that describe the material. However, in isotropic materials, that is, materials in which the material properties are independent of spatial direction, only two elastic moduli are required. Assuming that the body is stress free before deformation, Hooke's law for linear elastic isotropic materials takes the form

$$\sigma = 2\mu\varepsilon(u) + \lambda(\nabla \cdot u)I \tag{11.7}$$

where I is the 3×3 identity matrix. The elastic moduli μ and λ are the so called Lamé parameters, defined by

$$\mu = \frac{E}{2(1+\nu)}, \qquad \lambda = \frac{E\nu}{(1+\nu)(1-2\nu)} \tag{11.8}$$

where E is Young's elastic modulus, and ν is Poisson's ratio. Young's modulus describes the stiffness of the material, whereas Poisson's ratio describes the material's tendency to shrink its cross section when stretched. For homogeneous materials, E and ν are constant throughout the material.

Combining the equilibrium equation (11.3) with the constitutive relation (11.7) we get a system of two vector valued partial differential equations, in the unknowns σ and u.

11.1.3 Boundary Conditions

To obtain a unique solution u, (11.3) and (11.7) must be supplemented by boundary conditions, which can be of the two standard types, Dirichlet and Neumann. Dirichlet boundary conditions are constraints on the displacements u and take the form $u = g_D$, where g_D is given function. Often, $g_D = 0$, which corresponds to a situation where the material is clamped to the surrounding and unable to move. Neumann boundary conditions are constraints on the normal stress and take the form $\sigma \cdot n = g_N$, where n is the outward unit normal to the boundary, and g_N a given so-called traction load.

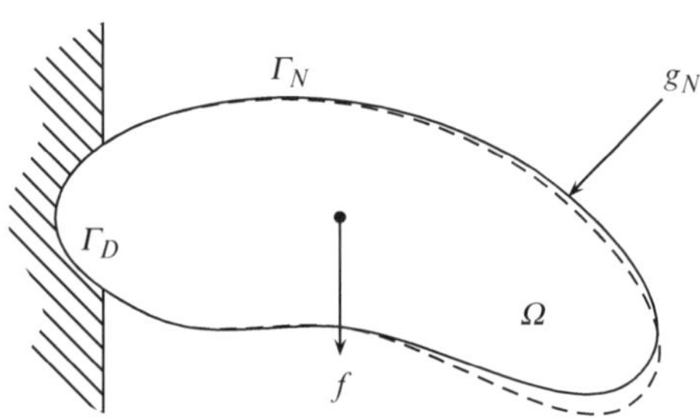

Fig. 11.1 Illustration of a clamped elastic material body Ω deforming under a body load f and a traction load g_N. The *solid line* shows the initial and the *dashed line* the deformed configuration

11.2 Linear Elastostatics

Thus, the basic problem of linear elastostatics is to find the stress tensor σ and the displacement vector u such that

$$-\nabla \cdot \sigma = f, \quad \text{in } \Omega \tag{11.9a}$$

$$\sigma = 2\mu\varepsilon(u) + \lambda(\nabla \cdot u)I, \quad \text{in } \Omega \tag{11.9b}$$

$$u = 0, \quad \text{on } \Gamma_D \tag{11.9c}$$

$$\sigma \cdot n = g_N, \quad \text{on } \Gamma_N \tag{11.9d}$$

where Γ_D and Γ_N are two boundary segments associated with the Dirichlet and Neumann boundary conditions, respectively. See Fig. 11.1.

11.2.1 Weak Form

To derive the weak form of (11.9), let V be the Hilbert space

$$V = \{v \in [H^1(\Omega)]^3 : v|_{\Gamma_D} = 0\} \tag{11.10}$$

That is, all sufficiently smooth displacement vectors vanishing on Γ_D.

Multiplying $f = -\nabla \cdot \sigma$ with a test function $v \in V$, and integrating by parts, we have

$$(f, v) = (-\nabla \cdot \sigma, v) \tag{11.11}$$

$$= \sum_{i,j=1}^{3} \left(-\frac{\partial \sigma_{ij}}{\partial x_j}, v_i\right) \tag{11.12}$$

$$= \sum_{i,j=1}^{3} -\left(\sigma_{ij}, n_j v_i\right)_{\partial\Omega} + \left(\sigma_{ij}, \frac{\partial v_i}{\partial x_j}\right) \tag{11.13}$$

Introducing the contraction operator:, defined by

$$A : B = \sum_{i,j=1}^{3} A_{ij} B_{ij} \tag{11.14}$$

for any two 3×3 matrices A and B we further have

$$-(\sigma \cdot n, v)_{\partial\Omega} + (\sigma : \nabla v) = (f, v) \tag{11.15}$$

where the entries of the 3×3 gradient matrix ∇v are given by $[\nabla v]_{ij} = \partial v_i / \partial v_j$. Using finally the Neumann boundary condition $\sigma \cdot n = g_N$ on Γ_N, and that $v = 0$ on Γ_D, we end up with

$$(\sigma : \nabla v) = (f, v) + (g_N, v)_{\Gamma_N}, \quad \forall v \in V \tag{11.16}$$

Actually, we can simplify this a bit more. If a 3×3 matrix A is symmetric and another 3×3 matrix B is anti-symmetric with zero diagonal, then $A : B = 0$. Now, recalling that any matrix can be decomposed into its symmetric and anti-symmetric part, viz., $A = (A + A^T)/2 + (A - A^T)/2$, it follows that

$$\sigma : \nabla v = \sigma : \frac{1}{2}(\nabla v + \nabla v^T) + \sigma : \frac{1}{2}(\nabla v - \nabla v^T) = \sigma : \epsilon(v) + 0 \tag{11.17}$$

This allows us to replace ∇v with $\epsilon(v)$ in (11.16), which yields

$$(\sigma(u) : \epsilon(v)) = (f, v) + (g_N, v)_{\Gamma_N}, \quad \forall v \in V \tag{11.18}$$

or, if we insert Hooke's law $\sigma = 2\mu\epsilon(u) + \lambda(\nabla \cdot u)I$, and use that $I : \varepsilon(v) = \nabla \cdot v$,

$$2\mu(\varepsilon(u) : \varepsilon(v)) + \lambda(\nabla \cdot u, \nabla \cdot v) = (f, v) + (g_N, v)_{\Gamma_N}, \quad \forall v \in V \tag{11.19}$$

Hence, the weak form of (11.9) reads: find $u \in V$ such that

$$a(u, v) = l(v) \quad \forall v \in V \tag{11.20}$$

where the bilinear from $a(\cdot, \cdot)$ and the linear form $l(\cdot)$ are defined by

$$a(u, v) = 2\mu(\varepsilon(u) : \varepsilon(v)) + \lambda(\nabla \cdot u, \nabla \cdot v) \tag{11.21}$$

$$l(v) = (f, v) + (g_N, v)_{\Gamma_N} \tag{11.22}$$

11.2.2 Existence and Uniqueness of the Solution

We next show that the weak form (11.20) fulfills the requirements of the Lax-Milgram lemma. In doing so, to measure the size of the various matrices, tensors, and vectors involved we define the following norms on V for any 3×3 matrix or tensor A, and any 3×1 vector b.

$$\|A\|_V^2 = \sum_{i,j=1}^{3} \|A_{ij}\|_{H^1(\Omega)}^2, \qquad \|b\|_V^2 = \sum_{i=1}^{3} \|b_i\|_{H^1(\Omega)}^2 \tag{11.23}$$

The semi-norms of A and b on V are defined analogously.

The continuity of $a(\cdot,\cdot)$ follows from the Cauchy-Schwarz inequality.

$$\begin{aligned} a(u,v) &= 2\mu(\varepsilon(u):\varepsilon(v)) + \lambda(\nabla\cdot u, \nabla\cdot v) && (11.24)\\ &\le 2\mu\|\varepsilon(u)\|\,\|\varepsilon(v)\| + \lambda\|\nabla\cdot u\|\,\|\nabla\cdot v\| && (11.25)\\ &\le C\|\nabla u\|\,\|\nabla v\| && (11.26)\\ &\le C\|u\|_V\|v\|_V && (11.27)\end{aligned}$$

The continuity of $l(\cdot)$ follows from the trace inequality

$$\|v\|_{\Gamma_N} \le C(\|\nabla v\| + \|v\|) \le C\|v\|_V \tag{11.28}$$

and, again, the Cauchy-Schwarz inequality.

$$\begin{aligned} l(v) &\le (f,v) + (g,v)_{\Gamma_N} && (11.29)\\ &\le \|f\|\,\|v\| + \|g\|_{\Gamma_N}\|v\|_{\Gamma_N} && (11.30)\\ &\le \|f\|\,\|v\|_V + \|g\|_{\Gamma_N}\|v\|_V && (11.31)\\ &\le C\|v\|_V && (11.32)\end{aligned}$$

To prove coercivity of $a(\cdot,\cdot)$ we need the following result.

Theorem 11.1 (Korn's Inequality). *There is a constant C such that*

$$C\|\nabla v\|^2 \le \|\varepsilon(v)\|^2 = \int_\Omega \sum_{i,j=1}^{3} \varepsilon_{ij}(v)\varepsilon_{ij}(v)\,dx \tag{11.33}$$

Proof. For simplicity, let us assume that $u = 0$ on the whole boundary $\partial\Omega$.

Straight forward calculation reveals that

$$\int_{\Omega} \sum_{i,j=1}^{3} \varepsilon_{ij}(v)\varepsilon_{ij}(v)\, dx = \int_{\Omega} \sum_{i,j=1}^{3} \frac{1}{2}\left(\frac{\partial v_i}{\partial x_j} + \frac{\partial v_j}{\partial x_i}\right) \frac{1}{2}\left(\frac{\partial v_i}{\partial x_j} + \frac{\partial v_j}{\partial x_i}\right) dx \tag{11.34}$$

$$= \frac{1}{4} \int_{\Omega} \sum_{i,j=1}^{3} \left(\frac{\partial v_i}{\partial x_j}\right)^2 + 2\frac{\partial v_i}{\partial x_j}\frac{\partial v_j}{\partial x_i} + \left(\frac{\partial v_j}{\partial x_i}\right)^2 dx \tag{11.35}$$

$$= \frac{1}{2}\|\nabla v\|^2 + \frac{1}{2} \sum_{i,j=1}^{3} \int_{\Omega} \frac{\partial v_i}{\partial x_j}\frac{\partial v_j}{\partial x_i}\, dx \tag{11.36}$$

Obviously, the claim follows if we can show that the last term is positive.

Using partial integration twice together with the boundary condition $v = 0$ on $\partial\Omega$, we have

$$\sum_{i,j=1}^{3} \int_{\Omega} \frac{\partial v_i}{\partial x_j}\frac{\partial v_j}{\partial x_i}\, dx = -\sum_{i,j=1}^{3} \int_{\Omega} v_i \frac{\partial^2 v_j}{\partial x_i \partial x_j}\, dx + \int_{\partial\Omega} n_j v_i \frac{\partial v_j}{\partial x_i}\, ds \tag{11.37}$$

$$= \sum_{i,j=1}^{3} \int_{\Omega} \frac{\partial v_i}{\partial x_i}\frac{\partial v_j}{\partial x_j}\, dx - \int_{\partial\Omega} n_i v_i \frac{\partial v_j}{\partial x_j}\, ds \tag{11.38}$$

$$= \int_{\Omega} \left(\sum_{i=1}^{3} \frac{\partial v_i}{\partial x_i}\right)\left(\sum_{j=1}^{3} \frac{\partial v_j}{\partial x_j}\right) dx \tag{11.39}$$

$$= \int_{\Omega} (\nabla \cdot v)^2\, dx \geq 0 \tag{11.40}$$

which concludes the proof. □

The coercivity of $a(\cdot,\cdot)$ now follows from Korn's inequality.

$$a(u,u) = 2\mu\|\varepsilon(u)\|^2 + \lambda\|\nabla \cdot u\|^2 \geq 2\mu\|\varepsilon(u)\|^2 \geq C\|\nabla u\|^2 \geq m\|v\|_V^2 \tag{11.41}$$

Thus, we conclude that the requirements for the Lax-Milgram lemma are fulfilled. Hence, there exist a unique solution $u \in V$ to the weak form (11.20).

11.2.3 Finite Element Approximation

From the Lax-Milgram lemma we know that the weak form (11.20) has a unique solution $u \in V$, which can be approximated with finite elements. To this end, we choose to approximate each component of u using continuous piecewise linears. Let $\mathcal{K} = \{K\}$ be a shape regular tetrahedral mesh on Ω, and let V_h be the polynomial space

$$V_h = \{v \in V : v|_K \in [P_1(K)]^3, \ \forall K \in \mathcal{K}\} \tag{11.42}$$

That is, all displacement vectors with continuous piecewise linear components vanishing on Γ_D.

Our finite element method takes the form: find $u_h \in V_h$, such that

$$a(u_h, v) = l(v), \quad \forall v \in V_h \tag{11.43}$$

11.2.4 A Priori Error Estimate

As always, we wish to assert the accuracy of the finite element solution u_h by estimating the error $e = u - u_h$. In doing so, we have the following a priori error estimate.

Theorem 11.2. *The finite element solution u_h, defined by* (11.43)*, satisfies the estimate*

$$\|\nabla e\| \leq Ch|u|_{H^2(\Omega)} \tag{11.44}$$

where C is constant independent of u, u_h, and the mesh size h.

Proof. Using the coercivity of $a(\cdot,\cdot)$ we have

$$m\|\nabla e\|^2 \leq a(e,e) = a(e, u - u_h) = a(e, u - \pi u + \pi u - u_h) = a(e, u - \pi u) \tag{11.45}$$

where we have added and subtracted an interpolant $\pi u \in V_h$ to u, and used that $a(e, \pi u) = 0$ by Galerkin orthogonality. Using also the continuity of $a(\cdot,\cdot)$ we further have

$$m\|\nabla e\|^2 \leq a(e, u - \pi u) \leq C\|\nabla e\|\,\|\nabla(u - \pi u)\| \tag{11.46}$$

Finally, using interpolation theory we have

$$\|\nabla(u - \pi u)\| \leq Ch|u|_{H^2(\Omega)} \tag{11.47}$$

Dividing by $\|\nabla e\|$ concludes the proof. □

11.2.5 Engineering Notation

To simplify the bookkeeping of the test and trial functions v and u_h and their components, it is customary to rewrite the bilinear form $a(u_h, v)$ as the product of a few matrices. The starting point is to write the six independent components of the stress tensor σ as a vector, viz.,

$$\sigma = \left[\sigma_{11}\ \sigma_{22}\ \sigma_{33}\ \sigma_{12}\ \sigma_{23}\ \sigma_{31}\right]^T \tag{11.48}$$

The strain tensor ε is similarly written as a vector, viz.,

$$\varepsilon = \left[\varepsilon_{11}\ \varepsilon_{22}\ \varepsilon_{33}\ 2\varepsilon_{12}\ 2\varepsilon_{23}\ 2\varepsilon_{31}\right]^T \tag{11.49}$$

Hooke's law (11.7) can then be written

$$\sigma = D\varepsilon \tag{11.50}$$

where the 6×6 matrix D is given by

$$D = \begin{bmatrix} \lambda + 2\mu & \lambda & \lambda & 0 & 0 & 0 \\ \lambda & \lambda + 2\mu & \lambda & 0 & 0 & 0 \\ \lambda & \lambda & \lambda + 2\mu & 0 & 0 & 0 \\ 0 & 0 & 0 & \mu & 0 & 0 \\ 0 & 0 & 0 & 0 & \mu & 0 \\ 0 & 0 & 0 & 0 & 0 & \mu \end{bmatrix} \tag{11.51}$$

in three-dimensions.

In two-dimensions one has to differ between so-called plane strain, defined by

$$\varepsilon_{13} = \varepsilon_{23} = \varepsilon_{31} = 0, \qquad \sigma_{33} = \nu(\sigma_{11} + \sigma_{22}) \tag{11.52}$$

and so-called plane stress, defined by

$$\sigma_{13} = \sigma_{23} = \sigma_{31} = 0, \qquad \varepsilon_{33} = -\frac{\nu}{E}(\sigma_{33} + \sigma_{22}) \tag{11.53}$$

Both cases can be handled by a constitutive law of the form $\sigma = D\varepsilon$, where

$$\sigma = \left[\sigma_{11}\ \sigma_{22}\ \sigma_{12}\right]^T, \qquad \varepsilon = \left[\varepsilon_{11}\ \varepsilon_{22}\ 2\varepsilon_{12}\right]^T \tag{11.54}$$

and

$$D = \begin{bmatrix} \lambda + 2\mu & \lambda & 0 \\ \lambda & \lambda + 2\mu & 0 \\ 0 & 0 & \mu \end{bmatrix} \tag{11.55}$$

for plane strain and

$$D = \frac{E}{1-\nu^2}\begin{bmatrix} 1 & \nu & 0 \\ \nu & 1 & 0 \\ 0 & 0 & (1-\nu)/2 \end{bmatrix} \tag{11.56}$$

for plane stress. We shall return to plane strain and stress shortly.

Now, the so-called engineering notation adopted so far allows us to write

$$\varepsilon : \sigma = \varepsilon^T \sigma = \varepsilon^T D \varepsilon \tag{11.57}$$

which implies

$$a(u_h, v) = \int_\Omega \varepsilon(v) : \sigma(u_h)\, dx = \int_\Omega \varepsilon^T(v)\sigma(u_h)\, dx = \int_\Omega \varepsilon^T(v) D \varepsilon(u_h)\, dx \tag{11.58}$$

Here, it is convenient to write the finite element ansatz $u_h \in V_h$ in matrix form, viz.,

$$u_h = \begin{bmatrix} u_1 \\ u_2 \\ u_3 \end{bmatrix}_h = \begin{bmatrix} \varphi_1 & 0 & 0 & \varphi_2 & 0 & 0 & \dots & \varphi_{n_i} & 0 & 0 \\ 0 & \varphi_1 & 0 & 0 & \varphi_2 & 0 & \dots & 0 & \varphi_{n_i} & 0 \\ 0 & 0 & \varphi_1 & 0 & 0 & \varphi_2 & \dots & 0 & 0 & \varphi_{n_i} \end{bmatrix} \begin{bmatrix} d_{11} \\ d_{12} \\ d_{13} \\ d_{21} \\ d_{22} \\ d_{23} \\ \vdots \\ d_{N1} \\ d_{N2} \\ d_{N3} \end{bmatrix} = \varphi d \tag{11.59}$$

where φ_i, $i = 1, 2, \dots, n_i$ are the hat basis functions, and d is a vector containing the nodal displacements. Because there are three displacements per node, d is of length $3n_i$, with n_i the number of interior nodes.

The strain field is linked to the displacements by (11.5). An alternative way of writing this is

$$\begin{bmatrix} \varepsilon_{11} \\ \varepsilon_{22} \\ \varepsilon_{33} \\ 2\varepsilon_{12} \\ 2\varepsilon_{23} \\ 2\varepsilon_{31} \end{bmatrix} = \begin{bmatrix} \partial/\partial x_1 & 0 & 0 \\ 0 & \partial/\partial x_2 & 0 \\ 0 & 0 & \partial/\partial x_3 \\ \partial/\partial x_2 & \partial/\partial x_1 & 0 \\ 0 & \partial/\partial x_3 & \partial/\partial x_2 \\ \partial/\partial x_3 & 0 & \partial/\partial x_1 \end{bmatrix} \begin{bmatrix} u_1 \\ u_2 \\ u_3 \end{bmatrix} \tag{11.60}$$

Introducing the strain matrix

$$B = \begin{bmatrix} \partial/\partial x_1 & 0 & 0 \\ 0 & \partial/\partial x_2 & 0 \\ 0 & 0 & \partial/\partial x_3 \\ \partial/\partial x_2 & \partial/\partial x_1 & 0 \\ 0 & \partial/\partial x_3 & \partial/\partial x_2 \\ \partial/\partial x_3 & 0 & \partial/\partial x_1 \end{bmatrix} \begin{bmatrix} \varphi_1 & 0 & 0 & \varphi_2 & 0 & 0 & \dots & \varphi_{n_i} & 0 & 0 \\ 0 & \varphi_1 & 0 & 0 & \varphi_2 & 0 & \dots & 0 & \varphi_{n_i} & 0 \\ 0 & 0 & \varphi_1 & 0 & 0 & \varphi_2 & \dots & 0 & 0 & \varphi_{n_i} \end{bmatrix} \tag{11.61}$$

we have the discrete strains and stresses

$$\varepsilon = Bd \tag{11.62}$$

$$\sigma = DBd \tag{11.63}$$

The linear system arising from the finite element method (11.43) can now be written in matrix form as

$$\left(\int_\Omega B^T DB \, dx \right) d = \int_\Omega \varphi^T f \, dx + \int_{\Gamma_N} \varphi^T g_N \, ds \tag{11.64}$$

or simply

$$Kd = F \tag{11.65}$$

where K is the $3n_i \times 3n_i$ stiffness matrix

$$K = \int_\Omega B^T DBdx \tag{11.66}$$

and F is the $3n_i \times 1$ load vector

$$F = \int_\Omega \varphi^T f \, dx + \int_{\Gamma_N} \varphi^T g_N \, ds \tag{11.67}$$

11.2.6 Computer Implementation

Although deformation is a genuine three-dimensional phenomenon it is sometimes possible to reduce to two dimensions. For example, say that we have a very slender structure oriented along the x_3-axis with length much greater than cross-section area. Then the strains associated with length (i.e., ε_{13}, ε_{23}, and ε_{33}) are small compared to the cross-sectional strains, since they are constrained by nearby material. In such a case, it suffice to consider a reduced two-dimensional elastic problem in the cross-section of the structure to deduce the deformation. The conditions that $u_3 = 0$ and that there is no variation with respect to x_3 in any quantity is called the state of plain strain. By analogy, the state of plane stress applies to structures, which are large, but thin, such as plates or shells, for instance.

Let us work through the details of writing a two-dimensional elastic finite element solver. To this end, let $\Omega \subset \mathbb{R}^2$ denote a two-dimensional domain in the x_1x_2-plane, and let $\mathcal{K} = \{K\}$ be a triangle mesh of Ω.

As usual, the stiffness matrix (11.66) and the load vector (11.67) can be assembled by summing integral contributions from each element. Consider therefore an element K with the three nodes N_i, $i = 1, 2, 3$. On K the element displacements u_h^K are given by

$$u_h^K = \begin{bmatrix} \varphi_1 & 0 & \varphi_2 & 0 & \varphi_3 & 0 \\ 0 & \varphi_1 & 0 & \varphi_2 & 0 & \varphi_3 \end{bmatrix} \begin{bmatrix} d_{11} \\ d_{12} \\ d_{21} \\ d_{22} \\ d_{31} \\ d_{32} \end{bmatrix} = \varphi^K d^K \tag{11.68}$$

where φ_i are the hat functions. Recall that these are given by $\varphi_i = a_i + b_i x_1 + c_i x_2$ where a_i, b_i, and c_i are determined from $\varphi_i(N_j) = \delta_{ij}$. Further, on K the element strains are given by

$$\varepsilon^K = \begin{bmatrix} \partial/\partial x_1 & 0 \\ 0 & \partial/\partial x_2 \\ \partial/\partial x_2 & \partial/\partial x_1 \end{bmatrix} u_h^K = \begin{bmatrix} b_1 & 0 & b_2 & 0 & b_3 & 0 \\ 0 & c_1 & 0 & c_2 & 0 & c_3 \\ c_1 & b_1 & c_2 & b_2 & c_3 & b_3 \end{bmatrix} d^K = B^K d^K \tag{11.69}$$

Note that the strain matrix B^K is constant and that all strains are constant on the element. Because the element strains are constant, so are also the element stresses $\sigma^K = D\varepsilon^K$.

Now, the element stiffness matrix is given by

$$K^K = \int_K {B^K}^T D B^K \, dx \tag{11.70}$$

which simplifies to $K^K = {B^K}^T D B^K |K|$, since the integrand is constant.

Writing a code for computing K^K is easy.

```
function KK = ElasticStiffness(x,y,mu,lambda)
% triangle area and gradients (b,c) of hat functions
[area,b,c]=HatGradients(x,y);
% elastic matrix
D=mu*[2 0 0; 0 2 0; 0 0 1]+lambda*[1 1 0; 1 1 0; 0 0 0];
% strain matrix
BK=[b(1)    0 b(2)     0 b(3)     0 ;
        0 c(1)     0 c(2)     0 c(3);
```

```
    c(1) b(1) c(2) b(2) c(3) b(3)];
% element stiffness matrix
KK=BK'*D*BK*area;
```

Input to this routine is the element node coordinates `x` and `y`, and the Lamé parameters `lambda` and `mu`. Output is the 6×6 element stiffness matrix `KK`.

Continuing, the element load vector is given by

$$F^K = \int_K \varphi^{K^T} f dx = \int_K \begin{bmatrix} \varphi_1 & 0 \\ 0 & \varphi_1 \\ \varphi_2 & 0 \\ 0 & \varphi_2 \\ \varphi_3 & 0 \\ 0 & \varphi_3 \end{bmatrix} \begin{bmatrix} f_1 \\ f_2 \end{bmatrix} dx \tag{11.71}$$

To evaluate these integrals without effort we can use the old trick of replacing f with its linear interpolant πf, and then integrate the interpolant. Recall that πf is defined on K by

$$\pi f = \begin{bmatrix} \pi f_1 \\ \pi f_2 \end{bmatrix} = \begin{bmatrix} \varphi_1 & 0 & \varphi_2 & 0 & \varphi_3 & 0 \\ 0 & \varphi_1 & 0 & \varphi_2 & 0 & \varphi_3 \end{bmatrix} \begin{bmatrix} f_{11} \\ f_{21} \\ f_{12} \\ f_{22} \\ f_{13} \\ f_{23} \end{bmatrix} = \varphi^{K^T} f^K \tag{11.72}$$

where $f_{ij} = f_i(N_j)$ are the nodal force values. This gives us

$$F^K = \int_K \varphi^{K^T} f dx \approx \int_K \begin{bmatrix} \varphi_1 & 0 \\ 0 & \varphi_1 \\ \varphi_2 & 0 \\ 0 & \varphi_2 \\ \varphi_3 & 0 \\ 0 & \varphi_3 \end{bmatrix} \begin{bmatrix} \varphi_1 & 0 & \varphi_2 & 0 & \varphi_3 & 0 \\ 0 & \varphi_1 & 0 & \varphi_2 & 0 & \varphi_3 \end{bmatrix} \begin{bmatrix} f_{11} \\ f_{21} \\ f_{12} \\ f_{22} \\ f_{13} \\ f_{23} \end{bmatrix} dx \tag{11.73}$$

$$= \int_K \begin{bmatrix} \varphi_1^2 & 0 & \varphi_2\varphi_1 & 0 & \varphi_3\varphi_1 & 0 \\ 0 & \varphi_1^2 & 0 & \varphi_2\varphi_1 & 0 & \varphi_3\varphi_1 \\ \varphi_1\varphi_2 & 0 & \varphi_2^2 & 0 & \varphi_3\varphi_2 & 0 \\ 0 & \varphi_1\varphi_2 & 0 & \varphi_2^2 & 0 & \varphi_3\varphi_2 \\ \varphi_1\varphi_3 & 0 & \varphi_2\varphi_3 & 0 & \varphi_3^2 & 0 \\ 0 & \varphi_1\varphi_3 & 0 & \varphi_2\varphi_3 & 0 & \varphi_3^2 \end{bmatrix} \begin{bmatrix} f_{11} \\ f_{21} \\ f_{12} \\ f_{22} \\ f_{13} \\ f_{23} \end{bmatrix} dx = M^K f^K \tag{11.74}$$

where M^K is the element mass matrix. Evaluating its integrals one finds that

$$M^K = \frac{1}{12}\begin{bmatrix} 2 & 0 & 1 & 0 & 1 & 0 \\ 0 & 2 & 0 & 1 & 0 & 1 \\ 1 & 0 & 2 & 0 & 1 & 0 \\ 0 & 1 & 0 & 2 & 0 & 1 \\ 1 & 0 & 1 & 0 & 2 & 0 \\ 0 & 1 & 0 & 1 & 0 & 2 \end{bmatrix} |K| \tag{11.75}$$

which immediately translates into MATLAB code.

```
function MK = ElasticMass(x,y)
area=polyarea(x,y);
MK=[2 0 1 0 1 0;
    0 2 0 1 0 1;
    1 0 2 0 1 0;
    0 1 0 2 0 1;
    1 0 1 0 2 0;
    0 1 0 1 0 2]*area/12;
```

Since the element load is approximately given by $F^K = M^K f^K$ on each element, it is straight forward to assemble the load vector F as the sum $F = \sum_K F^K$.

When performing the assembly of the global system of equations, we need to take into account that there are two unknowns, or, degrees of freedom, per node. This makes the insertion of element matrix contributions into the global system matrix a bit more trickier than usual. In order to add the local stiffness K_{ij}^K to its correct location in the global stiffness matrix K, we have to make a local-to-global map between the node numbers and the numbering of the displacement degrees of freedom. Actually, we have already set up this mapping when ordering the nodal displacements in the vector d. Recall that all odd vector entries d_{2i-1} have to do with the x_1-displacements, and that all even entries d_{2i} have to do with x_2-displacements. This is also true for the element displacement vector d^K. Thus, the two displacement components in node number i is mapped onto vector entries d_{2i-1} and d_{2i}, $i = 1, 2, \dots, n_i$, and the map between a node N_i and its degrees of freedom is consequently $i \mapsto (2i-1, 2i)$. For example, if element K has the nodes 3, 5 and 6, then the degrees of freedom is 5, 6, 9, 10, 11, and 12. From this it follows that the local stiffness K_{15}^K should be added to row 5 column 11 in the global stiffness matrix K.

Using the subroutines `ElasticStiffness` and `ElasticMass` we can now write a routine for assembling the global stiffness matrix K and the global load vector F. For future use let us also assemble the global mass matrix M.

```
function [K,M,F] = ElasticAssembler(p,e,t,lambda,mu,force)
ndof=2*size(p,2); % total number of degrees of freedom
K=sparse(ndof,ndof); % allocate stiffness matrix
M=sparse(ndof,ndof); % allocate mass matrix
F=zeros(ndof,1); % allocate load vector
```

```
dofs=zeros(6,1); % allocate element degrees of freedom
for i=1:size(t,2) % assemly loop over elements
    nodes=t(1:3,i); % element nodes
    x=p(1,nodes); y=p(2,nodes); % node coordinates
    dofs(2:2:end)=2*nodes; % element degrees of freedom
    dofs(1:2:end)=2*nodes-1;
    f=force(x,y); % evaluate force at nodes
    KK=ElasticStiffness(x,y,lambda,mu); % element stiffness
    MK=ElasticMass(x,y); % element mass
    fK=[f(1,1) f(2,1) f(1,2) f(2,2) f(1,3) f(2,3)]';
    FK=MK*fK; % element load
    K(dofs,dofs)=K(dofs,dofs)+KK; % add to stiffness matrix
    M(dofs,dofs)=M(dofs,dofs)+MK; % add to mass matrix
    F(dofs)=F(dofs)+FK; % add to load vector
end
```

Input is the usual point, edge, and connectivity matrices `p`, `e`, and `t`, the Lamé parameters `lambda`, and `mu`, and a function handle to a subroutine `Force` specifying the body force. For example,

```
function f = Force(x,y)
f=[35/13*y-35/13*y.^2+10/13*x-10/13*x.^2;
  -25/26*(-1+2*y).*(-1+2*x)];
```

Output is the global stiffness matrix `K`, the global mass matrix `M`, and the global load vector `F`.

The Lamé parameters λ and μ can conveniently be computed from the Young modulus E and the Poisson's ratio ν, which are the usual physical data available, with the following subroutine.

```
function [mu,lambda] = Enu2Lame(E,nu)
mu=E/(2*(1+nu));
lambda=E*nu/((1+nu)*(1-2*nu));
```

For the stiffness matrix to be invertible some boundary conditions must be enforced. Assuming a Dirichlet type boundary condition this can be done as usual by eliminating the known nodal displacements from the stiffness matrix and adding them to the load vector. For example, if we have the homogeneous Dirichlet boundary conditions on the whole boundary, then these can be enforced with the code

```
bdry=unique([e(1,:) e(2,:)]); % boundary nodes
fixed=[2*bdry-1 2*bdry]; % boundary degrees of freedom, DoFs
values=zeros(length(fixed),1); % zero boundary values
ndof=length(F); % total number of DoFs
free=setdiff([1:ndof],fixed); % free DoFs
F=F(free)-K(free,fixed)*values; % modify load for BC
K=K(free,free); % modify stiffness for BC
d=zeros(ndof,1); % nodal displacement vector
```

```
d(free)=K\F; % solve for free DoFs
d(fixed)=values; % insert known DoFs
```

The main routine for our linear elastic finite element solver is given below.

```
function ElasticSolver()
g=Rectg(0,0,1,1);
[p,e,t]=initmesh(g,'hmax',0.1);
E=1; nu=0.3;
[mu,lambda]=Enu2Lame(E,nu);
[K,M,F]=ElasticAssembler(p,e,t,mu,lambda,@Force);
bdry=unique([e(1,:) e(2,:)]);
fixed=[2*bdry-1 2*bdry];
values=zeros(length(fixed),1);
ndof=length(F);
free=setdiff([1:ndof],fixed);
F=F(free)-K(free,fixed)*values;
K=K(free,free);
d=zeros(ndof,1);
d(free)=K\F;
d(fixed)=values;
U=d(1:2:end); V=d(2:2:end);
figure(1), pdesurf(p,t,U), title('(u_h)_1')
figure(2), pdesurf(p,t,V), title('(u_h)_2')
```

11.2.6.1 Verifying the Energy Norm Convergence

Let us verify that our finite element solver is implemented correctly. By taking the logarithm of $\sqrt{a(e,e)} \leq Ch$, which is nothing but the a priori estimate of Theorem 11.2, we find that the error $e = u - u_h$ obeys

$$\log \sqrt{a(e,e)} \leq \log(Ch) = C + \log(h) \tag{11.76}$$

where C is a constant depending on D^2u. Recall that $\sqrt{a(\cdot,\cdot)}$ is the energy norm $|||\cdot|||$. From (11.76) it follows that if we make a plot of $\log(h)$ versus $\log(|||e|||)$, then we should asymptotically get a straight line with slope 1. However, to be able to compute e we need to know the exact solution u, and we shall therefore manufacture a problem with known solution. To this end, let $\Omega = [0,1]^2$, and let $u = [x_1(1-x_1)x_2(1-x_2), 0]$. This choice of u assures that $u = 0$ on the boundary $\partial\Omega$. Using u to first compute the strain tensor ε, and then the stress tensor σ, and finally $-\nabla \cdot \sigma$, we find that the force f equals

$$f = \begin{bmatrix} 35/13x_2 - 35/13x_2^2 + 10/13x_1 - 10/13x_1^2 \\ -25/26(-1+2x_2)(-1+2x_1) \end{bmatrix} \tag{11.77}$$

Table 11.1 Convergence of $|||e|||$ for a sequence of finer and finer meshes

h	$\sqrt{F^T d}$	$\|\|\|e\|\|\|$
0.1250	0.1372	0.0201
0.1125	0.1374	0.0187
0.1000	0.1377	0.0162
0.0875	0.1379	0.0146
0.0750	0.1381	0.0125
0.0625	0.1383	0.0103
0.0500	0.1384	0.0083
0.0375	0.1385	0.0061
0.0250	0.1386	0.0040
0.0125	0.1387	0.0020

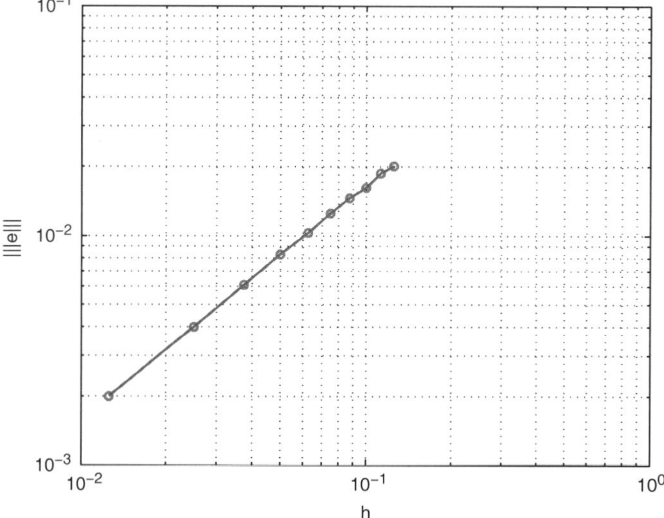

Fig. 11.2 Loglog plot of the mesh size h versus the error in the energy norm $|||e|||$

with $E = 1$ and $\nu = 0.3$. In the same way we also find that

$$a(u, u) = (\sigma(u) : \varepsilon(u)) = 1/52 \tag{11.78}$$

Next, to compute $a(e, e)$ we note that $a(e, e) = a(u, u) - a(u_h, u_h)$ by Galerkin orthogonality, and that $a(u_h, u_h)$ can be easily computed as $a(u_h, u_h) = d^T K d = F^T d$. Recording the mesh size h and the energy norm error $|||e|||$ for ten different uniform meshes we get the results shown in Table 11.1. In Fig. 11.2 we show a loglog plot of the obtained data. Looking at the plot we see that it is almost a straight line and by doing a linear least squares fit on the data we find that the slope of the line is 1.0104, which is close to the predicted value of 1, indeed.

11.2.7 A Posteriori Error Estimate

To formulate adaptive finite elements we wish to derive an posteriori estimate for the error $e = u - u_h$.

Starting from the coercivity of $a(\cdot,\cdot)$, and using Galerkin orthogonality $a(e,v) = 0$ for all $v \in V_h$ with v chosen as the interpolant $\pi e \in V_h$ of e, we have

$$C\|\nabla e\|^2 \leq a(e, e - \pi e) \tag{11.79}$$

$$= a(u, e - \pi e) - a(u_h, e - \pi e) \tag{11.80}$$

$$= (f, e - \pi e) - a(u_h, e - \pi e) \tag{11.81}$$

$$= \sum_{K\in\mathcal{K}} (f, e - \pi e)_K - (\sigma_h(u_h) : \varepsilon(e - \pi e))_K \tag{11.82}$$

$$= \sum_{K\in\mathcal{K}} (f, e - \pi e)_K + (\nabla\cdot\sigma_h, e - \pi e) - (\sigma_h \cdot n, e - \pi e)_{\partial K} \tag{11.83}$$

$$+ (g_N, e - \pi e)_{\partial K \cap \Gamma_N}$$

$$= \sum_{K\in\mathcal{K}} (f + \nabla\cdot\sigma_h, e - \pi e)_K - \left(\frac{1}{2}[\sigma_h \cdot n], e - \pi e\right)_{\partial K\setminus\partial\Omega} \tag{11.84}$$

$$+ (g_N - \sigma_h \cdot n, e - \pi e)_{\partial K \cap \Gamma_N}$$

Here, $[\sigma_h \cdot n]$ denotes the jump in the computed normal stress over the element boundaries. On the domain boundary Γ_N we obtain the term $(g_N, e - \pi e)_{\Gamma_N}$ when integrating by parts due to the Neumann traction boundary condition. This term does not vanish, since $e - \pi e$ is not zero on Γ_N. However, $e - \pi e$ is zero on Γ_D, since both u and u_h satisfies the Dirichlet boundary conditions.

Now, using the Cauchy-Schwarz inequality on each term in (11.84) we immediately get

$$\|\nabla e\|^2 \leq C \sum_{K\in\mathcal{K}} \|f + \nabla\cdot\sigma_h\|_K \|e - \pi e\|_K \tag{11.85}$$

$$\|\frac{1}{2}[\sigma_h \cdot n]\|_{\partial K\setminus\partial\Omega} \|e - \pi e\|_{\partial K\setminus\partial\Omega}$$

$$+ \|g_N - \sigma_h \cdot n\|_{\partial K\cap\Gamma_N} \|e - \pi e\|_{\partial K\cap\Gamma_N}$$

Next, recalling the Trace inequality $\|v\|_{\partial K} \leq C(h_K^{-1/2}\|v\|_K + h_K^{1/2}\|\nabla v\|_K)$, the interpolation estimate $\|v - \pi v\|_K \leq Ch_K\|\nabla v\|_K$, and the stability estimate $\|\nabla(\pi v)\| \leq C\|\nabla v\|$, we have

$$h_K^{1/2}\|e - \pi e\|_{\partial K} \leq C(\|e - \pi e\|_K + h_K\|\nabla(e - \pi e)\|_K) \leq h_K\|\nabla e\|_K \tag{11.86}$$

Using this result and the Cauchy-Schwarz inequality, again, we further have

$$\|\nabla e\|^2 \leq C \sum_{K\in\mathcal{K}} \Big(h_K\|f + \nabla\cdot\sigma_h\|_K + h_K^{1/2}\Big(\frac{1}{2}\|[\sigma_h\cdot n]\|_{\partial K\setminus\partial\Omega}$$

$$+ \|g_N - \sigma_h\cdot n\|_{\partial K\cap\Gamma_N}\Big)\Big)\|\nabla e\|_K \tag{11.87}$$

$$\leq C\left(\sum_{K\in\mathcal{K}} h_K^2\|f + \nabla\cdot\sigma_h\|_K^2 + h_K\left(\frac{1}{4}\|[\sigma_h\cdot n]\|^2_{\partial K\setminus\partial\Omega}\right.\right.$$

$$\left.\left. + \|g_N - \sigma_h\cdot n\|^2_{\partial K\cap\Gamma_N}\right)\right)^{1/2} \|\nabla e\| \tag{11.88}$$

Finally, dividing by $\|\nabla e\|$ we end up with

$$\|\nabla e\| \leq C\left(\sum_{K\in\mathcal{K}} h_K^2\|f + \nabla\cdot\sigma_h\|_K^2\right.$$

$$\left. + h_K\left(\frac{1}{4}\|[\sigma_h\cdot n]\|^2_{\partial K\setminus\partial\Omega} + \|g_N - \sigma_h\cdot n\|^2_{\partial K\cap\Gamma_N}\right)\right)^{1/2} \tag{11.89}$$

$$\leq C\sum_{K\in\mathcal{K}} h_K\|f + \nabla\cdot\sigma_h\|_K + h_K^{1/2}\left(\frac{1}{2}\|[\sigma_h\cdot n]\|_{\partial K\setminus\partial\Omega} + \|g_N - \sigma_h\cdot n\|_{\partial K\cap\Gamma_N}\right) \tag{11.90}$$

Hence, we have shown the following a posteriori error estimate.

Theorem 11.3. *The finite element solution u_h, defined by* (11.43)*, satisfies the estimate*

$$\|\nabla e\| \leq C\sum_{K\in\mathcal{K}} \eta_K \tag{11.91}$$

where the element residual η_K is the sum of the cell residual $R_K = h_K\|f + \nabla\cdot\sigma_h\|_K$ and the edge residual $r_K = h_K^{1/2}(\frac{1}{2}\|[\sigma_h\cdot n]\|_{\partial K\setminus\partial\Omega} + \|g_N - \sigma_h\cdot n\|_{\partial K\cap\Gamma_N})$.

Next we show how to compute the cell and edge residuals.

The cell residual is easy to compute with one point quadrature. Note that it simplifies to $R_K = \|f\|_K$ for a piecewise linear u_h.

```
function RK = CellResiduals(p,t,force)
nt=size(t,2); % number of elements
RK=zeros(nt,1); % allocate element residuals
for i=1:nt % loop over elements
  nodes=t(1:3,i); % nodes
```

```
  x=p(1,nodes); y=p(2,nodes); % node coordinates
  [area,ds]=Triutils(x,y); % area and side lengths
  f=force(mean(x),mean(y)); % force at element centroid
  h=max(ds); % local mesh size is max side length
  RK(i)=h*sqrt(dot(f,f)*area); % cell residual h|f|_K
end
```

Here, we use the following utility routine to compute the area, edge lengths, and outward unit normals on an element. Edge 1 is opposite node 1, edge 2 opposite node 2, etc.

```
function [area,ds,nx,ny] = Triutils(x,y)
area=polyarea(x,y); % triangle area
dx=[x(3)-x(2); x(1)-x(3); x(2)-x(1)];
dy=[y(2)-y(3); y(3)-y(1); y(1)-y(2)];
ds=sqrt(dx.*dx+dy.*dy); % side lengths
nx=-dy./ds; % outward unit normal components
ny=-dx./ds;
```

The edge residual is a little more complicated to compute than the cell residual, since it requires information about the element neighbors. A routine called `Tri2Tri` for computing element neighbors is given in the Appendix. Also, for simplicity, we do not compute the term stemming from the traction g_N.

```
function rK = EdgeResiduals(p,t,E,nu,U,V)
nt=size(t,2);
rK=zeros(nt,1); % allocate edge residuals
nbrs=Tri2Tri(p,t); % get element neighbours
[mu,lambda]=Enu2Lame(E,nu);
[ux,uy]=pdegrad(p,t,U); % gradient of U
[vx,vy]=pdegrad(p,t,V);
for i=1:nt
    nodes=t(1:3,i);
    x=p(1,nodes); y=p(2,nodes);
    r=0; % sum of edge residuals sqrt(h)|0.5[n.sigma]|_dK
    [area,ds,nx,ny]=Triutils(x,y);
    h=max(ds);
    for j=1:3 % loop over element edges
        n=nbrs(i,j); % element neighbour
        if n<0 % no neighbour
            continue; % don't compute on domain boundary!
                      % should compute sqrt(h)|g_N-n.sigma|_dK
        end
        Sp=Stress(mu,lambda,ux,uy,vx,vy,i); % stress on element i
        Sm=Stress(mu,lambda,ux,uy,vx,vy,n); % stress on neighbour
        jump=0.5*(Sm-Sp)*[nx(j); ny(j)]; % stress jump
        r=r+dot(jump,jump)*ds(j);
    end
    rK(i)=sqrt(h)*sqrt(r);
end
```

To compute the stress tensor on a given element we use the following subroutine.

```
function sigma = Stress(mu,lambda,ux,uy,vx,vy,i)
divu=ux(i)+vy(i); % div u
dudx=[ux(i) uy(i); vx(i) vy(i)]; % grad u
epsilon=(dudx+dudx')/2; % strain
sigma=2*mu*epsilon+lambda*divu*eye(2); % stress
```

11.2.7.1 Adaptive Mesh Refinement on a Rotated L-Shaped Domain

We illustrate the use of the element indicator η_K by adaptively solving a problem with a manufactured solution. The domain Ω is a rotated L-shaped polygon with vertex points $(-1,-1)$, $(0,0)$, $(-1,1)$, $(0,2)$, $(2,0)$, and $(0,-2)$. The solution u is known in polar coordinates (r,θ)

$$u_r(r,\theta) = \frac{1}{2\mu} r^\alpha ((c_2 - \alpha - 1)\cos((\alpha - 1)\theta) - (\alpha + 1)\cos((\alpha + 1)\theta)) \quad (11.92)$$

$$u_\theta(r,\theta) = \frac{1}{2\mu} r^\alpha ((\alpha + 1)\sin((\alpha + 1)\theta) + (c_2 + \alpha - 1)\sin((\alpha - 1)\theta)) \quad (11.93)$$

where the exponent α is the solution to the equation $\alpha\sin(2\omega)+\sin(2\omega\alpha) = 0$ with $\omega = 3\pi/4$, $c_1 = -\cos((\alpha+1)\omega)/\cos((\alpha-1)\omega)$, and $c_2 = 2(\lambda+2\mu)/(\lambda+\mu)$. This displacement field satisfies the linear elastic equations with $f = 0$ and $\Gamma_D = \partial\Omega$. In the computations we use $E = 1$ and $\nu = 0.3$.

By construction, the gradient of the solution u tends to infinity at the re-entrant corner of the domain. Thus, in order to capture this rapid growth of the gradient it is necessary to have a high density of nodes near this corner. Now, from the a posteriori error estimate we know that the gradient error can be controlled by using the element residuals η_K to select elements for refinement. Indeed, starting with the coarse mesh with ten elements and making ten adaptive refinement loops we obtain the mesh shown in Fig. 11.3. Clearly, the adaptivity has identified and resolved the region around the re-entrant corner. The computed displacement is shown in Fig. 11.4.

11.3 Linear Thermoelasticity

Heating or cooling of a material often leads to isotropic expansion or contraction. In such cases, it is common to assume that the total strain ε is the sum of the mechanical and the thermal strain ε_M and ε_T, respectively. The former is assumed to obey Hooke's law, while the latter is given by

$$\varepsilon_T = \alpha(T - T_0)I \quad (11.94)$$

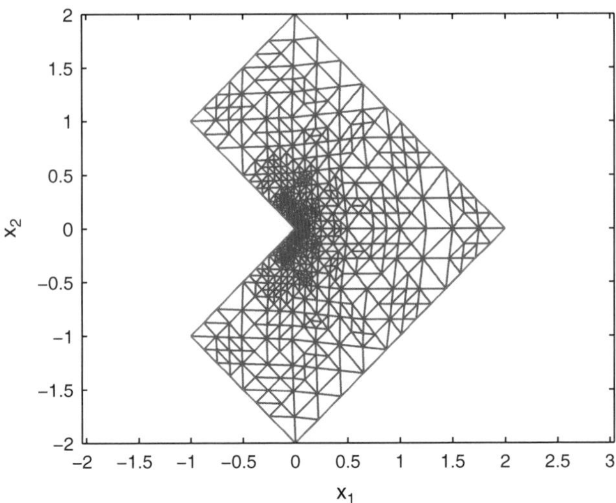

Fig. 11.3 Final mesh

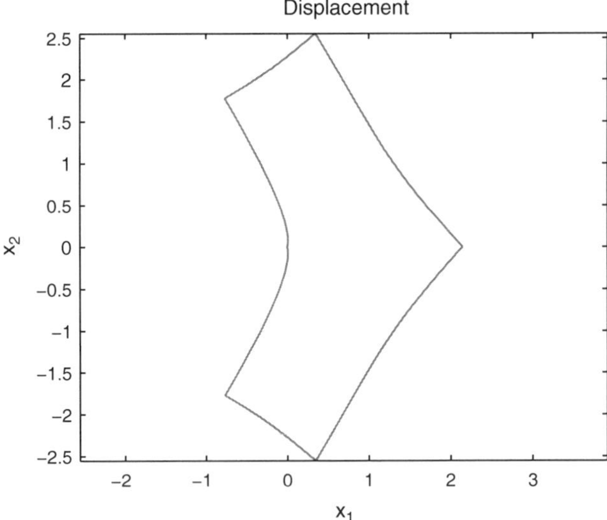

Fig. 11.4 Computed displacement of the rotated L-shaped domain

where α is the thermal expansion coefficient, T the temperature, and T_0 a reference temperature. As usual, I is the identity tensor.

These assumptions give rise to a generalized Hooke's law, relating stresses, temperature, and displacements, of the form

$$\sigma = 2\mu\varepsilon(u) + \lambda(\nabla \cdot u)I - \alpha(3\lambda + 2\mu)(T - T_0)I \tag{11.95}$$

The generalized Hooke's law can be combined with Cauchy's equilibrium equation to yield the equations of linear thermoelasticity. In weak form, assuming for simplicity $f = 0$, these equations reads: find $u \in V$ such that

$$\int_\Omega 2\mu\varepsilon(u) : \varepsilon(v) + \lambda(\nabla \cdot u)(\nabla \cdot v) dx = \int_\Omega \alpha(3\lambda+2\mu)(T-T_0)(\nabla \cdot v)\, dx, \quad \forall v \in V \tag{11.96}$$

From this we see that the thermal strains yield a load proportional to the temperature difference $T - T_0$.

Usually the temperature T is not available in closed form, but has to be computed by solving a heat transfer problem with finite elements. This is a simple, but very common, example of a so called multi-physics problem where two or more different types of physics are coupled together to describe the behavior of a system.

11.4 Linear Elastodynamics

We now consider the time dependent linear elasticity equations. Recall that Newton's second law, $F = ma$, says that the net force F acting on a particle equals the mass m of the particle times its acceleration a. Translated to the continuum setting this yields the equations of motion

$$\rho\ddot{u} = f + \nabla \cdot \sigma \tag{11.97}$$

where ρ is the density of the material and $\ddot{u}$ is the second derivative of the displacement u with respect to time t. To see the analogy between Newton's second law and (11.97) note that if we consider a small particle with volume dx inside a material body, then ρdx is precisely the mass of the particle, $\ddot{u}$ is its acceleration, and $(f + \nabla \cdot \sigma)dx$ is the net force acting on it. Moreover, $f dx$ is externally applied force, whereas $\nabla \cdot \sigma dx = \sigma \cdot n ds$ is internal stresses acting on the surface ds of dx with n the outward unit normal.

We can now write down the basic problem of linear elastodynamics, namely: find the time dependent symmetric stress tensor σ and the time dependent displacement vector u such that

$$\rho\ddot{u} - \nabla \cdot \sigma = f, \quad \text{in } \Omega \times I \tag{11.98a}$$

$$\sigma = 2\mu\varepsilon(u) + \lambda(\nabla \cdot u)I, \quad \text{in } \Omega \times I \tag{11.98b}$$

$$u = 0, \quad \text{on } \Gamma_D \times I \tag{11.98c}$$

$$\sigma \cdot n = 0, \quad \text{on } \Gamma_N \times I \tag{11.98d}$$

$$u = u_0, \quad \text{in } \Omega, \text{ for } t = 0 \tag{11.98e}$$

$$\dot{u} = v_0, \quad \text{in } \Omega, \text{ for } t = 0 \tag{11.98f}$$

where $I = (0, T]$ is the time interval, and u_0 and v_0 is a given initial displacement and velocity, respectively.

11.4.1 Modal Analysis

Since the equations of motion (11.98) resembles a wave equation it is natural to look for a solution in the form of a plane wave, that is,

$$u = z \sin(\omega t) \tag{11.99}$$

where z is a function independent of time t and ω a frequency. We note that both z and ω are unknown. Inserting this ansatz into $\rho\ddot{u} - \nabla \cdot \sigma(u) = f$, assuming $\rho = 1$ and $f = 0$, we obtain

$$-\nabla \cdot \sigma(z) = \omega^2 z \tag{11.100}$$

which we recognize as an eigenvalue problem for the pair (z, ω).

The weak form of (11.100) reads: find $(z, \omega) \in V \times \mathbb{R}$ such that

$$a(z, v) = \omega^2 (z, v), \quad \forall v \in V \tag{11.101}$$

The finite element approximation of (11.101) reads: find $(z_h, \omega_h) \in V_h \times \mathbb{R}$ such that

$$a(z_h, v) = \omega_h^2 (z_h, v), \quad \forall v \in V_h \tag{11.102}$$

The finite element method leads to the generalized algebraic eigenvalue problem

$$Kd = wMd \tag{11.103}$$

where K is the stiffness matrix, M the mass matrix, $w = \omega_h^2$ the eigenvalues, and d a vector containing the nodal values of the eigenmodes z_h.

The computation of eigenmodes and eigenvalues is important in engineering and is routinely performed in industry. This is typically done during the design process of a mechanical structure to identify resonance frequencies (i.e., eigenvalues). Indeed, resonance vibrations can cause the structure to wear out unreasonably fast or even fail due to fatigue.

11.4.1.1 Eigenvalues and Eigenmodes of a Steel Bracket

As a small numerical example we compute the ten smallest eigenvalues and eigenmodes of a freely vibrating steel bracket. The geometry and mesh is shown in

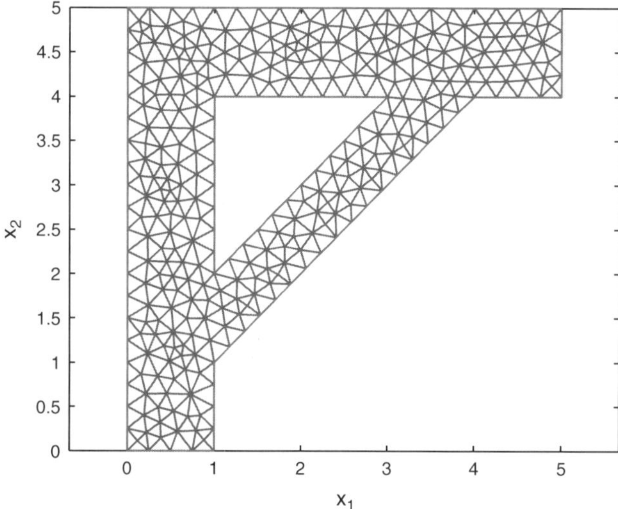

Fig. 11.5 Steel bracket and mesh

Table 11.2 The ten smallest eigenvalues w_i of the steel bracket

i	w_i
1	−0.0000
2	−0.0000
3	0.0000
4	0.0555
5	0.0732
6	0.1050
7	0.2411
8	0.2679
9	0.3972
10	0.4532

Fig. 11.5. A skeleton code for assembling the mass and stiffness matrix, and calling the MATLAB routine `eigs`, which computes eigenvalues and eigenmodes of sparse eigenvalue problems is given below.

```
function ElasticModalSolver()
[p,e,t]=initmesh(...); % create mesh
E=1; % Young modulus
nu=0.3; % Poisson ratio
[mu,lambda]=Enu2Lame(E,nu);
[K,M]=ElasticAssembler(p,e,t,mu,lambda,@Force);
[D,w]=eigs(K,M,10,'SM');
```

The computed eigenvalues are listed in Table 11.2, and in Fig. 11.6 we show the corresponding first, fourth, fifth, and eighth eigenmode. Note that the three lowest eigenvalues are zero. The corresponding eigenmodes are the rigid body modes.

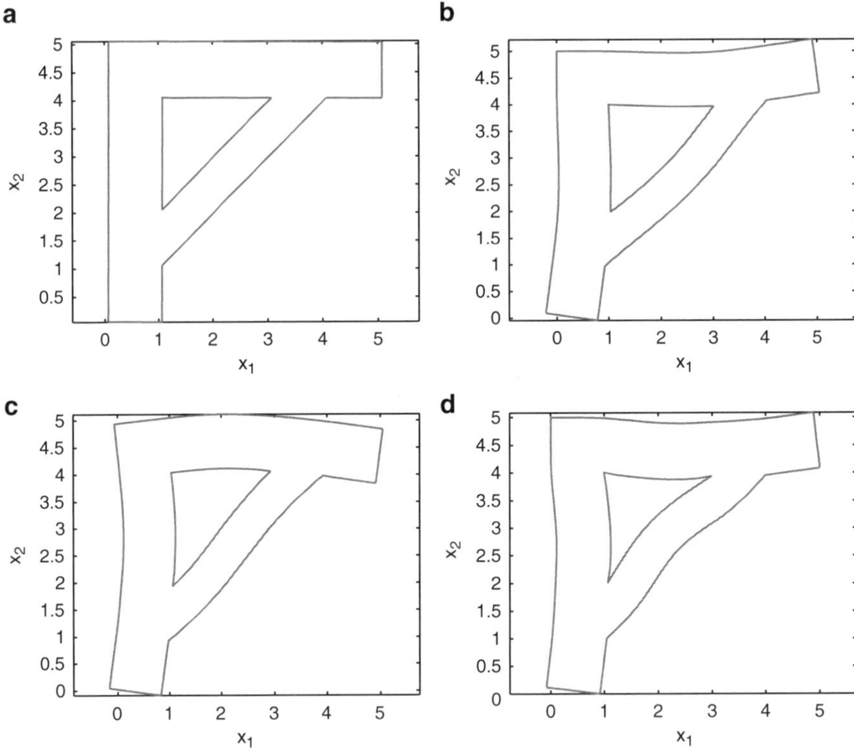

Fig. 11.6 Eigenmodes 1, 4, 5, and 8 of the steel bracket. (**a**) Mode 1. (**b**) Mode 4. (**c**) Mode 5. (**d**) Mode 8

These express two-dimensional translation and rotation, and cause no stress or strain in the bracket. Therefore, they belong to the kernel of the bilinear form $a(\cdot,\cdot)$, or equivalently, the null space of the stiffness matrix K.

11.4.2 Time Stepping

11.4.2.1 Rayleigh Damping

If the transient response of a loaded elastic body is important, then the equations of motion need to be solved for the time dependent stresses and displacements. In semi-discrete form these equations are given by the system of ODE

$$M\ddot{d}(t) + C\dot{d}(t) + Kd(t) = F(t) \tag{11.104}$$

where M is the mass matrix, K the stiffness matrix, and F the load vector, which may depend on time t. The time dependent vector $d = d(t)$ is the nodal values of the approximate displacement u_h. The matrix C is called a damping matrix and accounts for any dissipation of energy, for example, due to friction. A common choice is

$$C = \alpha M + \beta K \tag{11.105}$$

where α and β are parameters that are typically determined by experiments. This choice of C is called Rayleigh damping and means that the damping is the sum of viscous and solid damping (i.e., hysteresis).

11.4.2.2 Newmark's Method

The most popular algorithms for time stepping the equations of motion is based on the co-called Newmark family of methods. Newmark's method for advancing the displacement d_l from time level l to $l+1$ is defined by

$$\begin{aligned}\left(\frac{1}{\beta k^2}M + \frac{\gamma}{\beta k}C + K\right) d_{l+1} = F_{l+1} \qquad (11.106)\\ + M\left(\frac{1}{\beta k^2}u_l + \frac{1}{\gamma k}v_l + \left(\frac{1}{2\beta} - 1\right) a_l\right) + C\left(\frac{\gamma}{\beta k}u_l + \left(\frac{\gamma}{\beta} - 1\right) v_l\\ + \left(\frac{\gamma}{2\beta} - 1\right) k a_l\right)\end{aligned}$$

where k denotes the time step. Once the displacement d_{l+1} has been found the velocity v_l and acceleration a_l are advanced using the defining relations

$$v_{l+1} = v_l + k(1-\gamma)a_l + \gamma k a_{l+1} \tag{11.107}$$

$$a_{l+1} = \frac{1}{\beta k^2}(u_{l+1} - u_l) - \frac{1}{\beta k}v_l - \left(\frac{1}{2\beta} - 1\right) a_l \tag{11.108}$$

To start the time stepping scheme d_0, v_0 and a_0 are needed. The initial conditions d_0 and u_0 are usually given, as opposed to a_0, which must be computed from some form of equation, such as $Ma_0 = F_0 - Kd_0$, for instance.

In the Newmark method the parameters $0 \leq \beta \leq 1/2$ and $0 \leq \gamma \leq 1$ can be tuned to yield different numerical properties of the resulting time stepping scheme. The choice $\beta = 1/4$ and $\gamma = 1/2$ leads to both unconditional stability with respect to the size of the time step k, as well as second order accuracy in time. For any other value of γ than $1/2$ the time stepping scheme is only first order accurate. Moreover, with $\gamma > 1/2$ artificial damping occurs.

11.5 Thin Plates

In engineering special structures are commonly used, for instance, beams and plates. These are characterized by a special geometry of Ω which is very thin in certain directions. Beams are essentially one dimensional for beams and plates are two dimensional. To handle these structures efficiently special partial differential equations have been derived for certain load cases. In this short section we briefly describe a model for a thin plate and the finite element method.

A plate is a thin flat elastic object subjected to a load in the transversal direction, that is, the surface normal direction. The resulting transversal deflection may be modeled using so-called Kirchhoff-Love plate theory. As the thickness of the plate is assumed to be very thin compared to the other two dimensions it is sufficient to describe the deformation of the plate by the deflection of its mid-surface. Indeed, the following hypothesis is made for thin plates:

- Straight lines normal to the mid-surface of the plate remain straight and normal to the mid-surface after deformation.
- The thickness of the plate is constant during deformation.

While we will not give a derivation of the Kirchhoff-Love plate theory, we will below supply the resulting governing equations and derive the weak form of a thin plate problem.

11.5.1 Governing Equations

Consider a thin plate occupying a plane domain $\Omega \in \mathbb{R}^2$ under a transverse shear force load p with a clamped boundary $\partial\Omega$. We seek the (scalar) transversal deflection u of the plate. By demanding equilibrium of certain forces and moments it can be shown that u is governed by the following fourth-order problem, the so-called the Biharmonic equation,

$$\Delta^2 u = f, \quad \text{in } \Omega \tag{11.109a}$$

$$u = n \cdot \nabla u = 0, \quad \text{on } \partial\Omega \tag{11.109b}$$

with Δ^2 the operator

$$\Delta^2 = \frac{\partial^4}{\partial x_1^4} + 2\frac{\partial^4}{\partial x_1^2 \partial x_2^2} + \frac{\partial^4}{\partial x_2^4} \tag{11.110}$$

and $f = p/D$ with D the bending stiffness of the plate

$$D = \frac{Et^3}{12(1-\nu^2)} \tag{11.111}$$

where E is Young's modulus, ν Poisson's ratio, and t is the thickness of the plate.

We remark that the Biharmonic equation can be supplemented with different sets of boundary conditions. For example, for a plate rigidly clamped to the boundary we have $u = g_D$ and $n \cdot \nabla u = 0$, where g_D is a prescribed boundary displacement. Also, for a simply supported, or pinned, plate we have $u = g_D$ and $\Delta u = g_M$, where g_M is a given bending moment on the boundary.

11.5.1.1 Weak Form

To derive the weak form of the Biharmonic equation (11.109) we begin by introducing the Hilbert space

$$V = \{v \in H^2(\Omega) : v|_{\partial\Omega} = n \cdot \nabla v|_{\partial\Omega} = 0\} \tag{11.112}$$

Then, multiplying $f = \Delta^2 u$ by a test function $v \in V$ and integrating, using Green's formula twice, we end up with

$$(f, v) = (\Delta^2 u, v) \tag{11.113}$$

$$= (n \cdot \nabla \Delta u, v)_{\partial\Omega} - (\Delta u, n \cdot \nabla v)_{\partial\Omega} + (\Delta u, \Delta v) \tag{11.114}$$

$$= (\Delta u, \Delta v) \tag{11.115}$$

where we have used that $v|_{\partial\Omega} = 0$ and $n \cdot \nabla v|_{\partial\Omega} = 0$ to get rid of the boundary terms.

Thus, the weak form of (11.109) takes the form: find $u \in V$ such that

$$a(u, v) = l(v), \quad \forall v \in V \tag{11.116}$$

with $a(u, v) = (\Delta u, \Delta v)$ and $l(v) = (f, v)$.

11.5.1.2 The C^1 Continuity Requirement and the Morley Approximation

As indicated by the fact that the weak form (11.116) includes second order derivatives, this equation generally requires C^1 continuous finite elements to be solvable. An intuitive explanation for this necessity is that C^1 continuity is needed to ensure that the plate doesn't kink between elements, producing infinite curvature values (i.e., infinite Δu). This is a quite difficult requirement to meet, especially for triangular elements on unstructured meshes. Indeed, the only triangular finite element with full C^1 continuity is the Argyris element. However, since this is a costly element to use, much development has been done to derive non-conforming alternatives. For the Biharmonic equation the simplest such alternative element is the Morley element, which is neither C^1 nor C^0. Indeed, on a triangle mesh $\mathcal{K} = \{K\}$ the Morley space $V_h \not\subset V$ is given by

$$V_h = \{v \in L^2 : v|_K \in P_2(K)\ \forall K \in \mathcal{K}, \qquad (11.117)$$
$$v \text{ continuous at the vertices},$$
$$n \cdot \nabla v \text{ continuous at the edge mid-points}\}$$

In addition, to satisfy the homogeneous boundary conditions all degrees of freedom on the boundary are removed.

Using the Morley space, the finite element approximation of (11.116) takes the form: find $u_h \in V_h$ such that

$$a(u_h, v) = l(v), \quad \forall v \in V_h \qquad (11.118)$$

11.6 Further Reading

Finite element methods was originally developed for solid mechanics applications and during the years the literature on this topic has grown and is today quite extensive.

An introductory book on finite element methods for static linear elasticity is the one by Ottosen and Petersson [51], which covers also finite element analysis of trusses, beams, and plates. This is also covered by Šzabo and Babuška in [68] from a kind of historic and mathematical point of view. Advanced books on finite elements for dynamic and non-linear elasticity include the ones by Hughes [43], Bathe [8], Wriggers [73], and Zienkiewicz and Taylor [74], which covers many of the more complicated phenomenons of elasticity, such as large displacements, contact, and creep, for instance.

In this context we mention the book by Kwon and Bang [4], which contains many useful MATLAB implementations of the numerical methods used for elasticity and heat transfer problems.

11.7 Problems

Exercise 11.1. Given the stress field $\sigma_{11} = x_1 x_2$, $\sigma_{12} = (1 - x_2^2)/2$, and $\sigma_{22} = 0$. Determine if this corresponds to a state of equilibrium under a zero body force.

Exercise 11.2. Show the vector identity $2\nabla\cdot\varepsilon(v) = \Delta v + \nabla(\nabla\cdot v)$ with $v = [v_1,\ v_2]$.

Exercise 11.3. Use the previous result to rewrite (11.3) and(11.7) as the single equation $\mu\Delta u + (\lambda + \mu)\nabla(\nabla \cdot u) + f = 0$.

Exercise 11.4. Show that the strain tensor $\varepsilon(u)$ is zero under the deformation

$$u = \begin{bmatrix} a \\ b \end{bmatrix} + \begin{bmatrix} 0 & -\theta \\ \theta & 0 \end{bmatrix} \begin{bmatrix} x_1 \\ x_2 \end{bmatrix}$$

where a, b, and θ are constants. Can you give a physical interpretation of u, assuming that θ is small?

Exercise 11.5. Show that $\varepsilon(v) : I = \nabla \cdot v$.

Exercise 11.6. Verify that the requirements for the Lax-Milgram lemma are fulfilled by the weak form (11.20). For simplicity, you only have to consider the case of homogeneous Dirichlet boundary conditions $u = 0$ on the whole boundary $\partial\Omega$.

Exercise 11.7. Calculate the element stiffness matrix K^K by hand for the triangle with corners at $(0,0)$, $(3,1)$, and $(2,2)$. Assume that

$$D = \begin{bmatrix} 4 & 1 & 0 \\ 1 & 4 & 0 \\ 0 & 0 & 2 \end{bmatrix}$$

Verify that K^K has three zero eigenvalues. Can you explain why?

Exercise 11.8. Calculate by hand the element mass matrix M^K assuming a unit density on the reference triangle with vertices at origo, $(1,0)$, and $(0,1)$.

Exercise 11.9. A mesh of the square domain $\Omega = [-1,1]^2$ is obtained by typing `[p,e,t]=initmesh('squareg')`. Compute and plot the ten lowest eigenmodes on this domain. Assume elastic constants $\rho = 1$, $E = 1$, and $\nu = 0.3$. Test both clamped and stress free boundary conditions.

Exercise 11.10. Calculate the Morley basis functions on the triangle with vertices at origo, $(1,0)$, and $(1,1)$.

Chapter 12
Fluid Mechanics

Abstract In this chapter we study finite elements for incompressible fluids (i.e., most liquids and gases). We start by reviewing the governing equations of mass and momentum balance and derive the Navier-Stokes equations. Restricting attention to laminar flow we then introduce the Stokes system and formulate a finite element method for the velocity and pressure. We discuss the inf-sup condition as a necessary requirement for existence and uniqueness of the solution. Three types of inf-sup stable finite elements are presented. Uzawa's algorithm and the solution of saddle-point linear systems is briefly touched upon. The lid-driven cavity benchmark is studied numerically. Both a priori and a posteriori error estimates are derived using B-stability. Finally, we introduce Chorin's classical projection method as a simple numerical method to simulate time-dependent nearly turbulent fluid flow.

12.1 Governing Equations

12.1.1 Conservation of Mass

In classical physics mass can neither be destroyed nor created. This means that the mass of any small volume dx of matter (e.g., a fluid) can change over time only by flow in and out of the boundary ds. Letting u denote the flow velocity vector we immediately obtain the following mass balance equation for a fluid occupying the domain Ω.

$$(\dot{\rho}, 1) + (\rho, u \cdot n)_{\partial\Omega} = 0 \tag{12.1}$$

Here, ρ is the density of the fluid and n is the outward pointing unit normal on the boundary $\partial\Omega$. Because $dm = \rho dx$ is the mass of dx, the first term represents the rate of change of mass within the domain. Further, during the small time span dt a

M.G. Larson and F. Bengzon, *The Finite Element Method: Theory, Implementation, and Applications*, Texts in Computational Science and Engineering 10, DOI 10.1007/978-3-642-33287-6_12, © Springer-Verlag Berlin Heidelberg 2013

total volume of matter of $dm = \rho u \cdot n ds$ will flow out of the surface ds. Hence, the second term represents the rate of mass loss through the domain boundary. Now, using the divergence theorem on the surface integral we have

$$\dot{\rho} + \nabla \cdot (\rho u) = 0 \tag{12.2}$$

Assuming a constant density ρ, this simplifies to

$$\nabla \cdot u = 0 \tag{12.3}$$

Physically, this means that the volume of any small fluid particle dx does not change under deformation. Such fluids are said to be incompressible. Most everyday fluids (e.g., water) are incompressible to a very high degree.

12.1.2 Momentum Balance

Besides mass conservation a fluid also obeys conservation of momentum (i.e., Newton's second law). Recall that the momentum of a particle with mass m and velocity u is defined as the product $p = mu$, and that Newton's second law says that the rate of change of momentum equals the net force F acting on the particle, that is, $\dot{p} = F$.

Now, the momentum dp of a small volume of fluid dx is given by $dp = \rho u dx$, so taking into consideration the fact that momentum can be transported in and out of the boundary $\partial\Omega$ of the domain Ω we also have the following equation for momentum balance.

$$(\dot{\rho u}, 1) + (\rho u, u \cdot n)_{\partial\Omega} = F \tag{12.4}$$

Here, we can use our knowledge from mechanics to write the net force $F = (\nabla \cdot \sigma + f, 1)$ with σ the stress tensor of the fluid, and f a given body load, such as gravity, for instance. Using, again, the divergence theorem on the surface integral we arrive at

$$\partial_t \rho u + \nabla \cdot (\rho u u) = \nabla \cdot \sigma + f \tag{12.5}$$

Here, the right left side can be simplified by differentiating using the chain rule, viz.,

$$\partial_t \rho u + \nabla \cdot (\rho u u) = u\dot{\rho} + \rho\dot{u} + u\nabla \cdot (\rho u) + \rho(u \cdot \nabla)u \tag{12.6}$$

$$= \rho\dot{u} + \rho(u \cdot \nabla)u \tag{12.7}$$

where we have used conservation of mass (12.2) to eliminate the first and third term in the right hand side of (12.6). Hence, we end up with

$$\rho\dot{u} + \rho(u \cdot \nabla)u = \nabla \cdot \sigma + f \tag{12.8}$$

12.1.3 Incompressible Newtonian Fluids

The kinds of stresses acting on a fluid particle are of two types, namely:

- Internal stresses, σ_1, due to the fluid pressure.
- Viscous stresses, σ_2.

Internal stresses always arise when a fluid is brought into motion, since the pressure p is changed from that existing when the fluid is at rest. The corresponding stress tensor takes the form

$$\sigma_1 = -pI \tag{12.9}$$

with I the $d \times d$ identity tensor, with $d = 1, 2$, or 3 the space dimension.

Viscosity is a measure of the resistance of a fluid to being deformed by stresses. Indeed, it may be thought of as a friction caused by neighboring layers of fluid rubbing against each other. In reality, though, it is fluid molecules with different velocities that bump into each other. Viscosity is commonly perceived as the thickness of the fluid. Thus, water is thin, having a lower viscosity, while oil is thick having a higher viscosity. All real fluids have some resistance to stress, so a fluid with no resistance is called either inviscid or ideal.

Viscous stresses oppose deformation of neighboring fluid particles. Now, because a constant velocity field does not give rise to any relative movement between the fluid particles, it is reasonable to assume that the stress tensor σ_2 is related only to the velocity gradients ∇u. Clearly, the simplest assumption is that this relation is linear. Recalling that σ is symmetric yields

$$\sigma = \sigma_1 + \sigma_2 = -pI + \mu(\nabla u + \nabla u^T) \tag{12.10}$$

where the coefficient of proportionality μ is the viscosity of the fluid. Fluids obeying this constitutive law are called Newtonian.

Finally, inserting (12.10) into (12.8), using that $\nabla \cdot \sigma = \mu(\Delta u + \nabla(\nabla \cdot u)) - \nabla p$, and assuming $\nabla \cdot u = 0$, we obtain a set of partial differential equations for the velocity u and pressure p, namely,

$$\dot{u} + (u \cdot \nabla)u = \nu\Delta u - \frac{\nabla p}{\rho} + f \tag{12.11a}$$

$$\nabla \cdot u = 0 \tag{12.11b}$$

where $\nu = \mu/\rho$. These are the famous Navier-Stokes equations.

12.1.4 *Boundary- and Initial Conditions*

In order to yield a unique velocity-pressure pair (u, p) the Navier-Stokes equations must be supplemented by appropriate boundary conditions. The most common of these have names, such as:

- Slip, $u \cdot n = 0$.
- No-slip, $u = g_D$.
- Stress free, $\sigma \cdot n = 0$.
- Do-nothing, $n \cdot \nabla u - pn = 0$.

Slip and no-slip boundary conditions apply at a solid wall with normal n. Slip boundary conditions says that the fluid flow is parallel to the boundary (i.e., orthogonal to n). No-slip conditions prescribe that the velocity u agrees with a known vector g_D on the boundary (e.g., fluid flow near a moving wall with velocity g_D). Often $g_D = 0$ meaning that the fluid is at rest. Stress free and do-nothing boundary conditions are generally used on so-called outflow boundaries where the flow leaves the domain. Stress free boundary conditions model free flow into a large reservoir, while do-nothing boundary conditions are used to truncate very long channel like domains.

Due to the time derivative on the velocity, it is also necessary to specify initial conditions of the type $u(\cdot, t_0) = u_0$ with u_0 a given velocity at the initial time t_0.

12.2 The Stokes System

12.2.1 *The Stationary Stokes System*

Many applications involve laminar fluid flow, which means that the flow is slow and calm with essentially parallel streamlines. In such cases it is possible to omit the non-linear term $(u \cdot \nabla)u$, which governs inertial effects, from the Navier-Stokes equations (12.11). Omitting also any time dependence (i.e., assuming steady state), we end up with a set of linear stationary equations called the Stokes system. These are given by

$$-\Delta u + \nabla p = f, \quad \text{in } \Omega \tag{12.12a}$$

$$\nabla \cdot u = 0, \quad \text{in } \Omega \tag{12.12b}$$

$$u = g_D, \quad \text{on } \partial\Omega \tag{12.12c}$$

where f is a given body force, and g_D is given boundary data. For simplicity, we assume a unit viscosity $\nu = 1$.

Perhaps needless to say, the reason for studying the Stokes system is that it is much easier to analyze than the Navier-Stokes equations, but still provides a realistic model of many fluid flows.

Since only the gradient of the pressure p enters the equations, p is only determined up to an arbitrary constant called the hydrostatic pressure level. To determine this constant it is customary to require the pressure to have a zero mean value, that is,

$$(p, 1)/|\Omega| = 0 \tag{12.13}$$

This is a characteristic feature of all enclosed flows (i.e., flow problems with only no-slip boundary conditions).

As we shall see, for well-posedness, we require the boundary data g_D to satisfy $(g_D, n)_{\partial\Omega} = 0$.

In this context we remark that a nice thing with the do-nothing boundary condition is that it automatically fixes the hydrostatic pressure level.

12.2.2 Weak Form

In order to derive a weak form of the Stokes system (12.12) we need to introduce two function spaces V_g and Q for the velocity u and pressure p, respectively. To this end, let

$$V_g = \{v \in [H^1(\Omega)]^d : v|_{\partial\Omega} = g_D\} \tag{12.14}$$

$$Q = \{q \in L^2(\Omega) : (q, 1) = 0\} \tag{12.15}$$

We see that the pressure space Q is the subset of $L^2(\Omega)$ with zero mean value.

Now, multiplying the momentum equation $f = -\Delta u + \nabla p$ by a test vector $v \in V_0$ and integrating by parts, we have

$$(f, v) = (-\Delta u, v) + (\nabla p, v) \tag{12.16}$$

$$= (-n \cdot \nabla u, v)_{\partial\Omega} + (\nabla u : \nabla v) + (pn, v)_{\partial\Omega} - (p, \nabla \cdot v) \tag{12.17}$$

which, since $v = 0$ on $\partial\Omega$, simplifies to

$$(f, v) = (\nabla u : \nabla v) - (p, \nabla \cdot v) \tag{12.18}$$

Similarly, multiplying the incompressibility constraint $\nabla \cdot u = 0$ by a test function $q \in Q$ and integrating, we trivially have

$$(\nabla \cdot u, q) = 0 \tag{12.19}$$

One might ask why Q is the appropriate test space for the incompressibility constraint $\nabla \cdot u = 0$? After all, the functions in Q are somewhat peculiar since they all have a zero mean value. The reason is that it suffice to test against these functions, since the variational equation (12.19) is void anyway for q constant. To see this let c be a constant and recall that by assumption $(g_D, n)_{\partial\Omega} = 0$. Using integration by parts we then have

$$(\nabla \cdot u, c) = (u, nc)_{\partial\Omega} + (u, \nabla c) = c(g_D, n)_{\partial\Omega} = 0 \tag{12.20}$$

Summarizing, the variational formulation of (12.12) reads: find $u \in V_g$ and $p \in Q$ such that

$$a(u, v) + b(v, p) = l(v), \quad \forall v \in V_0 \tag{12.21a}$$

$$b(u, q) = 0, \qquad \forall q \in Q \tag{12.21b}$$

where we have introduced the linear forms

$$a(u, v) = (\nabla u : \nabla v) \tag{12.22}$$

$$b(u, q) = -(\nabla \cdot u, q) \tag{12.23}$$

$$l(v) = (f, v) \tag{12.24}$$

Notice that the sign of the incompressibility constraint (12.21b) can be chosen arbitrarily, since the right hand side is zero anyway. Often, $b(u, q) = 0$ is preferred over the perhaps more correct $-b(u, q) = 0$, since it gives a symmetric variational form. From a theoretical point of view it does not matter which sign is chosen. However, it can have a large impact on the numerics. Recall that symmetric matrices are often to be preferred when it comes to solving linear systems.

Equations of the form (12.21) are called saddle-point problems.

To avoid working with different trial and test spaces V_g and V_0 let us extend g_D from $\partial\Omega$ to Ω, and write $u = g_D + u_0$ with $u_0 \in V_0$ the new unknown. Then, the weak form for u_0 takes the form: find $u_0 \in V_0$ such that

$$a(u_0, v) + b(v, p) = l(v) - a(g_D, v), \quad \forall v \in V_0 \tag{12.25a}$$

$$b(u_0, q) = 0, \qquad \forall q \in Q \tag{12.25b}$$

Thus, for the purpose of analysis it suffice to consider only $g_D = 0$, since any other g_D can be studied by defining a new body force $\tilde{f}$ by $(\tilde{f}, v) = (f, v) - a(g_D, v)$ for all $v \in V_0$. Therefore, let us drop the subscripts and simply write $V_g = V_0 = V$.

There is a more compact way of writing the weak form (12.21), namely: find the solution pair $(u, p) \in V \times Q$ such that

$$B((u,p),(v,q)) = F((v,q)), \quad \forall (v,q) \in V \times Q \tag{12.26}$$

where the big linear forms $B(\cdot,\cdot)$ and $F(\cdot)$ are defined by

$$B((u,p),(v,q)) = a(u,v) + b(v,p) + b(u,q) \tag{12.27}$$

$$F((v,q)) = l(v) \tag{12.28}$$

That (12.26) is equivalent to (12.21a) and (12.21b) can be seen by choosing the test functions $(v,0)$ and $(0,q)$, respectively.

12.2.3 Equivalent Optimization Problem

The weak form (12.21) can be viewed as the constrained minimization problem

$$u = \min_{v \in V} F(v) \tag{12.29}$$

with

$$F(v) = \tfrac{1}{2}a(v,v) - l(v) \tag{12.30}$$

subject to the constraint

$$b(v,q) = 0 \tag{12.31}$$

with $q \in Q$. The minimum, u, is of course the velocity, whereas the pressure, p, is a Lagrange multiplier enforcing the constraint $\nabla \cdot u = 0$.

In the language of optimization finding the minimum, u, of the constrained optimization problem (12.29) amounts to finding saddle-points (u,p) of the unconstrained optimization problem

$$\min_{q \in Q} \max_{v \in V} L(v,q) \tag{12.32}$$

where $L(\cdot,\cdot)$ is the Lagrangian

$$L(v,q) = \tfrac{1}{2}a(v,v) - l(v) + b(v,q) \tag{12.33}$$

Indeed, demanding that the partial derivatives of $L(\cdot,\cdot)$ with respect to v and q vanish yields precisely (12.21a) and (12.21b). Of course, the saddle-point nature of the pair (u,p) is the reason why these equations are called a saddle-point problem.

12.2.4 The Continuous Inf-Sup Condition

Existence of a solution pair (u, p) to the variational equation (12.21) does not follow from the Lax-Milgram lemma alone, since it is impossible to establish coercivity of the big bilinear form $B((u, p), (v, q)) = a(u, v) + b(v, p) \pm b(u, q)$ on $V \times Q$. Attempting to do so gives $B((u, p), (u, p)) = a(u, u)$, since $b(u, p) = 0$. As p is missing, no lower bound on $B((u, p), (u, p))$ of the form $m(\|u\|_V + \|p\|_Q)$ can be obtained. Thus, the question of existence of u and p must be settled in another way.

On close examination it turns out that it is the existence of p that is difficult to deduce, but not so for u. Indeed, let $Z = \{v \in V : b(v, q) = 0, \ \forall q \in Q\}$ be the null space of $b(\cdot, \cdot)$, that is, the subspace of V containing all divergence free vectors. This is a Hilbert space with norm $\| \cdot \|_Z = \| \cdot \|_V$. On Z the variational equation (12.21a) reduces to: find $u \in Z$ such that

$$a(u, v) = l(v), \quad \forall v \in Z \tag{12.34}$$

Consequently, since $a(\cdot, \cdot)$ is continuous and coercive on Z, and $l(v)$ continuous on Z, we can simply invoke the Lax-Milgram lemma to show existence, and also uniqueness, of u.

Given $u \in Z$, we are next faced with the problem of trying to determine $p \in Q$ from

$$b(v, p) = (r, v), \quad \forall v \in V \tag{12.35}$$

where r is the weak residual, defined by $(r, v) = l(v) - a(u, v)$ for all $v \in V$. We note that due to (12.34) r is orthogonal to Z, and therefore a member of the orthogonal complement $Z^\perp = \{v \in V : (v, z) = 0, \ \forall z \in Z\}$.

The well-posedness of (12.35) is by no means obvious, since the trial and test spaces Q and V are more or less unrelated. Clearly, we can not just pick any combination of V and Q and hope to solve the equation. For solvability, there needs to be some kind of compatibility condition between Q, V, and $b(\cdot, \cdot)$. Indeed, (12.35) means solving $\nabla p = r$ for p, with r in $Z^\perp$. For this to be possible the range of the gradient operator ∇ must coincide with $Z^\perp$. Now, using functional analysis it can be shown that $Z^\perp$ is indeed the range of ∇ if $V = [H_0^1(\Omega)]^d$ and $Q = L_0^2(\Omega)$. The combination $V = [H_0^1(\Omega)]^d$ and $Q = L_0^2(\Omega)$ is therefore compatible in the sense that (12.35) is well-posed.

This is succinctly in the next theorem, alternatingly called the inf-sup, Babuška-Brezzi, or Ladyshenskaya-Babuška-Brezzi condition.

Theorem 12.1 (Inf-Sup Condition). *There exists a constant $\beta > 0$ such that*

$$\beta \|q\|_Q \le \sup_{v \in V} \frac{b(v, q)}{\|v\|_V}, \quad \forall q \in Q \tag{12.36}$$

Proof. The proof is complicated and outside the scope of this book.

The inf-sup condition can also be written

$$\beta \leq \inf_{q \in Q} \sup_{v \in V} \frac{|b(v,q)|}{\|v\|_V \|q\|_Q} \tag{12.37}$$

Hence, its name.

The inf-sup condition is a kind of coercivity for $b(\cdot,\cdot)$ on the spaces V and Q. In fact, it may be thought of as an abstract condition of the angle between V and Q.

Other equivalent statements of the inf-sup conditions include:

- The orthogonal complement to the null space of $\nabla\cdot$ is the range of ∇.
- The orthogonal complement to the null space of ∇ is the range of $\nabla\cdot$.
- For every $q \in Q$ there is a $v \in V$ such that $\nabla \cdot v = q$ with $\|v\|_V \leq C\|q\|_Q$.
- For every $z \in Z^\perp$ there is a $q \in Q$ such that $\nabla q = z$ with $\|q\|_Q \leq C\|z\|_V$.

Let us show how to use the inf-sup condition by proving the uniqueness of p. To this end, suppose that both p and $\tilde{p}$ satisfy (12.35). Then, by subtraction, we have

$$b(v, p - \tilde{p}) = 0, \quad \forall v \in V \tag{12.38}$$

Combining this with the inf-sup condition, we further have

$$\beta\|p - \tilde{p}\|_Q \leq \sup_{v \in V} \frac{b(v, p - \tilde{p})}{\|v\|_V} = 0 \tag{12.39}$$

from which it readily follows that $\|p - \tilde{p}\|_Q = 0$, or $p = \tilde{p}$.

Summarizing, the following existence and uniqueness result hold for saddle-point problems.

Theorem 12.2 (Brezzi). *Let V and Q be Hilbert spaces, and let $a(\cdot,\cdot)$ and $b(\cdot,\cdot)$ be continuous bilinear forms on $V \times V$ and $V \times Q$, respectively. Denote the kernel of $b(\cdot,\cdot)$ by $Z = \{v \in V : b(v,q) = 0, \forall q \in Q\}$. If $a(\cdot,\cdot)$ is coercive on Z, and if $b(\cdot,\cdot)$ satisfies the inf-sup condition*

$$\beta\|q\|_Q \leq \sup_{v \in V} \frac{b(v,q)}{\|v\|_V}, \quad \forall q \in Q \tag{12.40}$$

then there exist a unique solution $(u,p) \in V \times Q$ to the saddle-point problem (12.21).

Perhaps needless to say, continuity of $b(\cdot,\cdot)$ amounts to the inequality

$$b(v,q) \leq C\|v\|_V \|q\|_Q \tag{12.41}$$

which is easily shown using the Cauchy-Schwarz and the trivial inequality $\|\nabla \cdot v\| \leq \|\nabla v\|$.

12.2.5 Finite Element Approximation

In order to formulate a numerical method let V_h and Q_h be two spaces of piecewise polynomials that approximates V and Q.

The finite element approximation of (12.21) takes the form: find $u_h \in V_h$ and $p_h \in Q_h$ such that

$$a(u_h, v) + b(v, p_h) = l(v), \quad \forall v \in V_h \tag{12.42a}$$

$$b(u_h, q) = 0, \quad \forall q \in Q_h \tag{12.42b}$$

or, in more compact form: find $(u_h, p_h) \in V_h \times Q_h$ such that

$$B((u_h, p_h), (v, q)) = F((v, q)), \quad \forall (v, q) \in V_h \times Q_h \tag{12.43}$$

A finite element method which uses two spaces to approximate two different variables is commonly referred to as a mixed method.

Now, let $\{\varphi_i\}_1^n$ be a set of vector valued basis functions for V_h, and let $\{\chi_i\}_1^m$ be a set of scalar basis functions for Q_h. The finite element method (12.42) results in a linear system which can be written in block form

$$\begin{bmatrix} A & B^T \\ B & 0 \end{bmatrix} \begin{bmatrix} \xi \\ \psi \end{bmatrix} = \begin{bmatrix} b \\ 0 \end{bmatrix} \tag{12.44}$$

where A is the $n \times n$ stiffness matrix, and B is the $n \times m$ divergence matrix with entries

$$A_{ij} = a(\varphi_j, \varphi_i) \tag{12.45}$$

$$B_{ij} = b(\varphi_j, \chi_i) \tag{12.46}$$

Further, b is the $n \times 1$ load vector with entries $b_i = l(\varphi_i)$. Also, ξ and ψ are $n \times 1$ and $m \times 1$ vectors containing the unknown degrees of freedom of $u_h = \sum_{j=1}^n \xi_j \varphi_j$ and $p_h = \sum_{j=1}^m \psi_j \chi_j$, respectively.

Saddle-point linear systems, such as (12.44), are known to be notoriously difficult to solve due to the all zero $m \times m$ lower diagonal block.

12.2.6 The Discrete Inf-Sup Condition

So far we have not said anything more specific about the finite element spaces V_h and Q_h. In fact, we do not even know if the finite element solution (u_h, p_h) is well defined. To assert this we must make sure that the linear system (12.44) can be solved. This is equivalent to establishing a discrete inf-sup condition on V_h and Q_h.

More precisely, there must exist a constant $\gamma > 0$ such that

$$\gamma \|q\|_Q \leq \sup_{v \in V_h} \frac{b(v,q)}{\|v\|_V}, \quad \forall q \in Q_h \tag{12.47}$$

or, using matrix notation,

$$\gamma(\theta^T M \theta)^{1/2} \leq \max_{\varpi \in \mathbb{R}^n, \varpi \neq 0} \frac{\theta^T B \varpi}{(\varpi^T A \varpi)^{1/2}}, \quad \forall \theta \in \mathbb{R}^m, \theta \neq 1 \tag{12.48}$$

where A is the stiffness matrix, B the divergence matrix, and M the $m \times m$ pressure mass matrix with entries $M_{ij} = (\chi_j, \chi_i)$. We emphasize that it is a non-trivial task to show this, because even if we know that the inf-sup condition is satisfied on the continuous spaces V and Q, it need not hold on the discrete spaces V_h and Q_h, not even if the inclusions $V_h \subset V$ and $Q_h \subset Q$ hold. The condition $\theta \neq 1$ means that θ can not be any constant vector except the zero vector, which is necessary for the zero mean property.

That continuous inf-sup stability does not imply discrete ditto can be understood by considering the discrete incompressibility constraint, that is, finding $u_h \in V_h$ such that $b(u_h, q) = 0$ for all $q \in Q_h$. Choosing Q_h as a big subspace of Q means that there are many test functions q. As each test function puts a restriction on $\nabla \cdot u_h$, in the form of a moment, there is a risk of over determining u_h if V_h is a small subspace of V. This is know as a locking phenomenon. What the discrete inf-sup condition does is that it asserts that there is a good balance between the number of velocity and pressure degrees of freedom.

All the same, if the discrete inf-sup condition does hold, then the discrete pressure p_h exists and is unique. This can be shown by doing block elimination on the $(n+m) \times (n+m)$ linear system (12.44). From the first row we have $\xi = A^{-1}(b - B^T \psi)$. Plugging this into the second row, $B\xi = 0$, and rearranging terms we obtain the $m \times m$ linear system

$$BA^{-1}B^T \psi = BA^{-1}b \tag{12.49}$$

for the pressure degrees of freedom ψ. For this to make sense the matrix $S = BA^{-1}B^T$, which is called the Schur complement, must be invertible. To show this, we recall that since A is SPD it has the Cholesky factorization $A = LL^T$, implying $A^{-1} = L^{-T}L^{-1}$. Thus, $S = BL^{-T}L^{-1}B^T$, from which we see that S is symmetric. Further, due to the discrete inf-sup condition (12.48), for all $\theta \in \mathbb{R}^m$ with $\theta \neq 1$, we have

$$0 < \gamma \leq \max_{\varpi \neq 0} \frac{\theta^T B \varpi}{(\varpi^T A \varpi)^{1/2} (\theta^T M \theta)^{1/2}} \tag{12.50}$$

$$= \frac{1}{(\theta^T M\theta)^{1/2}} \max_{w=L^T\varpi,\varpi\neq 0} \frac{\theta^T BL^{-T} w}{(\varpi^T LL^T \varpi)^{1/2}} \tag{12.51}$$

$$= \frac{1}{(\theta^T M\theta)^{1/2}} \max_{w\neq 0} \frac{w^T L^{-1} B^T \theta}{(w^T w)^{1/2}} \tag{12.52}$$

Here, the maximum is attained for $w = L^{-1} B^T \theta$, yielding

$$0 < \gamma \leq \frac{((L^{-1}B^T\theta)^T(L^{-1}B^T\theta))^{1/2}}{(\theta^T M\theta)^{1/2}} = \frac{(\theta^T BL^{-T}L^{-1}B^T\theta)^{1/2}}{(\theta^T M\theta)^{1/2}} = \frac{(\theta^T S\theta)^{1/2}}{(\theta^T M\theta)^{1/2}} \tag{12.53}$$

This shows that S is SPD and, therefore, invertible.

Once ψ has been found, ξ can be retrieved by solving the $n \times n$ linear system

$$A\xi = b - B^T\psi \tag{12.54}$$

This shows that also the discrete velocity u_h exists and is unique.

The Schur complement S is not useful from a practical point of view, since it is a full matrix requiring $m \times m$ floating point numbers to store in a computer. This has among other things motivated development of iterative methods for solving saddle-point linear systems.

12.2.6.1 The Uzawa Algorithm

A simple and robust, but slow, way of iteratively solving the saddle-point linear system (12.44) is the Uzawa algorithm, which is defined by the following iteration scheme:

Algorithm 28 The Uzawa algorithm

1: Set $\xi^{(0)} = \psi^{(0)} = 0$.
2: Choose a relaxation parameter $\tau > 0$, and a preconditioner M for the Schur complement $S = BA^{-1}B^T$.
3: **for** $k = 1, 2, \dots$ until convergence **do**
4: Solve

$$A\xi^{(k)} = b - B^T\psi^{(k-1)} \tag{12.55}$$

5: Set

$$\psi^{(k)} = \psi^{(k-1)} + \tau M^{-1} B\xi^{(k)} \tag{12.56}$$

6: **end for**

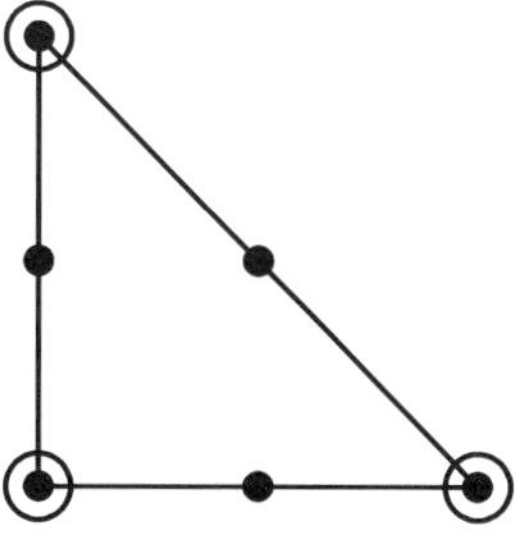

Fig. 12.1 Velocity • and pressure ○ nodes for the Taylor-Hood element

In this algorithm, substituting the first equation into the second yields

$$\psi^{(k)} = \psi^{(k-1)} + \tau M^{-1} B A^{-1} (b - B^T \psi^{(k-1)}) \tag{12.57}$$

which is nothing but a preconditioned Richardson iteration on the linear system (12.49)

Often, the preconditioner M is chosen as the $m \times m$ pressure mass matrix, which is due to the fact that the eigenvalues of this matrix and the Schur complement S has the same sign and order of magnitude. Thus, in a way, M resembles S. Recall that if $M = S$, then the algorithm would converge in a single iteration.

We remark that the relaxation parameter $\tau > 0$ must obey $0 < \tau < 2\nu$, with ν the viscosity parameter, for convergence.

12.2.7 Three Inf-Sup Stable Finite Elements

As we have seen, it is important to choose the finite element spaces so that the discrete inf-sup condition is satisfied. We now present three finite elements that has this property.

12.2.7.1 The Taylor-Hood Element

The Taylor-Hood finite element is the standard finite element for simulating incompressible fluid flow, since it gives a good approximation of both velocity and pressure, and since it is not too numerically costly to use. The element consists of a continuous piecewise quadratic approximation of each velocity component combined with a continuous piecewise linear approximation of the pressure. That is, the velocity space is $V_h = \{v \in [C^0(\Omega)]^d : v|_K \in [P^2(K)]^d\}$, and the pressure space $Q_h = \{v \in C^0(\Omega) : v|_K \in P^1(K)\}$. Figure 12.1 shows the position of the velocity and pressure nodes on a triangle element K.

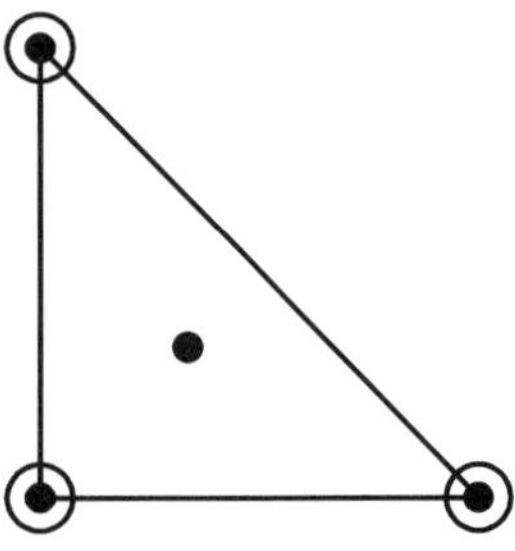

Fig. 12.2 Velocity • and pressure ○ nodes for the MINI element

12.2.7.2 The MINI Element

The MINI element is the simplest inf-sup stable element. It consists of a continuous piecewise linear approximation for each velocity component as well as for the pressure. However, on each element the velocity space is enriched by cubic bubble functions of the form

$$\varphi_{\text{bubble}} = \varphi_1 \varphi_2 \varphi_3 \tag{12.58}$$

where φ_i, $i = 1, 2, 3$, are the usual hat functions. More precisely, the velocity space is given by $V_h = \{v \in [C^0(\Omega)]^d : v|_K \in [P^1(K)]^d \bigoplus [B(K)]^d\}$, where $B(K) = \text{span}\{\varphi_{\text{bubble}}\}$ is the space of bubble functions on element K. Perhaps needless to say, the bubble function has earned its name from the fact that it has the shape of a bubble. By construction φ_{bubble} vanishes on the boundary ∂K, which is important as it allows all bubble functions to be eliminated from the saddle-point linear system (12.44) before attempting to invert it. The MINI element has become popular because it is easy to implement. However, it is also known for giving a poor approximation of the pressure. The velocity and pressure nodes on a triangle element K are shown in Fig. 12.2.

12.2.7.3 The Non-conforming $P^1 - P^0$ Element

The non-conforming $P^1 - P^0$ element amounts to approximating the velocity components with Crouzeix-Raviart functions and the pressure with piecewise constants. This element has the desirable property of being able to yield a finite element solution that is exactly divergence free on each element. As we shall see shortly, it is also fairly easy to implement. The velocity and pressure node locations are shown in Fig. 12.3.

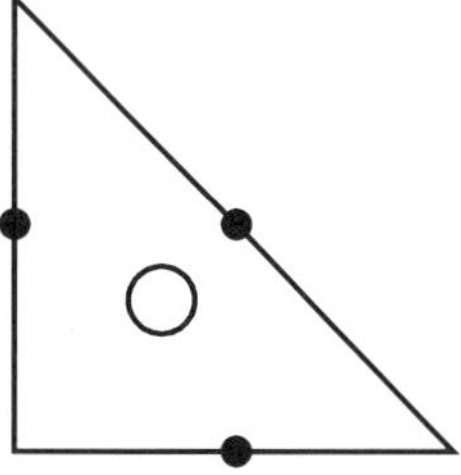

Fig. 12.3 Velocity • and pressure ○ nodes for the non-conforming $P^1 - P^0$ element

12.2.8 Asserting the Inf-Sup Condition

12.2.8.1 Fortin's Trick

There are a few ways to prove that the inf-sup condition holds for a particular finite element. One is Fortin's trick where one seeks to construct an interpolation operator Π from V to V_h satisfying

$$b(v - \Pi v, q) = 0, \quad \forall v \in V,\ q \in Q_h \tag{12.59}$$

and

$$\|\Pi v\|_V \leq C \|v\|_V \tag{12.60}$$

If there is such an operator then the inf-sup condition holds.

To show that the inf-sup condition follows if such an interpolation operator exists, we observe that (12.59) and (12.60) implies

$$\beta \|q\|_Q \leq \sup_{v \in V} \frac{b(v,q)}{\|v\|_V} = \sup_{v \in V} \frac{b(\Pi v,q)}{\|v\|_V} \leq C \sup_{v \in V} \frac{b(\Pi v,q)}{\|\Pi v\|_V} \tag{12.61}$$

Hence, we have

$$\beta \|q\|_Q \leq C \sup_{v \in V_h} \frac{b(v,q)}{\|v\|_V}, \quad \forall q \in Q_h \tag{12.62}$$

which is the discrete inf-sup condition.

Let us use Fortin's trick to show that the non-conforming $P^1 - P^0$ element is inf-sup stable. In this case, the discrete velocity and pressure space is given by $V_h = \{v \in [L_2(\Omega)]^2 : v \in [P_1(K)]^2,\ \forall K \in \mathcal{K}, ([v_i], 1)_{\partial K} = 0, i = 1,2\}$ and $Q_h = \{q \in L_0^2(\Omega) : q \in P_0(K),\ \forall K \in \mathcal{K}\}$, respectively. We consider the piecewise linear interpolant $\Pi v = [\Pi v_i]_{i=1}^2 \in V_h$, defined on each triangle K by

$$\Pi v_i(m_j) = (v_i, 1)_{E_j}/|E_j|, \quad j = 1,2,3 \tag{12.63}$$

where m_j is the mid-point of triangle edge E_j. In other words, component i of Πv on K is given by

$$\Pi v_i = \sum_{j=1}^{3} \frac{(v_i, 1)_{E_j}}{|E_j|} S_j^{CR} \tag{12.64}$$

with S_j^{CR} the Crouzeix-Raviart shape functions.

Now, using the divergence theorem, the definition of Πv, and that the integral of the linear function Πv over edge E_j is given by the value $\Pi v(m_j)|E_j|$ at the edge mid-point m_j, we have

$$(\nabla \cdot v, 1)_K = (v, n)_{\partial K} \tag{12.65}$$

$$= \sum_{j=1}^{3} (v, n)_{E_j} \tag{12.66}$$

$$= \sum_{j=1}^{3} \Pi v(m_j) \cdot n|E_j| \tag{12.67}$$

$$= \sum_{j=1}^{3} (\Pi v, n)_{E_j} \tag{12.68}$$

$$= (\Pi v, n)_{\partial K} \tag{12.69}$$

$$= (\nabla \cdot \Pi v, 1)_K \tag{12.70}$$

where n is the outward unit normal on ∂K. Thus, we have shown that $b(\Pi v, q) = b(v, q)$.

We must also show that $\|\Pi v\|_V \leq C\|v\|_V$. To this end, since the V norm is blind to constants, let us write $v_i = \hat{v}_i + \bar{v}_i$ with $\bar{v}_i = \Pi v_i(m_0)$, that is, $\bar{v}_i$ is the value of Π_i at edge mid-point m_0. As a consequence, $\hat{v}$ vanishes at m_0. Further, since constants are interpolated exactly by Π we have $\Pi v = \Pi(\hat{v} + \bar{v}) = \Pi\hat{v} + \bar{v}$. Also, since $\Pi\hat{v}$ is a Crouzeix-Raviart function, it can be written $\Pi\hat{v}_i = \sum_{j=1}^{3}(\hat{v}_j, 1)_{E_j} S_j^{CR}/|E_j|$. Using these observations, we have

$$\|\nabla \Pi v_i\|_K = \|\nabla \Pi \hat{v}_i\|_K \tag{12.71}$$

$$\leq C \max_j |(\hat{v}_j, 1)_{E_j}| \tag{12.72}$$

$$\leq C\|\hat{v}_i\|_{\partial K} \tag{12.73}$$

$$\leq C(\|\hat{v}_i\|_K + \|\nabla \hat{v}_i\|_K) \tag{12.74}$$

$$\leq C\|\nabla \hat{v}_i\|_K \tag{12.75}$$

$$= C\|\nabla v_i\|_K \tag{12.76}$$

Here, we have used the Trace inequality, and the Poincaré inequality, which is applicable since $\hat{v}_i$ vanishes at m_0. Summing this result over the elements K and the space dimension $i = 1, 2$ we obtain $\|\Pi v\|_V \leq C\|v\|_V$.

12.2.9 A Priori Error Estimate

For the big bilinear form $B(\cdot, \cdot)$ there holds a kind of inf-sup condition called B-stability, which is convenient to work with. In particular, for deriving error estimates. By definition, B-stability is expressed by the inequality

$$\sup_{(v,q)\in V\times Q} \frac{B((u, p), (v, q))}{\|(v, q)\|_{V\times Q}} \geq C\|(u, p)\|_{V\times Q} \tag{12.77}$$

where the product norm $\|(v, q)\|^2_{V\times Q} = \|v\|^2_V + \|q\|^2_Q$.

B-stability is the same as if there for every pair $(u, p) \in V \times Q$ exists another pair $(v, q) \in V \times Q$ satisfying

$$B((u, p), (v, q)) \geq C\|(u, p)\|^2_{V\times Q} \tag{12.78}$$

with

$$\|(v, q)\|_{V\times Q} \leq C\|(u, p)\|_{V\times Q} \tag{12.79}$$

To show B-stability we begin by observing that due to the inf-sup condition (12.36) for every $p \in Q$ there exist a $w \in V$ such that $b(w, p) \geq \kappa\|p\|^2_Q$ with $\|w\|_V \leq \|p\|_Q$. Choosing then $v = \alpha u + \beta w$ and $q = -\alpha p$ with $\alpha = 1 + \kappa^{-2}$ and $\beta = 2\kappa^{-1}$ we have

$$B((u, p), (v, q)) = a(u, \alpha u + \beta w) + b(\alpha u + \beta w, p) + b(u, -\alpha p) \tag{12.80}$$

$$= \alpha a(u, u) + \beta a(u, w) + \beta b(w, p) \tag{12.81}$$

$$\geq \alpha\|u\|^2_V - \tfrac{\beta}{2\kappa}\|u\|^2_V - \tfrac{\beta\kappa}{2}\|w\|^2_V + \beta\kappa\|p\|^2_Q \tag{12.82}$$

$$\geq (\alpha - \tfrac{\beta}{2\kappa})\|u\|^2_V + (-\tfrac{\beta\kappa}{2} + \beta\kappa)\|p\|^2_Q \tag{12.83}$$

$$= \|(u, p)\|^2_{V\times Q} \tag{12.84}$$

where we have used the arithmetic-geometric mean inequality $2a \cdot b \geq -(a \cdot a/\kappa + \kappa b \cdot b)$, which holds for all vectors a and b, scalars $\kappa > 0$, and all scalar products.

Further, using the Triangle inequality we also have

$$\|(v, q)\|^2_{V\times Q} = \|\alpha u + \beta w\|^2_V + \| - \alpha p\|^2_Q \tag{12.85}$$

$$\leq \alpha^2\|u\|_V^2 + \beta^2\|w\|_V^2 + \alpha^2\|p\|_Q^2 \tag{12.86}$$

$$\leq \alpha^2\|u\|_V^2 + (\beta^2 + \alpha^2)\|p\|_Q^2 \tag{12.87}$$

$$\leq C\|(u, p)\|_{V\times Q}^2 \tag{12.88}$$

Hence, combining the above results we have showed the B-stability property.

B-stability also holds on V_h and Q_h provided the finite element employed is inf-sup stable. This follows from a similar line of reasoning as shown above. In this case, by subtracting the weak form (12.26) from the finite element method (12.43) we obtain the Galerkin orthogonality

$$B((u - u_h, p - p_h), (v, q)) = 0, \quad \forall (v, q) \in V_h \times Q_h \tag{12.89}$$

Not surprising, since both $a(\cdot,\cdot)$ and $b(\cdot,\cdot)$ are continuous, so is $B(\cdot,\cdot)$. That is, we have

$$B((u, p), (v, q)) \leq C\|(u, p)\|_{V\times Q}\|(v, q)\|_{V\times Q} \tag{12.90}$$

Now, regarding the velocity and pressure error $u - u_h$ and $p - p_h$ we find by a simple application of the Triangle inequality

$$\|(u - u_h, p - p_h)\|_{V\times Q}^2 = \|u - u_h\|_V^2 + \|p - p_h\|_Q^2 \tag{12.91}$$

$$= \|u - v + v - u_h\|_V^2 + \|p - q + q - p_h\|_Q^2 \tag{12.92}$$

$$\leq \|u - v\|_V^2 + \|p - q\|_Q^2 + \|v - u_h\|_V^2 + \|q - p_h\|_Q^2 \tag{12.93}$$

where we have added and subtracted arbitrary functions v and q from V_h and Q_h, respectively. We may interpret each of the first two terms in (12.93) as a measure of how close a discrete function may come to the solution. By contrast, each of the last two terms in (12.93) measure of how close the finite element solution may come to the best approximation. The first two terms can be estimated using interpolation results. Thus, to turn this into an a priori error estimate we need bounds for the last two terms. To this end, we use the B-stability (12.77) and the Galerkin orthogonality (12.89) to obtain

$$C\|(v - u_h, q - p_h)\|_{V\times Q} \leq \sup_{(w,s)\in V_h\times Q_h} \frac{B((v - u_h, q - p_h), (w, s))}{\|(w, s)\|_{V\times Q}} \tag{12.94}$$

$$= \sup_{(w,s)\in V_h\times Q_h} \frac{B((v - u + u - u_h, q - p + p - p_h), (w, s))}{\|(w, s)\|_{V\times Q}} \tag{12.95}$$

$$= \sup_{(w,s)\in V_h\times Q_h} \frac{B((v-u,q-p),(w,s))}{\|(w,s)\|_{V\times Q}} \tag{12.96}$$

$$\leq \sup_{(w,s)\in V_h\times Q_h} \frac{C\|(v-u,q-p)\|_{V\times Q}\|(w,s)\|_{V\times Q}}{\|(w,s)\|_{V\times Q}} \tag{12.97}$$

$$\leq C\|(v-u,q-p)\|_{V\times Q} \tag{12.98}$$

where we have used the continuity of $B(\cdot,\cdot)$ in the second last line.

Thus we have shown the best approximation result

$$\|(u-u_h,p-p_h)\|_{V\times Q} \leq C\|(u-v,p-q)\|_{V\times Q}, \quad \forall (v,q)\in V_h\times Q_h \tag{12.99}$$

As usual, depending on the type of finite element employed and its interpolation properties the term $\|(u-v,p-q)\|_{V\times Q}$ can be further estimated in terms of u, p, and the mesh size h. For example, assuming a Taylor-Hood approximation and choosing v and q as the corresponding interpolants of u and p, and using standard interpolation estimates, we have the a following a priori error estimate provided that the solution is sufficiently regular.

Theorem 12.3. *For a Taylor-Hood finite element approximation* (u_h, p_h) *there holds the estimate*

$$\|u-u_h\|_V + \|p-p_h\|_Q \leq Ch^2(|u|_{H^3(\Omega)} + |p|_{H^2(\Omega)}) \tag{12.100}$$

In this context we remark that B-stability can be proven more generally than we have done here (i.e., for a general saddle-point problem). In fact, it is a property inherited by the big bilinear form $B(\cdot,\cdot)$ as soon as the requirements for Brezzi's Theorem 12.2 are fulfilled. We shall return to show this later on.

12.2.10 Computer Implementation

12.2.10.1 The Lid-Driven Cavity

Let us implement a Stokes solver to simulate a classical benchmark problem called the lid-driven cavity. The setup is very simple. A square cavity $\Omega = [-1,1]^2$ is filled with a viscous incompressible fluid. No-slip boundary conditions apply on all four sides of the cavity. On the bottom and walls $u_1 = u_2 = 0$, while $u_1 = 1$ and $u_2 = 0$ on the lid. This creates a swirling flow inside the cavity. There is no body load. We wish to compute the velocity field u and pressure distribution p.

Let us write our solver based on the non-conforming $P^1 - P^0$ element. Recall that this amounts to approximating both u_1 and u_2 with Crouzeix-Raviart functions and p with piecewise constants on a triangle mesh. To this end, we introduce the following basis for the velocity space V_h.

$$\{\varphi_i\}_{i=1}^{2n_e} = \left\{ \begin{bmatrix} S_1^{CR} \\ 0 \end{bmatrix}, \begin{bmatrix} S_2^{CR} \\ 0 \end{bmatrix}, \dots, \begin{bmatrix} S_{n_e}^{CR} \\ 0 \end{bmatrix}, \begin{bmatrix} 0 \\ S_1^{CR} \end{bmatrix}, \begin{bmatrix} 0 \\ S_2^{CR} \end{bmatrix}, \dots, \begin{bmatrix} 0 \\ S_{n_e}^{CR} \end{bmatrix} \right\} \tag{12.101}$$

where S_i^{CR}, $i = 1, \dots, n_e$, are the Crouzeix-Raviart shape functions and n_e the number of triangle edges. A basis for the pressure space Q_h is trivially given by $\{\chi_i\}_{i=1}^{n_t}$ with χ_i the characteristic function on triangle i and n_t the number of triangles. That is, $\chi_i = 1$ on triangle i and zero otherwise.

With this choice of bases the saddle-point linear system (12.44) can be written in block form

$$\begin{bmatrix} A_{11} & 0 & B_1^T \\ 0 & A_{22} & B_2^T \\ B_1 & B_2 & 0 \end{bmatrix} \begin{bmatrix} \xi_1 \\ \xi_2 \\ \psi \end{bmatrix} = \begin{bmatrix} b_1 \\ b_2 \\ 0 \end{bmatrix} \tag{12.102}$$

where the matrix and vector entries are given by

$$(A_{ss})_{ij} = (\nabla S_j^{CR}, \nabla S_i^{CR}), \quad i, j = 1, 2, \dots, n_e \tag{12.103}$$

$$(B_s)_{ij} = -(\partial_{x_s} S_j^{CR}, \chi_i), \quad i = 1, 2, \dots, n_t, \ j = 1, 2, \dots, n_e \tag{12.104}$$

$$(b_s)_i = (f_s, S_i^{CR}), \quad i = 1, 2, \dots, n_e \tag{12.105}$$

with $s = 1, 2$.

The assembly of the global matrices A_{ss} and B_s is done as usual by looping over the elements and summing the local matrices from each element into the appropriate places. In doing so, all element matrices are easy to evaluate since their integrands are constant. The 3×3 element matrix $(A_{11}^K)_{ij} = (\nabla S_j^{CR}, \nabla S_i^{CR})_K$ can be computed as

```
A11K=(Sx*Sx'+Sy*Sy')*area;
```

where Sx and Sy are the derivatives of the Crouzeix-Raviart shape functions output from the routine CRGradients. Further, the 1×3 element matrices $(B_1^K)_{ij} = -(\partial_{x_1} S_j^{CR}, \chi_i)_K$ and $(B_2^K)_{ij} = -(\partial_{x_2} S_j^{CR}, \chi_i)_K$ can be computed as

```
B1K=-Sx'*area;
B2K=-Sy'*area;
```

The numbering of the edges is done by reusing our routine Tri2Edge, which outputs a $n_t \times 3$ matrix t2e containing numbers for the edges in each triangle.

Putting things together we have the following routine for assembling the matrices A_{11}, B_1 and B_2.

```
function [A11,B1,B2,areas] = NCAssembler(p,t2e,t)
nt=size(t,2);
```

```
ne=max(t2e(:));
A11=sparse(ne,ne);
B1=sparse(nt,ne);
B2=sparse(nt,ne);
areas=zeros(nt,1);
for i=1:nt
  vertex=t(1:3,i);
  x=p(1,vertex);
  y=p(2,vertex);
  [area,Sx,Sy]=CRGradients(x,y);
  edges=t2e(i,:);
  A11(edges,edges)=A11(edges,edges)+(Sx*Sx'+Sy*Sy')*area;
  B1(i,edges)=-Sx'*area;
  B2(i,edges)=-Sy'*area;
  areas(i)=area;
end
```

Starting to write the main routine, we have

```
function NCStokesSolver()
[p,e,t]=initmesh('squareg'); % mesh square [-1,1]^2
t2e=Tri2Edge(p,t); % triangle-to-edge adjacency
nt=size(t,2); % number of triangles
ne=max(t2e(:)); % number of edges
[A11,B1,B2,areas]=NCAssembler(p,t2e,t); % assemble
nu=0.1; % viscosity parameter
LHS=[nu*A11 sparse(ne,ne) B1';
     sparse(ne,ne) nu*A11 B2';
     B1 B2 sparse(nt,nt)]; % LHS matrix
rhs=zeros(2*ne+nt,1); % RHS vector
```

where `LHS` and `rhs` is the left hand side matrix and right hand side vector of the linear system (12.102), respectively. Should we attempt to invert `LHS` we would find that it is singular. This is of course due to the fact that we have neither enforced any boundary conditions on the velocity nor a zero mean on the pressure.

In the discrete setting zero mean value on p_h means that

$$(p_h, 1) = \sum_{K=1}^{n_t} \psi_K (\chi_i, 1)_K = a^T \psi = 0 \tag{12.106}$$

where a is the vector `areas`. To enforce this constraint we augment the saddle-point linear system with this equation together with a Lagrangian multiplier μ to get

$$\begin{bmatrix} A_{11} & 0 & B_1 & 0 \\ 0 & A_{11} & B_2 & 0 \\ B_1^T & B_2^T & 0 & a \\ 0 & 0 & a^T & 0 \end{bmatrix} \begin{bmatrix} \xi_1 \\ \xi_2 \\ \psi \\ \mu \end{bmatrix} = \begin{bmatrix} b_1 \\ b_2 \\ 0 \\ 0 \end{bmatrix} \tag{12.107}$$

The code to do this looks like

```
last=[zeros(2*ne,1); areas]; % last row and column
LHS=[LHS last; last' 0];
rhs=[rhs; 0];
```

The last thing we need to do is to enforce the no-slip boundary condition $u = g$. We do this as explained by first writing $u_h = g_h + u_0$ with g_h the Crouzeix-Raviart interpolant of g, then shrinking the matrix LHS and the vector `rhs` according to (12.25), and finally solving for u_0. The setting up of g_h is a bit messy since it requires the computation of edge mid-points along with corresponding edge numbers. We can compute this information using the routine `EdgeMidPoints` listed below. The values of g_h can then be found by looping over the edges and evaluating g on each edge. We have

```
[xmid,ymid,edges] = EdgeMidPoints(p,t2e,t);
fixed=[]; % fixed nodes
gvals=[]; % nodal values of g
for i=1:length(edges) % loop over edges
  n=edges(i); % edge (ie. node) number
  x=xmid(i); % x-coordinate of edge mid-point
  y=ymid(i); % y-
  if (x<-0.99 | x>0.99 | y<-0.99 | y>0.99) % boundary
    fixed=[fixed; n; n+ne]; % fix velocity nodes
    u=0; v=0; % bc values
    if (y>0.99), u=1; end % u=1,v=0 on lid
    gvals=[gvals; u; v];
  end
end
```

To shrink the matrix and vector and solve the resulting linear system we type

```
neq=2*ne+nt+1; % number of equations
free=setdiff([1:neq],fixed);
rhs=rhs(free)-LHS(free,fixed)*gvals; % shrink vector
LHS=LHS(free,free); % shrink matrix
sol=zeros(neq,1); % allocate solution
sol(fixed)=gvals; % insert no-slip values
sol(free)=LHS\rhs; % solve linear system
```

Finally, we make plots of the velocity and pressure

```
U=sol(1:ne); V=sol(1+ne:2*ne); P=sol(2*ne+1:2*ne+nt);
figure(1), pdesurf(p,t,P')
figure(2), quiver(xmid,ymid,U',V')
```

Running this code we get the velocity and pressure of Figs. 12.4 and 12.5. As expected the velocity glyphs show a swirling fluid due to the moving lid. The pressure distribution shows a high pressure in the upper right corner of the cavity,

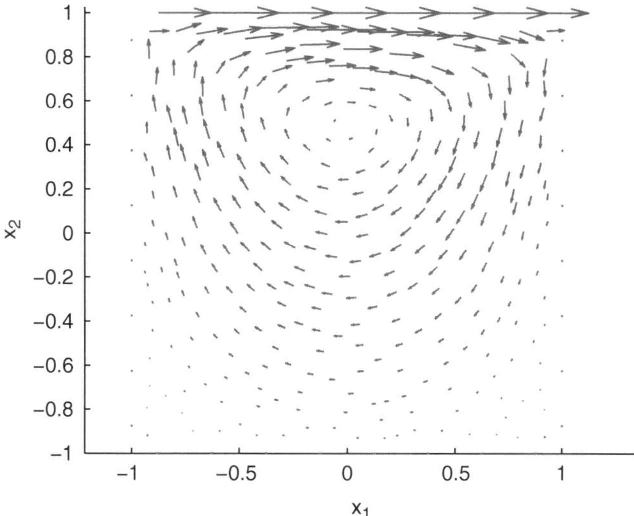

Fig. 12.4 Glyphs of velocity u_h in the cavity

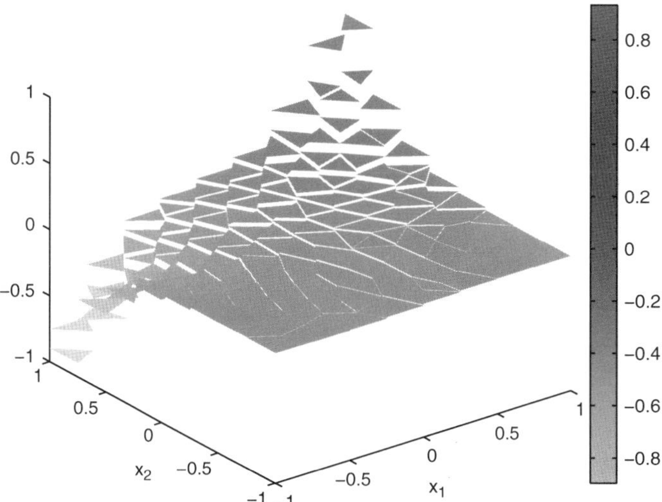

Fig. 12.5 Pressure distribution p_h in the cavity

where the fluid collides with the right wall. Similarly, a low pressure is visible in the upper left corner, where the fluid is swept away from the left wall by the moving lid.

To check our implementation we can make a test and compute the null space of the matrix $B^T = [B_1 \ B_2]^T$, which is the discrete gradient operator $-\nabla$. Recall that this operator determines the pressure p_h and should therefore have a null space consisting of the single vector 1, or a scaled copy of this vector. This is the discrete hydrostatic pressure mode, which we eliminated by adding the zero mean value constraint for p_h. The B^T matrix can be extracted from the saddle-point linear system. In doing so we must remember that the matrix LHS shrunk when we removed the boundary conditions. The null space is computed using the null command.

```
nfix=length(fixed);
n=2*ne-nfix; % number of free velocity nodes
Bt=LHS(1:n,n+1:n+nt); % extract B'
nsp=null(full(Bt)) % compute null space of B'
```

Indeed, the result of executing these lines is the vector nsp, which is a constant times the vector 1. This is a necessary condition for a finite element to be inf-sup stable. Of course it does not prove inf-sup stability, but it is one way of testing the code. Other ways to validate the code include computing the eigenvalues of the Schur complement, which should be positive, or checking that the Lagrange multiplier is close to the machine precision.

12.2.11 A Posteriori Error Estimate

We next show how to derive a basic a posteriori error estimate.

From the Galerkin orthogonality (12.89) we have

$$B((u-u_h,p-p_h),(v,q)) = B((u-u_h,p-p_h),(v,q)) - B((u-u_h,p-p_h),(\pi v,0)) \tag{12.108}$$

$$= B((u-u_h,p-p_h),(v-\pi v,q)) \tag{12.109}$$

since $\pi v \in V_h$ and the zero function is a member of Q_h. Rewriting this, we further have

$$B((u-u_h,p-p_h),(v-\pi v,q)) = F((v-\pi v,q)) - B((u_h,p_h),(v-\pi v,q)) \tag{12.110}$$

$$= l(v-\pi v) - a(u_h,v-\pi v) - b(v-\pi v,p_h) \tag{12.111}$$

$$- b(u_h,q)$$

$$= \sum_{K\in\mathcal{K}} (f+\Delta u_h - \nabla p_h, v-\pi v)_K \tag{12.112}$$

$$- \tfrac{1}{2}([n \cdot \nabla u_h - p_h n], v - \pi v)_{\partial K \setminus \partial \Omega}$$
$$+ (\nabla \cdot u_h, q)_K$$

where we have broken the integrals into a sum over the elements K in the mesh $\mathcal{K}$, and integrated by parts. As usual, the brackets $[\cdot]$ denote the jump of the enclosed quantity between adjacent elements. Note that if p_h is continuous, then $[p_h] = 0$. Further, denoting the sum (12.112) by S and estimating each term in it using the Cauchy-Schwartz inequality, and recalling the interpolation estimates $\|v - \pi v\|_K \leq C h_K \|\nabla v\|_K$, and $\|v - \pi v\|_{\partial K} \leq C h_K^{1/2} \|\nabla v\|_K$, we have

$$S \leq \sum_{K \in \mathcal{K}} \|f + \Delta u_h - \nabla p_h\|_K \|v - \pi v\|_K \tag{12.113}$$
$$+ \tfrac{1}{2} \|[n \cdot \nabla u_h - p_h n]\|_{\partial K \setminus \partial \Omega} \|v - \pi v\|_{\partial K \setminus \partial \Omega}$$
$$+ \|\nabla \cdot u_h\|_K \|q\|_K$$
$$\leq C \sum_{K \in \mathcal{K}} h_K \|f + \Delta u_h - \nabla p_h\|_K \|\nabla v\|_K \tag{12.114}$$
$$+ \tfrac{1}{2} h_K^{1/2} \|[n \cdot \nabla u_h - p_h n]\|_{\partial K \setminus \partial \Omega} \|\nabla v\|_K$$
$$+ \|\nabla \cdot u_h\|_K \|q\|_K$$
$$\leq \left(\sum_{K \in \mathcal{K}} h_K^2 \|f + \Delta u_h - \nabla p_h\|_K^2 + \tfrac{1}{4} h_K \|[n \cdot \nabla u_h - p_h n]\|_{\partial K \setminus \partial \Omega}^2 + \|\nabla \cdot u_h\|_K^2 \right)^{1/2}$$
$$(\|\nabla v\| + \|q\|) \tag{12.115}$$

Introducing the element residual η_K, defined by

$$\eta_K = h_K \|f + \Delta u_h - \nabla p_h\|_K + \tfrac{1}{2} h_K^{1/2} \|[n \cdot \nabla u_h - p_h n]\|_{\partial K \setminus \partial \Omega} + \|\nabla \cdot u_h\|_K \tag{12.116}$$

we thus have

$$B((u - u_h, p - p_h), (v, q)) \leq C \left(\sum_{K \in \mathcal{K}} \eta_K \right) \|(v, q)\|_{V \times Q} \tag{12.117}$$

Finally, continuous B-stability implies

$$C \|(u - u_h, p - p_h)\|_{V \times Q} \leq \sup_{(v,q) \in V \times Q} \frac{B((v - u_h, q - p_h), (v, q))}{\|(v, q)\|_{V \times Q}} \tag{12.118}$$

$$\leq C \sup_{(v,q)\in V\times Q} \frac{\left(\sum_{K\in\mathcal{K}} \eta_K\right) \|(v,q)\|_{V\times Q}}{\|(v,q)\|_{V\times Q}} \tag{12.119}$$

$$\leq C \sum_{K\in\mathcal{K}} \eta_K \tag{12.120}$$

Hence, we have shown the following a posteriori error estimate.

Theorem 12.4. *The finite element solution* (u_h, p_h) *satisfies the estimate*

$$\|(u-u_h, p-p_h)\|_{V\times Q} \leq C \sum_{K\in\mathcal{K}} \eta_K \tag{12.121}$$

where the element residual η_K *is the sum of the cell residual* $R_K = h_K\|f + \Delta u_h - \nabla p_h\|_K + \|\nabla \cdot u_h\|_K$ *and the edge residual* $r_K = \frac{1}{2}h_K^{1/2}\|[n\cdot\nabla u_h - p_h n]\|_{\partial K\setminus\partial\Omega}$.

12.2.12 Stabilized Finite Element Methods

12.2.12.1 Spurious Pressure Modes

Unfortunately, the choice of equal order Lagrange spaces for both velocity and pressure does not satisfy the inf-sup condition. For example, it is not possible to use piecewise linears for both u_h and p_h. This is unfortunate since these elements are very simple to implement in a computer. Attempting to do so yields a gradient matrix B^T with a too large null space (i.e., not only the constant vector 1). The basis vectors for this larger null space represent unphysical pressures, called spurious or parasitic pressure modes, which pollute p_h. Typically, these parasites are highly oscillatory and renders the pressure useless from a practical point of view as it varies abruptly from element to element. In fact, neighbouring elements often have opposite values. This is called the checkerboard syndrome.

The explanation for the occurrence of spurious modes is that with equal order interpolation the discrete inf-sup constant γ becomes implicitly proportional to the mesh size h. This is a hidden dependency in the sense that there is no way to control the size of γ. In particular, we can not prevent it from getting too small. Indeed, as h tends to zero γ also tends to zero, and numerical stability is lost. This is particularly problematic from the point of view of mesh refinement.

12.2.12.2 Galerkin Least Squares Stabilization

One way of mitigating the effects of spurious pressure modes is to use some sort of Galerkin Least Squares (GLS) stabilization. This is also a common way to circumvent the inf-sup condition.

The most widespread GLS method, which is suitable for approximation of both the pressure and velocity using equal order Lagrange elements, amounts to replacing the continuous linear forms $B(\cdot,\cdot)$ and $F(\cdot)$ in the finite element method (12.43) by the discrete forms $B_h(\cdot,\cdot)$ and $F_h(\cdot)$, defined by

$$B_h((u_h,p_h),(v,q)) = B((u,p),(v,q)) + \delta \sum_{K\in\mathcal{K}} (-\Delta u_h + \nabla p_h, -\Delta v + \nabla q)_K \tag{12.122}$$

$$F_h((v,q)) = F((v,q)) + \delta \sum_{K\in\mathcal{K}} (f, -\Delta v + \nabla q)_K \tag{12.123}$$

where $B((u,p),(v,q))$ now is the non-symmetric version of the big bilinear form, that is,

$$B((u,p),(v,q)) = a(u,v) + b(v,p) - b(u,q) \tag{12.124}$$

and δ is a stabilization parameter to be chosen suitably.

In the special case with a continuous piecewise linear approximation of both the velocity and pressure the terms Δu_h and Δv drop out, and we are left with

$$B_h((u,p),(v,q)) = B((u,p),(v,q)) + \delta(\nabla p_h, \nabla q) \tag{12.125}$$

$$F_h((v,q)) = F((v,q)) + \delta(f, \nabla q) \tag{12.126}$$

This stabilization corresponds to relaxing the incompressibility constraint by adding the term $-\delta\Delta p$ to the equation $\nabla\cdot u = 0$. Roughly speaking the GLS stabilization term $\delta(\nabla p_h, \nabla q)$ makes the discrete pressure p_h smoother. In particular, it penalizes large gradients of p_h, which is important to prevent spurious pressure modes to occur. Setting $v = u$ and $q = p$ we note that

$$B((u,p),(u,p)) = \|\nabla u\|^2 + \delta\|\nabla p\|^2 \tag{12.127}$$

since we are using the non-symmetric form. Hence, existence and uniqueness of u_h and p_h follow from the Lax-Milgram lemma without the imposition of any inf-sup condition. The coercivity constant α will depend on the stabilization parameter δ and, thus, still vanish as δ tend to zero. However, this dependency is explicit and can be controlled by choosing δ properly. In particular, α can be prevented from getting too small. In doing so, we want δ to be large enough for numerical stability, but as small as possible not to perturb the solution too much, which leads to $\delta = Ch^2$ with h the mesh size and C a constant of moderate size. With this choice of stabilization parameter it is also possible to establish inf-sup stability, with inf-sup constant independent of h, of the method.

We remark that more elaborate GLS methods do exist, for example, incorporating stabilization of conservation of mass.

12.2.12.3 Macroelement Techniques

An old and maybe somewhat primitive way to prevent spurious pressure modes is to use a macroelement technique. This amounts to clustering a set of adjacent elements to form a so-called macroelement. The rationale for doing so is that internal velocity degrees of freedom are created within the macroelement thereby enlarging the velocity space. The pressure space is, however, not made larger. Indeed, generally, more velocity than pressure degrees of freedom is necessary for inf-sup stability. In a way, all macroelements are reminiscent of the MINI element in which bubble functions is added to the piecewise linears to make the velocity space larger. Of course, instead of enlarging the velocity space it is possible to restrict the pressure space. Indeed, this is a common way to make the popular $Q^1 - Q^0$ element inf-sup stable. This element is defined on quadrilaterals with a bilinear approximation of each velocity component and a constant approximation of the pressure. Groups of four adjacent quadrilaterals are created to form the macroelement giving effectively a nine node velocity element with the pressure being the average of the four pressure values. This averaging is often referred to as a filtering of the pressure. Filtering smooths the pressure and hinders the rapidly oscillating spurious modes to occur. An advantage of the filtering is that it is simple to implement in the sense that it does not change the assembly procedure, as it can be applied at the end of each step of an iterative method, such as the Uzawa algorithm for instance, when solving the saddle-point linear system (12.44) arising from finite element discretization.

12.3 The Navier-Stokes Equations

Having studied some of the basic features of incompressible fluid flow, let us now consider the Navier-Stokes equations, which is a time-dependent, non-linear, generally convection dominated, saddle-point problem. Indeed, the Navier-Stokes equations are so complex that still basic properties like existence, uniqueness, and stability, of the solution are not settled. However, due to their wide range of use the numerical study of the Navier-Stokes equations has grown into a discipline of its own called computational fluid dynamics (CFD). CFD is a vast field involving continuum mechanics, thermodynamics, mathematics, and computer science. The applications are many and range from optimizing the mix of air and fuel in turbine engines to predicting the stresses in the walls of human blood vessels. The grand theoretical challenge for CFD is the understanding of turbulence. Turbulence is the highly chaotic flow pattern exhibited by a rapidly moving fluid with low viscosity. Think of the irregular plume of smoke rising from a cigarette, for example. Physically, turbulence is caused by a combination of dissipation of energy into heat at the microscopic level, with a large transport of momentum at the macroscopic level. The basic measure of the tendency for a fluid to develop a turbulent flow is the dimensionless Reynolds number, defined as $\mathrm{Re} = UL/\nu$, where ν is the viscosity and U and L is a representative velocity and length scale, respectively. A high Reynolds number implies a turbulent flow, while a low implies a steady state laminar flow. Because turbulence occurs on all length scales, down to the smallest so-called Kolmogorov scale $\mathrm{Re}^{-3/4}$, it is very difficult to simulate using finite elements on a coarse mesh. This is further complicated by the fact that turbulent flows are highly convective, which requires stabilization of the corresponding finite element methods with subsequent potential loss of accuracy. To remedy this substantial efforts have been made to model the effect of turbulence on the small scales by statistical means and derive additional terms supplementing the original equations. This has lead to turbulence models, which hope to account for turbulence effects on average. For example, in the simplest case this amounts to changing the viscosity ν to $\nu + \nu_T$, where ν_T is a variable eddy viscosity depending on the magnitude of the local velocity gradients. This is the frequently used so-called Smagorinsky turbulence model. Obviously, there is much more to say on this matter, but we shall not attempt to do so here. Suffice to say that many important fluid mechanic applications are somewhere in between laminar and turbulent, which make them amenable to simulation.

For completeness, recall that the Navier-Stokes equations take the form

$$\dot{u} + (u \cdot \nabla)u + \nabla p - \nu \Delta u = f, \quad \text{in } \Omega \times J \tag{12.128a}$$

$$\nabla \cdot u = 0, \quad \text{in } \Omega \times J \tag{12.128b}$$

$$u = g_D, \quad \text{in } \Gamma_D \times J \tag{12.128c}$$

$$\nu n \cdot \nabla u - pn = 0, \quad \text{in } \Gamma_N \times J \tag{12.128d}$$

$$u = u_0, \quad \text{in } \Omega, \text{ for } t = 0 \tag{12.128e}$$

where ν is viscosity, u and p the sought velocity and pressure, and f a given body force. As usual, we assume that the boundary $\partial\Omega$ is divided into two parts Γ_D and Γ_N associated with the no-slip and the do-nothing boundary conditions (12.128c) and (12.128d), respectively. Typically, Ω is a channel and Γ_D denotes either the rigid walls of the channel, with $g_D = 0$, or the inflow region, with g_D the inflow velocity profile, while Γ_N denotes the outlet with the boundary condition $\nu n \cdot \nabla u - pn = 0$. The velocity at time $t = 0$ is given by the initial condition u_0 and $J = (0, T]$ is the time interval with final time T.

12.3.1 The Fractional Step Method

12.3.1.1 Time Discretization

There are many ways to derive a numerical method for the Navier-Stokes equations and it is certainly not easy to know which one is the best. Several questions may arise, for example:

- Should we use implicit or explicit time stepping?
- Do we need to use an inf-sup stable element?

- Is Newton's method necessary to cope with the non-linear term $u \cdot \nabla u$?
- Must the finite element method be GLS stabilized for dealing with any high convection?

Needless to say, each answer to these questions has its own pros and cons regarding the accuracy and computational cost of the resulting numerical method. As usual, the application at hand has to be considered, and a balance has to be struck.

In the following, we shall favor computational speed and present a simple method due to Chorin [17] called the fractional step method, for discretizing the Navier-Stokes equations. The basic idea is as follows.

Discretizing the momentum equation (12.128a) in time using the forward Euler method we have the time stepping scheme

$$\frac{u_{l+1} - u_l}{k_l} + (u_l \cdot \nabla)u_l + \nabla p_l - \nu \Delta u_l = f_l \tag{12.129}$$

where k_l is the timestep and the subscript l indicates the time level. Now, adding and subtracting a tentative velocity u_* in the discrete time derivative $k_l^{-1}(u_{l+1} - u_l)$ we further have

$$\frac{u_{l+1} - u_* + u_* - u_l}{k_l} + (u_l \cdot \nabla)u_l + \nabla p_l - \nu \Delta u_l = f_l \tag{12.130}$$

Obviously, this equation holds if

$$\frac{u_* - u_l}{k_l} = -(u_l \cdot \nabla)u_l + \nu \Delta u_l + f_l \tag{12.131}$$

and

$$\frac{u_{l+1} - u_*}{k_l} = -\nabla p_l \tag{12.132}$$

hold simultaneously.

The decomposition of (12.128a) into (12.131) and (12.132) is called operator splitting. The rationale is that we get a decoupling of the diffusion and convection of the velocity, and the pressure acting to enforce the incompressibility constraint. Thus, assuming we know u_l, we can compute u_* from (12.131) separately without having to worry about the pressure. However, to determine also the pressure we take the divergence of (12.132), yielding

$$\nabla \cdot \frac{u_{l+1} - u_*}{k_l} = -\nabla \cdot (\nabla p_l) \tag{12.133}$$

Now, since we desire $\nabla \cdot u_{l+1} = 0$ this reduces to

$$-\nabla \cdot \frac{u_*}{k_l} = -\Delta p_l \tag{12.134}$$

It follows that the pressure p_l can be determined from a Poisson type equation. In fact, (12.134) is frequently referred to as the pressure Poisson equation (PPE). Thus, given u_* we can solve (12.134) to get a pressure p_l, which makes the next velocity

u_{l+1} divergence free. As p_l is manufactured from the tentative velocity u_*, it is not the actual pressure p, but at best a first order approximation in time.

The actual computation of u_{l+1} is done by reusing (12.132), but now in the form

$$u_{l+1} = u_* - k_l \nabla p_l \tag{12.135}$$

This line of reasoning leads us to the following algorithm:

Algorithm 29 The fractional step method

1: Given an initial condition $u_0 = 0$.
2: **for** $l = 1, 2, \dots, m$ **do**
3: Compute the tentative velocity u_* from

$$\frac{u_* - u_l}{k_l} = -(u_l \cdot \nabla)u_l + \nu \Delta u_l + f_l \tag{12.136}$$

4: Solve the pressure Poisson equation

$$-\nabla \cdot u_* = -k_l \Delta p_l \tag{12.137}$$

for p_l.
5: Update the velocity

$$u_{l+1} = u_* - k_l \nabla p_l \tag{12.138}$$

6: **end for**

The boundary conditions for u_* and p_l are not obvious and has been the source of some controversy. The simplest choice is to put the no-slip boundary conditions (12.128d) on u_*, and the natural boundary condition $n \cdot \nabla p_l = 0$ on p_l. The exception is at the outflow, where the do-nothing boundary condition (12.128e) is imposed term by term by assuming $n \cdot \nabla u_l = 0$ and $p_l = 0$. This generally means that u_{l+1} will not satisfy the velocity boundary conditions other than in a vague sense. The source of controversy is the natural boundary condition on the pressure, which is unphysical and leads to a poor quality of both p_l and u_{l+1} near the boundary. This has raised questions of the validity of the projection method. Numerous methods have been suggested to remedy this with, at least, partial success.

12.3.1.2 Space Discretization

Next, in order to obtain a fully discrete method we apply finite elements to each of the three Eqs. (12.136)–(12.138). We use continuous piecewise linears to approximate all occurring velocities and pressures. Using matrix notation, the discrete counterpart of Algorithm 29 takes the form:

- Compute the discrete tentative velocity ξ_* from

$$\begin{bmatrix} \xi_1 \\ \xi_2 \end{bmatrix}_* = \begin{bmatrix} \xi_1 \\ \xi_2 \end{bmatrix}_l - k_l \begin{bmatrix} M & 0 \\ 0 & M \end{bmatrix}^{-1} \left(\begin{bmatrix} C + \nu A & 0 \\ 0 & C + \nu A \end{bmatrix} \begin{bmatrix} \xi_1 \\ \xi_2 \end{bmatrix} - \begin{bmatrix} b_1 \\ b_2 \end{bmatrix} \right)_l \tag{12.139}$$

- Solve the discrete pressure Poisson equation

$$k_l A \psi_l = - \begin{bmatrix} B_1 & B_2 \end{bmatrix} \begin{bmatrix} \xi_1 \\ \xi_2 \end{bmatrix}_* \tag{12.140}$$

  for ψ_l.
- Update the discrete velocity

$$\begin{bmatrix} \xi_1 \\ \xi_2 \end{bmatrix}_{l+1} = \begin{bmatrix} \xi_1 \\ \xi_2 \end{bmatrix}_* - k_l \begin{bmatrix} M & 0 \\ 0 & M \end{bmatrix}^{-1} \begin{bmatrix} B_1 \\ B_2 \end{bmatrix} \psi_l \tag{12.141}$$

Here, A is the $n_p \times n_p$ stiffness matrix with n_p the number of nodes. Similarly, M is the $n_p \times n_p$ mass matrix, which for computational efficiency can be lumped. Further, C_l is the $n_p \times n_p$ convection matrix with convection field $u_l = [u_1, u_2]_l$. Because C_l depends on the current velocity it must be reassembled at each timestep l. Furthermore, B_1 and B_2 are the $n_p \times n_p$ x_1- and x_2-differentiation matrices, that is, convection matrices arising from convection fields $[1, 0]$ and $[0, 1]$, respectively. As usual, the $n_p \times 1$ load vectors b_1 and b_2 contain contributions from any body force. Finally, ξ_1, ξ_2, and ψ denote the $n_p \times 1$ vectors of nodal velocity and pressure values. Of course, these equations have to be adjusted for boundary conditions.

The time step of the presented numerical method is limited by the use of the forward Euler scheme. Indeed, for numerical stability it is necessary that the time step k_l is of magnitude h/U for convection dominated flow with $\nu < Uh$, and h^2/ν for diffusion dominated flow with $\nu \geq Uh$.

12.3.2 Computer Implementation

12.3.2.1 The DFG Benchmark

The practical implementation of the fractional step method is straight forward. As test problem we use the so-called DFG benchmark, which is a two-dimensional channel flow around a cylinder. The channel is rectangular with length 2.2 and height 0.41. At $(0.2, 0.2)$ is a cut out circle with diameter 0.1. The fluid has viscosity $\nu = 0.001$ and unit density. On the upper and lower wall and on the cylinder a zero no-slip boundary condition is prescribed. A parabolic inflow profile with maximum velocity $U_{\max} = 0.3$ is prescribed on the left wall,

$$u_1 = \frac{4U_{\max} y(0.41 - y)}{0.41^2}, \quad u_2 = 0 \tag{12.142}$$

The boundary conditions on the right wall is of do-nothing type, since this is the outflow. There are no body forces. Zero initial conditions are assumed.

The channel geometry is output from the routine DFGg listed in the Appendix. We start writing our solver by calling this routine, creating the mesh, and extracting the number of nodes and the node coordinates from the point matrix p.

```
function NSChorinSolver()
channel=DFGg();
[p,e,t]=initmesh(channel,'hmax',0.25);
np=size(p,2);
x=p(1,:); y=p(2,:);
```

The zero boundary condition on the pressure is most easily enforced by adding large weights, say 10^6, to the diagonal entries of A corresponding to nodes on the outflow. This penalizes any deviation from zero of the pressure in these nodes. It is convenient to store the weights in a diagonal matrix R, which can be built with the following lines of code.

```
out=find(x>2.199); % nodes on outflow
wgts=zeros(np,1); % big weights
wgts(out)=1.e+6;
R=spdiags(wgts,0,np,np); % diagonal penalty matrix
```

Moreover, the boundary conditions on the velocity can be be enforced a little simpler than usual due to the explicit time stepping. In each time step we can simply zero out any current value of the no-slip nodes and replace with zero. However, to do so we need two vectors mask and g to identify nodes with no-slip boundary conditions and to store the corresponding nodal value. These are computed as follows.

```
in =find(x<0.001); % nodes on inflow
bnd=unique([e(1,:) e(2,:)]); % all nodes on boundary
bnd=setdiff(bnd,out); % remove outflow nodes
mask=ones(np,1); % a mask to identify no-slip nodes
mask(bnd)=0; % set mask for no-slip nodes to zero
x=x(in); % x-coordinate of nodes on inflow
y=y(in); % y-
Umax=0.3; % maximum inflow velocity
g=zeros(np,1); % no-slip values
g(in)=4*Umax*y.*(0.41-y)/0.41^2; % inflow profile
```

The assembly of all matrices M, A, C, and B_s is easy to do by using the built-in routine assema for A and M, and our own ConvectionAssembler2D for B_s and C. To speed up the computation we lump the mass matrix M. Thus, we have

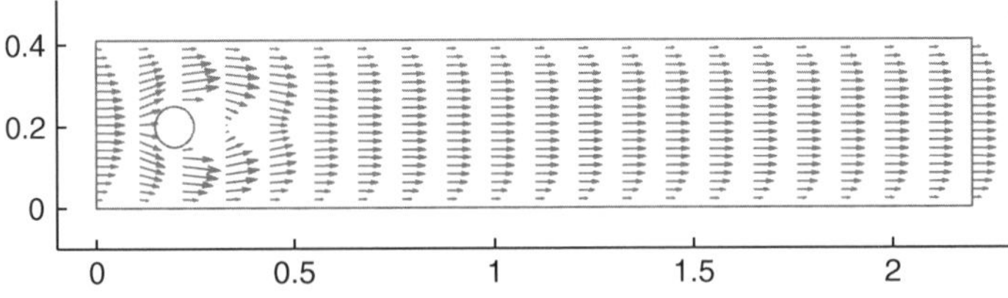

Fig. 12.6 Velocity glyphs for the DFG benchmark (Re = 20)

```
[A,unused,M]=assema(p,t,1,0,1);
Bx=ConvectionAssembler2D(p,t,ones(np,1),zeros(np,1));
By=ConvectionAssembler2D(p,t,zeros(np,1),ones(np,1));
```

Using these data structures the actual time loop with the projection scheme can be very compactly written, viz.,

```
dt=0.01;  % time step
nu=0.001; % viscosity
V=zeros(np,1); % x-velocity
U=zeros(np,1); % y-
for l=1:100
  % assemble convection matrix
  C=ConvectionAssembler2D(p,t,U,V);
  % compute tentative velocity
  U=U-dt*(nu*A+C)*U./M;
  V=V-dt*(nu*A+C)*V./M;
  % enforce no-slip BC
  U=U.*mask+g;
  V=V.*mask;
  % solve PPE
  P=(A+R)\-(Bx*U+By*V)/dt;
  % update velocity
  U=U-dt*(Bx*P)./M;
  V=V-dt*(By*P)./M;
  pdeplot(p,e,t,'flowdata',[U V]),axis equal,pause(.1)
end
```

The setup gives a Reynolds number of Re $=$ 20 with the characteristic velocity $U = \frac{2}{3}U_{\max} = 0.2$ the mean of the parabolic profile and $L = 0.1$ the cylinder diameter. This is a low Reynolds number and we expect to see a laminar flow. Running the code and simulating the flow during one second we obtain the results of Figs. 12.6–12.8. Due to the low Reynolds number a steady state flow has evolved and from the glyphs plot we see that it is, indeed, laminar. As we might have anticipated the pressure isocontours shows a high pressure in front of the cylinder and a low

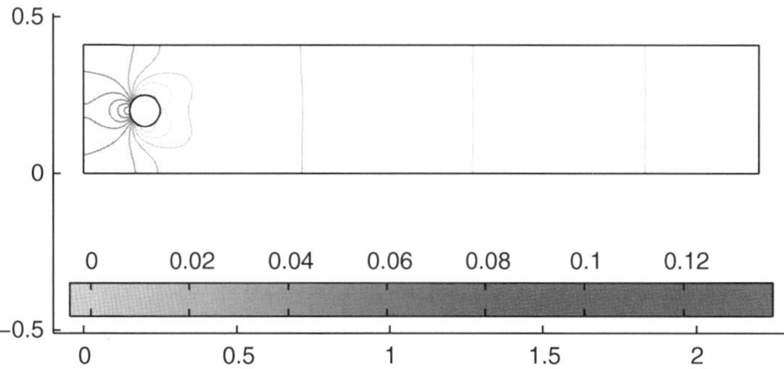

Fig. 12.7 Isocontours of the pressure (Re = 20)

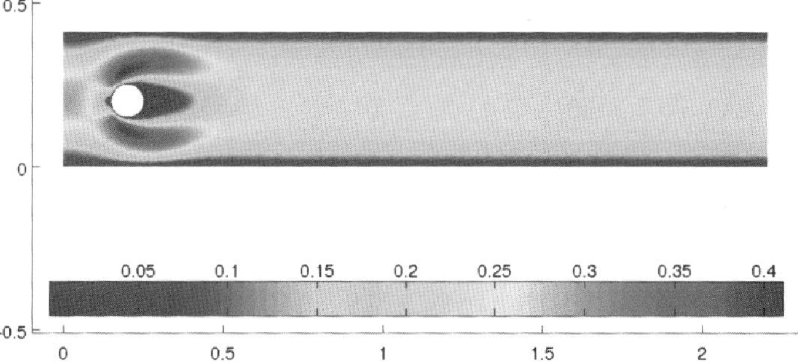

Fig. 12.8 Magnitude of the velocity (Re = 20)

pressure behind it. In this region we also see a small wake with recirculating flow forming. This is typical for incompressible fluid flow.

12.4 Further Reading

The finite element literature concerning fluids is relatively rich. Good books on fluid mechanics in general and numerical methods in particular include those of Chorin and Marsden [18], and Turek [71], Gresho and Sani [58], and Zienkiewicz, Taylor and Nithiarasu [75]. The book by Brezzi and Fortin [31] contains a detailed account of the theory for mixed finite elements and saddle-point problems, along with a list of inf-sup stable finite elements. This can also be found in the book by Braess [15]. Fast iterative solvers for solving the linear systems arising from saddle-point problems is the topic of the book by Elman, Silverster, and Wathen [26]. Stabilized GLS methods for the Stokes system are analyzed by Franca, Hughes and Stenberg in [38]. An overview of finite elements for Stokes equations is given by Boffi, Brezzi and Fortin in [12]. Books concerning only the Navier-Stokes equations include those by Girault and Raviart [34], and Temam [69].

12.5 Problems

Exercise 12.1. Formulate a finite element approximation of the Stokes equations using the Taylor-Hood element. In particular, deduce the entries of the saddle-point linear system, resulting from finite element discretization.

Exercise 12.2. Modify `NCStokesSolver` and solve the following problem called the colliding flow problem. The domain is the square $\Omega = [-1, 1]^2$, with no-slip boundary conditions on the whole boundary $\partial\Omega$ given by the benchmark solution

$$u_1 = 20x_1x_2^3, \quad u_2 = 5x_1^4 - 5x_2^4, \quad p = 60x_1^2x_2 - 20x_2^3$$

which satisfies the Stokes equations with zero body force, and zero mean pressure.

Exercise 12.3. Write a routine, `Uzawa`, for solving the saddle-point linear system in `NCStokesSolver` using the Uzawa algorithm. The calling syntax should be `sol(free)=Uzawa(A,Bt,b,areas);`. The relevant matrices and vectors can be extracted from `LHS` and `rhs` by typing

```
A=LHS(1:n,1:n); Bt=LHS(1:n,n+1:n+nt); b=rhs(1:n);
```

where `n` is the number of free velocity nodes. Note that the zero mean pressure condition must be enforced at each iteration k. That is, the constant vector 1 must be filtered out of $\psi^{(k)}$. This can be done by setting

$$\psi^{(k)} = \psi^{(k)} - (a^T\psi^{(k)})/(a^T 1)1$$

at the end of each iteration. Here, a is the `areas` vector. For simplicity, let $M = \text{diag}(a)$.

Exercise 12.4. How would Chorin's projection method look with Euler backward time stepping? What is the pros and cons of this as compared to Euler forward time stepping?

Exercise 12.5. Run a sequence of simulations on the DFG benchmark with varying viscosity from $\nu = 0.1$ to 0.005. In each run make 1,000 timesteps using $k_l = 0.01$. Study the transition from laminar to almost turbulent flow as you decrease ν. Make plots of the velocity and pressure.

Exercise 12.6. Simulate the Lid-Driven cavity problem using the fractional step method with viscosity $\nu = 0.1$ and 0.005. To fix the pressure you can set $p = 0$

in the point $(-1, 0)$. Make plots of the velocity magnitude. Can you say something about the effect of the non-linear term $u(\cdot\nabla u)$?

Exercise 12.7. Since the Navier-Stokes equations are non-linear it is possible to use Newton's method to solve them. This requires the linearization of the 3×1 vector $[-\nu\Delta u + (u \cdot \nabla u) + \nabla p, \nabla \cdot u]^T$. Do this by first setting $u_i = u_i^0 + \delta u_i$, $i = 1, 2$, and $p = p^0 + \delta p$. Then, discard all terms proportional to δ^2.

Exercise 12.8. Show that the GLS stabilized big linear form $B((u, p), (v, q)) = (\nabla u : \nabla v) - (p, \nabla \cdot v) + (\nabla \cdot u, q) + \delta(\nabla p, \nabla q)$, with $\delta > 0$ a stabilization parameter, is continuous and coercive on a suitable pair of velocity and pressure space (V, Q).

Chapter 13
Electromagnetics

Abstract In this chapter we briefly study finite elements for electromagnetic applications. We start off by recapitulating Maxwell's equations and look at some special cases of these, including the time harmonic and steady state case. Without further ado we then discretize the time harmonic electric wave equation using Nédélec edge elements. The computer implementation of the resulting finite element method is discussed in detail, and a simple application involving scattering from a perfectly conducting cylinder is studied numerically. Next, we introduce the magnetostatic potential equation as model problem. Using this equation we study the basic properties of the curl-curl operator $\nabla \times \nabla\times$, which frequently occurs in electromagnetic problems. In connection to this we also discuss the Helmholtz decomposition and its importance for characterizing the Hilbert space $H(\text{curl}; \Omega)$. The concept of a gauge is also discussed. For the mathematical analysis we reuse the theory of saddle-point problems and prove existence and uniqueness of the solution as well as derive both a priori and a posteriori error estimates.

13.1 Governing Equations

Electromagnetism is the area of science describing how stationary and moving charges affect each other. Loosely speaking stationary charges give rise to electric vector fields, whereas moving charges (i.e., current) give rise to magnetic vector fields. The electric vector field is commonly denoted by E, and the magnetic vector field by B. These fields extend infinitely in space and time and models the force felt by a small particle of charge q traveling with velocity v. This so-called Lorentz force is given by

$$F = q(E + v \times B) \tag{13.1}$$

and is used to define the E and B vectors at any given point and time.

M.G. Larson and F. Bengzon, *The Finite Element Method: Theory, Implementation, and Applications*, Texts in Computational Science and Engineering 10,
DOI 10.1007/978-3-642-33287-6_13,

13.1.1 Constitutive Equations

Certain media called dielectric materials have charges trapped inside their atomic structure. Under the influence of an externally applied electric field E these charges are slightly dislocated and form dipoles. The resulting so-called dipole moment polarizes the material. This polarization can be thought of as a frozen-in electric field, and is described by a polarization field P. Because it may be hard to know anything about P beforehand, it is desirable to try to separate the effects of free and bound charges within the material. This can sometimes be done by working with the electric displacement field D, defined by $D = \epsilon_0 E + P$ with ϵ_0 the electric permittivity of free space (i.e., vacuum). In linear, isotropic, homogeneous materials $P = \epsilon_0 \chi E$, where χ is the electric susceptibility indicating the degree of polarization of the dielectric material. This yields the constitutive relation

$$D = \epsilon E \tag{13.2}$$

where $\epsilon = \epsilon_0(1 + \chi)$. In free space ϵ is a positive scalar, but for a general linear, anisotropic, inhomogeneous medium in d dimensions it is a $d \times d$ positive definite tensor with entries that depend on space, but not time. For non-linear materials D is a more complicated function of E.

The D field has units charge per meters squared. It can be interpreted as a measure of the number of electrical force lines passing trough a given surface. Indeed, D is also called the electric flux field. By contrast, E has units of force per charge, and is called the electric field intensity.

Many electrically important materials are dielectrics. For example, many plastics used to make capacitors, and, also, nearly all insulators.

By analogy with the pair D and E, the magnetic flux density, or magnetic induction, field B, has a closely related cousin called the magnetic field strength H, which in linear materials obeys the constitutive relation

$$B = \mu H \tag{13.3}$$

where μ generally is a $d \times d$ positive definite tensor called the magnetic permeability.

We next study the laws of nature for the fields E, D, H, and B, that is, Gauss' laws, Ampère's law, and Faraday's law.

13.1.2 Gauss' Laws

Gauss' two laws are perhaps the simplest and most easily grasped fundamental physical principles governing electromagnetic phenomena.

The first Gauss' law states that the electrical flux through any closed surface $\partial\Omega$ is equal to the total net charge Q enclosed by the surface. In the language of mathematics this translates into

$$(D, n)_{\partial\Omega} = Q \tag{13.4}$$

where n is the outward unit normal on the surface. We can turn this surface integral into a volume integral by making use of the divergence theorem. In doing so, we have

$$(\nabla \cdot D, 1) = Q \tag{13.5}$$

By introducing a charge density q, defined by $(q, 1) = Q$, we have

$$(\nabla \cdot D - q, 1) = 0 \tag{13.6}$$

Finally, observing that the surface $\partial\Omega$ is arbitrary we conclude that

$$\nabla \cdot D = q \tag{13.7}$$

A similar argument for the magnetic flux leads to

$$\nabla \cdot B = 0 \tag{13.8}$$

The right hand side being zero since there are no magnetic monopoles (i.e., electrons). Because the magnetic field is solenoidal, all magnetic field lines must always be closed. This is Gauss' second law.

13.1.3 Ampère's Law

Ampère's law states that the flow of electric current gives rise to a magnetic field. More precisely, the circulation of the magnetic field H around a closed curve ∂S is equal to the amount of electric current J passing through the enclosed surface S. See Fig. 13.1. From the mathematical point of view this is equivalent to

$$(H, t)_{\partial S} = J \tag{13.9}$$

where t is the unit tangent vector on ∂S. To mold this integral into a differential we recall Stokes' theorem $(H, t)_{\partial S} = (\nabla \times H, n)_S$, with n the unit normal to S, from which it readily follows that

$$\nabla \times H = j \tag{13.10}$$

where j is the current density, defined by $(j, n)_S = J$,

The current density j can be written as the sum

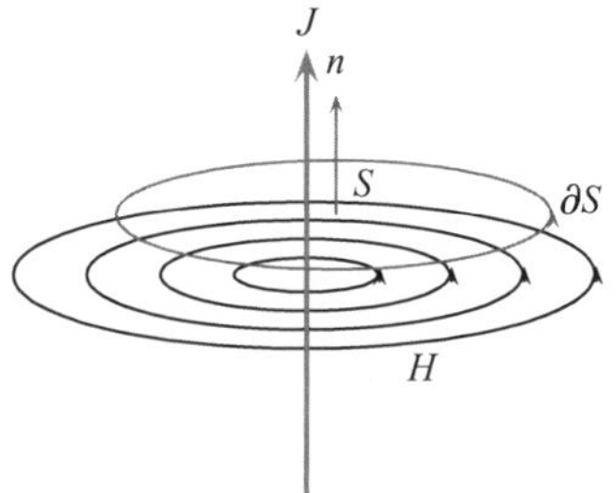

Fig. 13.1 Magnetic field H around a conducting wire with current J. Ampère's law states that the circulation of H around a closed curve ∂S equals the current through the enclosed surface S

$$j = j_a + \sigma E \tag{13.11}$$

where j_a is externally applied, or impressed, current, and where the so-called eddy current σE reflects the fact that the electric field E may itself give rise to current. The eddy current is a generalization of Ohm's famous law, which says that current equals resistance times voltage. This term is particularly important when dealing with metallic materials. The parameter σ is the electric conductivity of the material. In free space $\sigma = 0$.

Although (13.10) is valid at steady state, the current density j must be modified according to $j + \dot{D}$ to account for any varying electrical field. The so-called displacement current $\dot{D}$ is needed for consistency with conservation of charge. With this modification we arrive at

$$\nabla \times H = j + \dot{D} \tag{13.12}$$

13.1.4 Faraday's Law

Faraday's law says that a moving magnetic field generates a companion electrical field. In particular, the induced voltage in any closed circuit is equal to the time rate of change of the magnetic flux through the circuit. See Fig. 13.2. Mathematically, this is the same as

$$\nabla \times E = -\dot{B} \tag{13.13}$$

13.1.5 Maxwell's Equations

The collection of Gauss', Ampère's, and Faraday's laws are know as Maxwell's equations, and provide the fundamental mathematical model of electromagnetic interaction. Maxwell's equations are given by

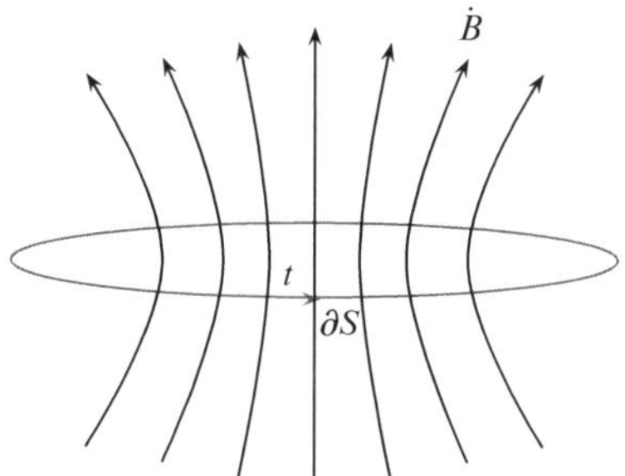

Fig. 13.2 An electric circuit ∂S in a time dependent magnetic field B. Faraday's law states that the induced voltage in the circuit equals the negative time rate of magnetic flux through the enclosed surface S, that is, $(E, t)_{\partial s} = -(\dot{B}, n)_S$

$$\nabla \times E = -\dot{B} \tag{13.14a}$$

$$\nabla \times H = j + \dot{D} \tag{13.14b}$$

$$\nabla \cdot D = q \tag{13.14c}$$

$$\nabla \cdot B = 0 \tag{13.14d}$$

By definition, current is moving charges, and therefore the current density j and the charge density q are related through a continuity equation, which takes the form

$$\nabla \cdot j = -\dot{q} \tag{13.15}$$

Of the four partial differential equations comprising Maxwell's equations, only three of them are independent, namely (13.14a), (13.14b), and (13.14c). Alternatively, (13.14a), and (13.14b) can be used together with the continuity equation (13.15) to obtain a well posed problem.

13.1.6 Initial- and Boundary Conditions

On the interface between two conducting media Maxwell's equations give rise to a set of continuity conditions for E, D, H, and B, given by

$$[E] \times n = 0 \tag{13.16}$$

$$[H] \times n = j_s \tag{13.17}$$

$$[D] \cdot n = q_s \tag{13.18}$$

$$[B] \cdot n = 0 \tag{13.19}$$

Here, the brackets $[\cdot]$ denote the jump across the boundary, and n is a unit normal. The letters j_s and q_s denote any external surface charge and surface current densities on the boundary surface, respectively.

These conditions mean that the tangential component of E is continuous across the interface, and that the difference of the tangential component of H is equal to the surface current density. By contrast, the difference of the normal components of D is equal to the surface charge density, and the normal components of B is continuous.

In the special, but common, case that the interface is a perfect conductor unable to sustain any electromagnetic fields the above interface conditions reduce to the boundary conditions

$$E \times n = 0 \tag{13.20}$$

$$H \times n = j_s \tag{13.21}$$

$$D \cdot n = q_s \tag{13.22}$$

$$B \cdot n = 0 \tag{13.23}$$

Of course, since the Maxwell's equations (13.14) are time dependent they must be supplemented by proper initial conditions to be solvable. In doing so, care must be take so that (13.14c) and (13.14d) are fulfilled by the initial conditions.

Finite element methods have only rarely been applied to the full set of Maxwell's equations without rewriting them in one way or another. The reason being either to adapt the equations to the application at hand or to limit the complexity or cost of the numerical method. We shall therefore next look at a few ways of simplifying Maxwell's equations under certain assumptions.

13.2 The Electric and Magnetic Wave Equations

To avoid working with two unknowns it is possible to eliminate either E (or D) or H (or B) from Maxwell's equations. The choice of unknown to eliminate is usually based on the boundary conditions prescribed. Elimination of H, say, is done by taking the curl of Faraday's law (13.14a) while differentiating Ampère's law (13.14b) with respect to time. This yields the two equations $\nabla \times \mu^{-1} \nabla \times E = -\nabla \times \dot{H}$ and $\nabla \times \dot{H} = \dot{j} + \epsilon \ddot{E}$, which can be combined to the single equation for E

$$\nabla \times \mu^{-1} \nabla \times E + \epsilon \ddot{E} = -\dot{j} \tag{13.24}$$

This is the so-called electric wave equation.

Taking the divergence of both sides of (13.24) and using the fact that the divergence of a curl is always zero, and the continuity equation (13.15), it is easy to show that a solution E also satisfies Gauss's law $\nabla \cdot D = q$. Thus, the electric wave equation is in a sense self contained.

There is a similar wave equation for H.

13.3 The Time Harmonic Assumption

Electrical devices are often powered by alternating current, which calls for a study of Maxwell's equations assuming periodically varying electric and magnetic fields of the form

$$E = \hat{E}e^{i\omega t}, \qquad H = \hat{H}e^{i\omega t} \tag{13.25}$$

where i is the imaginary unit with $i^2 = -1$, t is time, and $\omega > 0$ is a given frequency. The so-called phasors $\hat{E}$ and $\hat{H}$ are assumed to be independent of time, but complex-valued. Substituting this ansatz into Maxwell's equations, assuming j and q are also varying like $e^{i\omega t}$, we immediately obtain the time harmonic form of Maxwell's equations

$$\nabla \times \hat{E} = -i\omega\mu\hat{H} \tag{13.26a}$$

$$\nabla \times \hat{H} = \hat{j} + i\omega\epsilon\hat{E} \tag{13.26b}$$

$$\nabla \cdot \hat{B} = 0 \tag{13.26c}$$

$$\nabla \cdot \hat{D} = \hat{q} \tag{13.26d}$$

Here, since the time t has been eliminated by the time harmonic ansatz and replaced by the frequency ω we say that the equations are posed in the frequency domain, as opposed to the time domain.

13.4 Electro- and Magnetostatics

In the steady state all time derivatives vanish and Maxwell's four equations decouple into two systems with two equations each. One system governs the electric field E, (or D), whereas the other system governs the magnetic field H (or B). The electric system is given by

$$\nabla \times E = 0 \tag{13.27a}$$

$$\nabla \cdot D = q \tag{13.27b}$$

and the magnetic by

$$\nabla \times H = j \tag{13.28a}$$

$$\nabla \cdot B = 0 \tag{13.28b}$$

We remark that if the current density j contains the eddy current σE we may still get a coupling between the above two systems.

13.5 The Time Harmonic Electric Wave Equation

Substituting the time harmonic ansatz $E = \hat{E}e^{i\omega t}$ into the electric wave equation (13.24) assuming $j = j_a + \sigma E$ with $j_a = \hat{j}_a e^{i\omega t}$ we obtain the time harmonic electric wave equation

$$\nabla \times \mu^{-1}\nabla \times \hat{E} + (i\sigma\omega - \epsilon\omega^2)\hat{E} = -i\omega\hat{j}_a \tag{13.29}$$

for the phasor $\hat{E}$.

Perhaps needless to say, the solution $\hat{E}$ has wave characteristics.

In case the current density j is zero, (13.29) reduces to the so-called Maxwell eigenvalue problem

$$\nabla \times \mu^{-1}\nabla \times \hat{E} + \kappa^2\hat{E} = 0 \tag{13.30}$$

for $\hat{E}$ and $\kappa^2 = (i\sigma\omega - \epsilon\omega^2)$.

The time harmonic electric wave equation is arguably one of the most important equations in computational electromagnetics. There are two main reasons for this. First, it frequently occurs in real-world applications, such as antenna radiation or power loss in electrical motors, for instance. Second, it offers a very rich set of mathematical problems, both from the numerical and analytical point of view.

13.5.1 Boundary Conditions

There are two common types of boundary conditions for the time harmonic electric wave equation (13.29), namely:

- Dirichlet type boundary conditions of the form $n \times \hat{E} = \hat{E}_D$, with $\hat{E}_D = 0$ implying a perfect electric conductor.
- Neumann type boundary conditions of the form $\mu^{-1}(\nabla \times \hat{E}) \times n = -i\omega\hat{j}_N$, which is equivalent to $\hat{H} \times n = \hat{j}_N$. This boundary condition is typically used on antennas, since $\hat{j}_N$ can be interpreted as a surface current density. The homogeneous case $\hat{j}_N = 0$ implies a perfect magnetic conductor.

Other types boundary conditions include different kinds of so-called radiation boundary conditions, which loosely speaking says that the electric field $\hat{E}$ should vanish at infinity. These are the Silver-Müeller and Sommerfeld boundary conditions, which apply in infinite or unbounded domains. In principle, $\hat{E}$ is supposed to extend infinitely in space. However, in simulations, it is necessary to truncate the computational domain and introduce artificial boundary conditions. This leads to so-called absorbing boundary conditions (ABC). ABC s dampen the electrical waves near the boundary to avoid reflection of these back into the domain. The most famous type of ABC is called the perfectly matched layer (PML).

13.5.2 Green's Formula Revisited

The frequent occurrence of the curl $\nabla\times$ calls for special attention to this operator. Recall that the curl of a vector $v = [v_i]_{i=1}^d$ is defined differently depending on the space dimension d. For $d = 2$ the curl of v is the number $\nabla \times v = \partial_{x_1} v_2 - \partial_{x_2} v_1$. For $d = 3$ it is the vector $\nabla \times v = [\partial_{x_2} v_3 - \partial_{x_3} v_2, \partial_{x_3} v_1 - \partial_{x_1} v_3, \partial_{x_1} v_2 - \partial_{x_2} v_1]$. Also, the curl of a scalar s is sometimes defined as the vector $\nabla \times s = [\partial_{x_2} s, -\partial_{x_1} s]$.

With these definitions the following variant of Green's, or integration by parts, formula holds.

$$(\nabla \times u, v) = (u, \nabla \times v) - (u \times n, v)_{\partial\Omega} \tag{13.31}$$

The boundary term $-(u \times n, v)_{\partial\Omega}$ can also be written $(u, v \times n)_{\partial\Omega}$, or $(v \times u, n)_{\partial\Omega}$.

In this context we remark that if u and v are complex-valued functions, then we have to interpret the usual L^2 inner product (u, v) as $\int_\Omega u\bar{v}\,dx$, with $\bar{v}$ the complex conjugate of v. As a consequence, the L^2-norm $\|v\|^2 = (v, v) = \int_\Omega v\bar{v}\,dx$. Recall that the complex conjugate of a complex number $z = a + ib$ is $\bar{z} = a - ib$.

13.5.3 Weak Form

Let us derive the weak form of the time harmonic electric wave equation. To this end, we assume that boundary conditions are given by $\hat{E} \times n = 0$. Further, to ease the notation, let $\kappa^2 = (i\sigma\omega - \epsilon\omega^2)$, and $\hat{f} = -i\omega\hat{j}$. Then, multiplying, (13.29) with a sufficiently smooth complex-valued vector v, satisfying $v \times n = 0$ on the boundary, and integrating by parts, we have

$$\int_\Omega \hat{f}\cdot\bar{v}\,dx = \int_\Omega (\nabla \times \mu^{-1}\nabla \times \hat{E})\cdot\bar{v}\,dx + \kappa^2 \int_\Omega \hat{E}\cdot\bar{v}\,dx \tag{13.32}$$

$$= \int_\Omega \mu^{-1}(\nabla \times \hat{E})\cdot(\nabla \times \bar{v})\,dx + \int_{\partial\Omega} \mu^{-1}(\nabla \times \hat{E})\cdot(\bar{v} \times n)\,ds + \kappa^2 \int_\Omega \hat{E}\cdot\bar{v}\,dx \tag{13.33}$$

$$= \int_\Omega \mu^{-1}(\nabla \times \hat{E})\cdot(\nabla \times \bar{v})\,dx + \kappa^2 \int_\Omega \hat{E}\cdot\bar{v}\,dx \tag{13.34}$$

For the occurring integrals to be well defined it suffice that the integrands $\hat{E}$ and v have bounded norm of their curl in $L^2(\Omega)$, that is, belong to the space $H(\text{curl};\Omega)$, defined by

$$H(\text{curl};\Omega) = \{v \in [L^2(\Omega)]^d : \nabla \times v \in [L^2(\Omega)]^d\} \tag{13.35}$$

Further, to satisfy the Dirichlet boundary condition $\hat{E}$ and v also need to reside in the homogeneous space $H_0(\text{curl};\Omega) = \{v \in H(\text{curl};\Omega) : n \times v|_{\partial\Omega} = 0\}$.

Both $H(\text{curl}; \Omega)$ and $H_0(\text{curl}; \Omega)$ are Hilbert spaces. Their norm is the same and defined by

$$\|v\|_{H(\text{curl};\Omega)} = \|v\| + \|\nabla \times v\| \tag{13.36}$$

Thus, the weak form of (13.29) reads: find $E \in H_0(\text{curl}; \Omega)$ such that

$$a(\hat{E}, v) = l(v), \quad \forall v \in H_0(\text{curl}; \Omega) \tag{13.37}$$

where we have introduced the linear forms

$$a(\hat{E}, v) = \int_\Omega \mu^{-1}(\nabla \times \hat{E}) \cdot (\nabla \times \bar{v})\, dx + \kappa^2 \int_\Omega \hat{E} \cdot \bar{v}\, dx \tag{13.38}$$

$$l(v) = \int_\Omega \hat{f} \cdot \bar{v}\, dx \tag{13.39}$$

Formally, $\hat{E}$ should also satisfy Gauss' law, $\nabla \cdot (\epsilon \hat{E}) = \hat{q}$, but this is already built into (13.37). By combining Gauss's law with the continuity equation $\nabla \cdot \hat{j} = -i\omega\hat{q}$ and Ohm's law $\hat{j} = \hat{j}_a + \sigma \hat{E}$ we obtain $\nabla \cdot ((i\omega\epsilon - \sigma)\hat{E}) = \nabla \cdot \hat{j}_a$. But this is also what we get by setting $v = \nabla\phi$ with $\phi \in H_0^1(\Omega)$ and integrating by parts in (13.37). Thus, Gauss' law is automatically satisfied in a weak sense. In the following we shall assume that σ and ϵ are constants and that $\nabla \cdot \hat{j}_a = \hat{q} = 0$, which implies that $\nabla \cdot \hat{E} = 0$.

The question of existence of a solution $\hat{E}$ to (13.37) is a quite complicated matter. This has to do with the fact that the principal part of the linear form $a(\cdot, \cdot)$ has a large null space, and that $a(\cdot, \cdot)$ is indefinite for large values of ω. If $\sigma > 0$ then it can be shown that $a(\cdot, \cdot)$ is coercive and a variant of the Lax-Milgram lemma for complex-valued spaces can be used to show that $\hat{E}$ exist. However, if $\sigma = 0$ then a result from functional analysis called the Fredholm alternative must be used to assert existence of $\hat{E}$. Loosely speaking, Fredholm says that the operator $(I + \lambda T)$ is invertible up to a set of singular values λ provided that the operator T is compact. Now, in strong form (13.37) can be written $(I + \kappa^2 K^{-1})\hat{E} = K^{-1}\hat{f}$ with $K = \nabla \times \mu^{-1} \nabla \times$ so we are done by showing that $T = K^{-1}$ is compact. Suffice to say that this is indeed possible to do with some work. The singular values $\lambda = \kappa^2$ are called the Maxwell eigenvalues, and can formally be found by solving (13.30). In real-world applications, care must always be taken so that κ is not chosen as, or close to, a Maxwell eigenvalue, since this can cause resonance effects.

13.5.4 Finite Element Approximation

To design a robust and accurate finite element method we need to select a finite element which has good ability to approximate vectors in $H_0(\text{curl}; \Omega)$. In doing so, the natural choice is to use a member of the Nédélec finite element family. Recall that on a triangle mesh $\mathcal{K}$ the simplest of these consists of vectors which on every element K are linear in each component and tangential continuous across the element edges. Indeed, the tangent continuity is a characteristic feature of $H_0(\text{curl}; \Omega)$. The formal definition of the Nédélec space of lowest order is given by

$$V_h = \{v \in H(\text{curl}; \Omega) : v|_K = \begin{bmatrix} a_1 \\ a_2 \end{bmatrix} + b \begin{bmatrix} x_2 \\ -x_1 \end{bmatrix}, \forall K \in \mathcal{K}\} \tag{13.40}$$

Because we do not have a full linear polynomial in each component of v, the degree of the Nédélec element is somewhere between 0 and 1.

With this choice of discrete space V_h the finite element approximation of (13.37) takes the form: find $\hat{E}_h \in V_h$ such that

$$a(\hat{E}_h, v) = l(v), \quad \forall v \in V_h \tag{13.41}$$

A basis for V_h is given by the set of Nédélec shape functions $\{S_i^{ND}\}_{i=1}^{n_e}$, with n_e the number of triangle edges in the mesh. Using this basis we can write the unknown electric field $\hat{E}_h = \sum_{j=1}^{n_e} \xi_j S_j^{ND}$ for some coefficients ξ_j to be determined. As usual, these are given by the entries of the $n_e \times 1$ solution vector ξ to the linear system

$$A\xi = b \tag{13.42}$$

where the entries of the $n_e \times n_e$ stiffness matrix A and the $n_e \times 1$ load vector b are given by

$$A_{ij} = a(S_j^{ND}, S_i^{ND}), \quad i, j = 1, 2, \dots, n_e \tag{13.43}$$

$$b_i = l(S_i^{ND}), \quad i = 1, 2, \dots, n_e \tag{13.44}$$

In this context we point out that there are actually a good number of reasons for using edge elements to discretize (13.37), besides the fact that the Nédélec space is a subspace of $H(\text{curl}; \Omega)$. First, Nédélec vectors are divergence free within the interior of each element, so there is reason to believe that the equation $\nabla \cdot \hat{E} = 0$ is very nearly satisfied by $\hat{E}_h$. However, this does not imply that $\hat{E}_h$ is globally divergence free, since its normal component may jump across element edges. Second, if the material is such that ϵ is discontinuous, it is known that the tangent component of $\hat{E}$ is continuous, whereas the normal component is discontinuous. Indeed, this feature is possible to capture with edge elements, which have tangent, but not normal continuity. Third, edge elements fits the boundary conditions $n \times \hat{E} = \hat{E}_D$ perfectly in the sense that the degrees of freedom are exactly the curl of the shape functions for the edge elements.

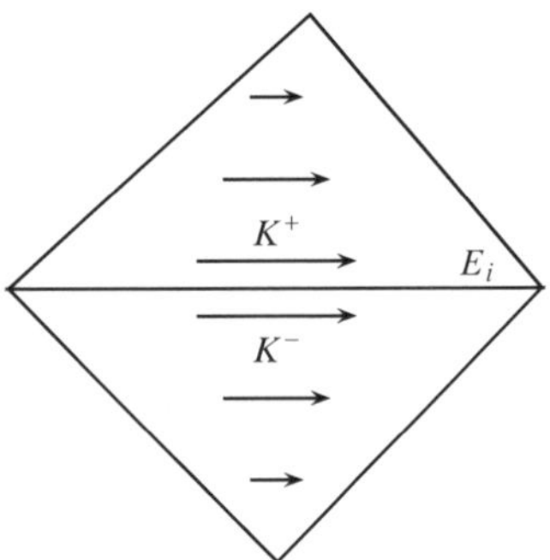

Fig. 13.3 The vector fields making up the Nédélec shape function S_i^{ND} must point in the same direction on edge E_i

13.5.5 Computer Implementation

The implementation of edge elements is essentially the same as for any other solver. The linear system resulting from the finite element discretization is assembled by adding local contributions from each element to the global stiffness matrix A and load vector b. In doing so, we must remember that the degrees of freedom for the edge elements are located on the element edges, which must be numbered. However, the edges must also be oriented in order for the shape functions to be truly tangent continuous. Indeed, the two vector fields making up shape function S_i^{ND} on the two neighbouring elements K^+ and K^- sharing edge E_i must point in the same direction. An illustration of this is shown in Fig. 13.3.

The edge orientation is easy as it amounts to assigning a sign σ to each edge on each element, signifying if the shapes need to be negated or not. On an element, if the element neighbor on the other side of an edge has a larger number than the element, then the edge has sign $\sigma = -1$, and otherwise $\sigma = +1$. This can be concisely coded. The element neighbours are found by calling `neighbors=Tri2Tri(p,t)`. The signs are then found by comparing all element numbers against the numbers of their neighbours. This can be done with the somewhat cryptic code

```
signs=2*((neighbors'<[1:nt; 1:nt; 1:nt])-1/2)
```

where `nt=size(t,2)` is the number of elements. The sign of edge `i` in element `j` is given by `signs(i,j)`. Each shape function must be multiplied by its sign prior to usage to be correct.

To compute the 3×1 element load vector b^K we note that, by definition, $S_i^{ND} = \sigma_i |E_i|(\varphi_j \nabla\varphi_k - \varphi_k \nabla\varphi_j) = |E_i|(\varphi_j [b_k, c_k]^T - \varphi_k [b_j, c_j]^T)$, where σ_i denotes the sign of edge E_i in element K. Also, the integral of any hat function φ_i over K is $|K|/3$. Assuming tacitly that the vector $\hat{f}$ is constant over K, we thus have the entries $b_i^K = (\hat{f}, S_i^{ND})_K = \sigma_i \hat{f} \cdot [b_k - b_j, c_k - c_j]^T |K|/3$, with cyclic permutation of the indices $\{i, j, k\}$ over $\{1, 2, 3\}$.

Continuing, entry A_{ij}^K of the 3×3 element stiffness matrix A^K is the sum of the curl-curl term $W_{ij}^K = (\mu^{-1}\nabla \times S_j^{ND}, \nabla \times S_i^{ND})_K$ and the mass term $M_{ij}^K = \kappa^2(S_j^{ND}, S_i^{ND})_K$. Recalling that $\nabla \times S_i^{ND} = \sigma_i|E_i|/|K|$ we compute without effort $W_{ij}^K = \mu^{-1}\sigma_i\sigma_j|E_i||E_j|/|K|$. By contrast, M_{ij}^K requires some labor to compute. To this end, let us first define $f_{ij} = b_ib_j + c_ic_j$. Then, recalling that $(\varphi_i, \varphi_j)_K = (\delta_{ij} + 1)|K|/12$ with δ_{ij} the Kronecker symbol, we have for the mass with, say, $i = j = 1$,

$$M_{11}^K = \kappa^2(S_1^{ND}, S_1^{ND})_K \tag{13.45}$$

$$= \kappa^2\sigma_1^2|E_1|^2(\varphi_2\nabla\varphi_3 - \varphi_3\nabla\varphi_2, \varphi_2\nabla\varphi_3 - \varphi_3\nabla\varphi_2)_K \tag{13.46}$$

$$= \kappa^2\sigma_1^2|E_1|^2(\varphi_2^2|\nabla\varphi_3|^2 - 2\varphi_2\varphi_3\nabla\varphi_3 \cdot \nabla\varphi_2 + \varphi_3^2|\nabla\varphi_2|^2, 1)_K \tag{13.47}$$

$$= \kappa^2\sigma_1^2|E_1|^2(f_{33} - f_{23} + f_{22})|K|/6 \tag{13.48}$$

The other masses can be computed similarly. In doing so, we have

$$M_{12}^K = \kappa^2\sigma_1\sigma_2|E_1||E_2|(f_{31} - f_{33} - 2f_{21} + f_{23})|K|/12 \tag{13.49}$$

$$M_{13}^K = \kappa^2\sigma_1\sigma_3|E_1||E_3|(f_{32} - 2f_{31} - f_{22} + f_{21})|K|/12 \tag{13.50}$$

$$M_{22}^K = \kappa^2\sigma_2^2|E_2|^2(f_{11} - f_{13} + f_{33})|K|/6 \tag{13.51}$$

$$M_{23}^K = \kappa^2\sigma_2\sigma_3|E_1||E_3|(f_{12} - f_{11} - 2f_{32} + f_{31})|K|/12 \tag{13.52}$$

$$M_{33}^K = \kappa^2\sigma_3^2|E_3|^2(f_{22} - f_{12} + f_{11})|K|/6 \tag{13.53}$$

Obviously, M^K is symmetric.

A routine for computing A^K and b^K is listed below.

```
function [AK,bK] = RotRot(x,y,sigmas,mu,kappa,fhat);
[area,b,c]=HatGradients(x,y);
f=b*b'+c*c';
len=zeros(3,1); % edge lengths
len(1)=sqrt((x(3)-x(2))^2+(y(3)-y(2))^2);
len(2)=sqrt((x(1)-x(3))^2+(y(1)-y(3))^2);
len(3)=sqrt((x(2)-x(1))^2+(y(2)-y(1))^2);
len=len.*sigmas; % edge lengths times signs
m11=(f(3,3)-f(2,3)+f(2,2))/6;
m22=(f(1,1)-f(1,3)+f(3,3))/6;
m33=(f(2,2)-f(1,2)+f(1,1))/6;
m12=(f(3,1)-f(3,3)-2*f(2,1)+f(2,3))/12;
m13=(f(3,2)-2*f(3,1)-f(2,2)+f(2,1))/12;
m23=(f(1,2)-f(1,1)-2*f(3,2)+f(3,1))/12;
MK=[m11 m12 m13; m12 m22 m23; m13 m23 m33]*area;
WK=ones(3)/area;
```

```
AK=(WK/mu+kappa^2*MK).*(len*len');
bK=zeros(3,1);
bK(1)=dot(fhat,[b(3); c(3)]-[b(2); c(2)]);
bK(2)=dot(fhat,[b(1); c(1)]-[b(3); c(3)]);
bK(3)=dot(fhat,[b(2); c(2)]-[b(1); c(1)]);
bK=bK.*len*area/3;
```

Input to this routine is the triangle vertex coordinates `x` and `y`, the signs of the triangle edges `signs`, the material parameters `mu` and `kappa`, and the current density vector `fhat`. Output is the element stiffness matrix `AK` and the element load vector `bK`.

13.5.5.1 Electromagnetic Scattering from a Cylinder

Scattering simulations are one of the most common tasks in computational electromagnetics. It finds applications in radar and remote sensing for instance. The basic problem setup is very simple. An incident electric wave hits a conducting object and is scattered. The task is to compute the scattered electric field. By measuring the amount of electric power reflected back by the scattered field (i.e., wave), the radar cross section of the scatterer can be determined.

Let us write a code to simulate the electric field scattered by a cylinder. This amounts to solving the time harmonic electric wave equation for the scattered field E^s within the square $S = \{x : -5 \leq x_1, x_2 \leq 5\}$ with the cut out cylinder $C = \{x : x_1^2 + x_2^2 < 1\}$. The cylinder C is perfectly conducting, and the surrounding cavity $\Omega = S \setminus C$ is free space, with $\epsilon = \mu = 1$. The current density $j = 0$.

It is customary to write the total electric field as the sum of the scattered and incident field, viz.

$$\hat{E} = \hat{E}^i + \hat{E}^s \tag{13.54}$$

Inserting this ansatz into (13.24) yields

$$\nabla \times \mu^{-1} \nabla \times (E^i + E^s) + \kappa^2 (E^i + E^s) = 0 \tag{13.55}$$

As $\hat{E}^i$ is known only $\hat{E}^s$ needs to be solved for. Let us choose $\hat{E}^i$ as

$$\hat{E}^i = \begin{bmatrix} 1 \\ 0 \end{bmatrix} e^{-i\omega x_2} \tag{13.56}$$

with cyclic frequency $\omega = 2\pi$. This represents a plane wave with unit wave length traveling in the x_2-direction. Moreover, with our choice of material parameters ϵ, μ, and $\sigma = 0$, it is a simple matter to check that E^i satisfies $\nabla \times \mu^{-1} \nabla \times E^i + \kappa^2 E^i = 0$, which reduces (13.55) to simply

$$\nabla \times \mu^{-1} \nabla \times E^s + \kappa^2 E^s = 0 \tag{13.57}$$

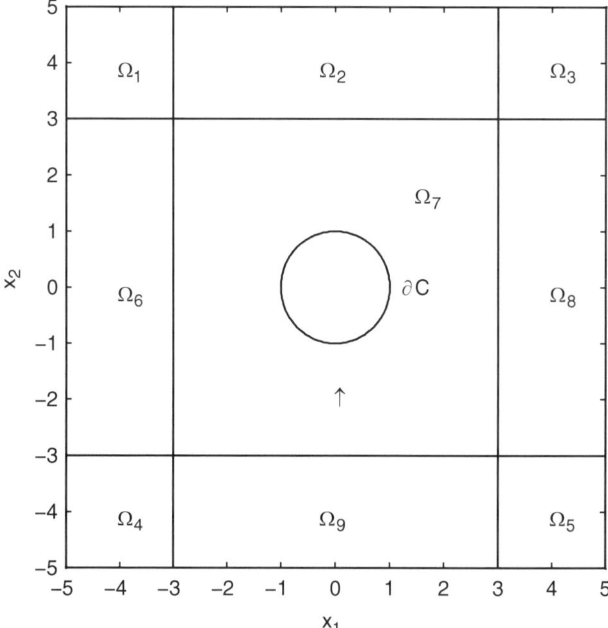

Fig. 13.4 Geometry with scattering cylinder and PML, and subdomain numbers

The boundary condition for E^s on the perfect conducting cylinder surface ∂C is $\hat{E} \times n = 0$, which implies

$$\hat{E}^s \times n = -\hat{E}^i \times n, \quad \text{on } \partial C \tag{13.58}$$

To avoid reflection of the scattered wave at the outer boundary of the square back into the cavity we use an absorbing boundary condition in the form of a perfectly matched layer (PML). This amounts to increasing the conductivity σ in a strip of width, say, 2, near the boundary. This non-physical conductivity is assumed to have a polynomial profile in the absorbing layer, being zero in the cavity. Indeed, referring to Fig. 13.4, which shows the subdomain numbering from `pdeinit`, we choose

$$\sigma = 2^{-n}\sigma_0 \begin{cases} (|x_1| - 3)^n, \text{ if } x \in \Omega_6 \cup \Omega_8 \\ (|x_2| - 3)^n, \text{ if } x \in \Omega_2 \cup \Omega_9 \\ (|x_1| - 3)^n + (|x_2| - 3)^n, \text{ if } x \in \Omega_1 \cup \Omega_3 \cup \Omega_4 \cup \Omega_5 \end{cases} \tag{13.59}$$

where the parameter $n = 4$ is the degree of absorption, and $\sigma_0 = 1$ is the maximum conductivity in the absorbing layer. The PML region is terminated with a perfectly conducting boundary condition $\hat{E}^s \times n = 0$. Of course, due to the PML it is only within the cavity subdomain Ω_7 that we can expect the solution $\hat{E}^s$ to be accurate.

The main routine is listed below.

```
function MaxwellSolver()
dl=Scatterg();
[p,dummy,t]=initmesh(dl,'hmax',0.1);
neighbors = Tri2Tri(p,t);
e = Tri2Edge(p,t);
ne = max(e(:));
nt = size(t,2);
A = sparse(ne,ne);
b = zeros(ne,1);
omega = 2*pi/1;
mu = 1;
epsilon = 1;
signs=2*((neighbors'<[1:nt; 1:nt; 1:nt])-1/2)
for i=1:size(t,2);
    nodes = t(1:3,i);
    x = p(1,nodes);
    y = p(2,nodes);
    edges = e(i,:);
    sigma = 0;
    sd=t(4,i);
    if sd==7 % cavity
        %
    elseif sd==2 | sd==9 % up down pml
        sigma=5*(abs(mean(y))-3)^4;
    elseif sd==6 | sd==8 % left right pml
        sigma=5*(abs(mean(x))-3)^4;
    else
        sigma=5*(abs(mean(x))-3)^4+(abs(mean(y))-3)^4;
    end
    kappa=sqrt(sqrt(-1)*sigma*omega-epsilon*omega^2);
    [AE,FE] = RotRot(x,y,signs(:,i),mu,kappa,[0,0]);
    A(edges,edges) = A(edges,edges) + AE;
    b(edges) = b(edges) + FE;
end
[xmid,ymid,edges] = EdgeMidPoints(p,e,t);
fixed=[]; % fixed nodes
gvals=[];
hold on
for i=1:ne % loop over edges
  r=edges(i); % edge or node number
  x=xmid(i); % node x-coordinate
  y=ymid(i); %      y-
  if (sqrt(x^2+y^2)<1.001) % cylinder
```

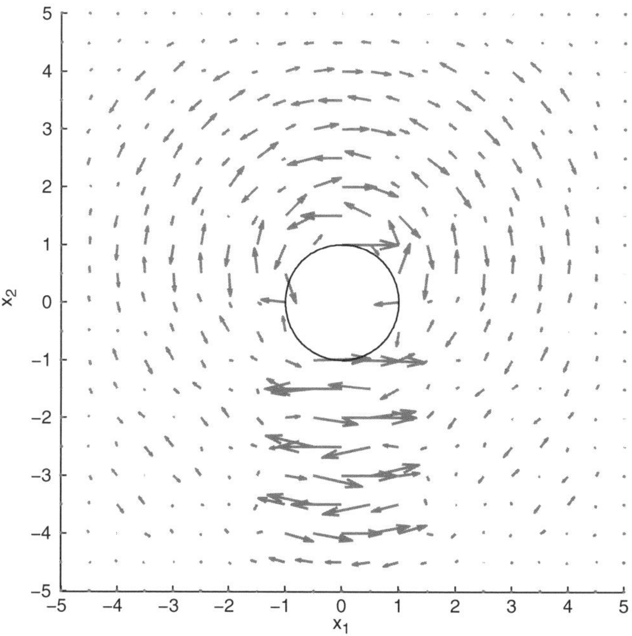

Fig. 13.5 Glyphs of the scattered electric field $\Re \hat{E}_h^s$

```
        normal = -[x; y]/sqrt(x^2+y^2);
        fixed = [fixed; r];
        gvals = [gvals; normal(2)*exp(omega*y*sqrt(-1))];
    end
    if (abs(x)>4.999 | abs(y)>4.999) % pml
        fixed = [fixed; r];
        gvals = [gvals; 0];
    end
end
free=setdiff([1:ne],fixed);
b=b(free)-A(free,fixed)*gvals;
A=A(free,free);
xi=zeros(ne,1);
xi(fixed)=gvals;
xi(free)=A\b;
```

Running the code we get the scattered electric field $\hat{E}_h^s$ shown in Fig. 13.5. Figures 13.6 and 13.7 show the components of $\hat{E}_h^s$. Only the real part of $\hat{E}_h^s$ is plotted. Note the unit wave length, the complex interference pattern, and the damping of the wave in the PML region.

Fig. 13.6 Iso-contours of the component $\Re \hat{E}^s_{h,1}$

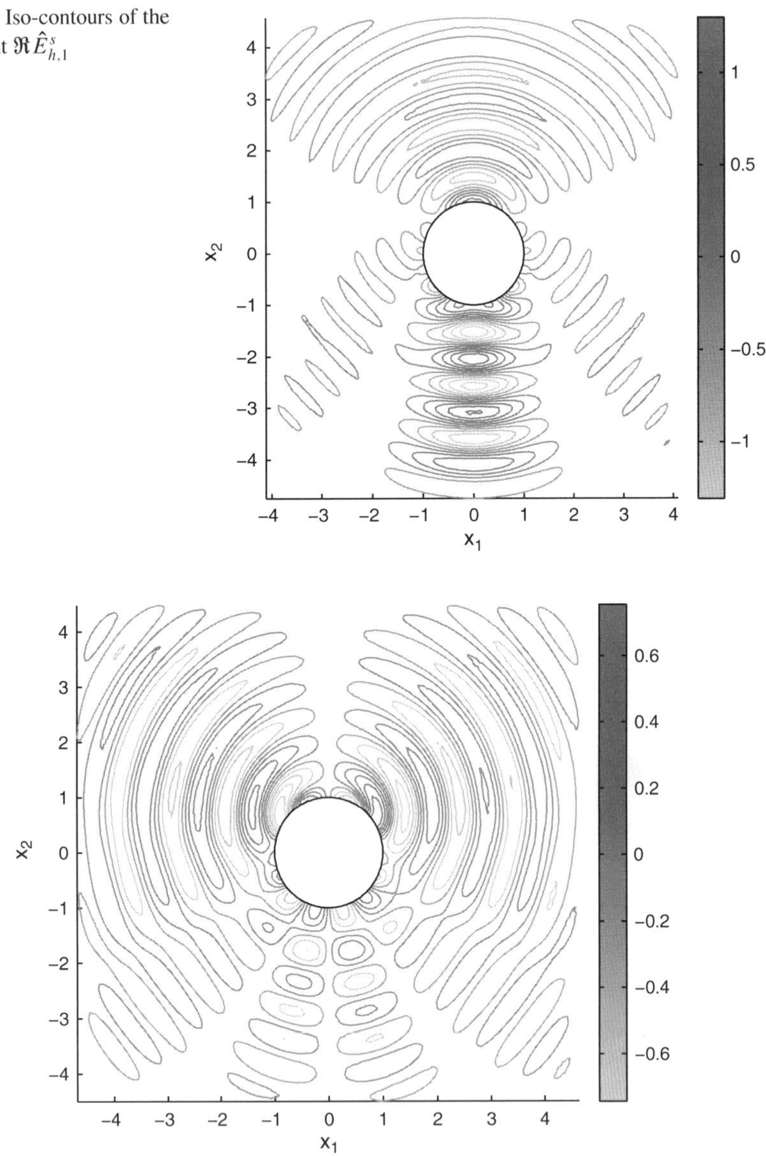

Fig. 13.7 Iso-contours of the component $\Re \hat{E}^s_{h,2}$

13.6 The Magnetostatic Potential Equation

Having familiarized ourselves somewhat with the practical aspects of computational electromagnetics, let us now turn to study this area from a more theoretical point of view.

Returning to Maxwell's equations, in the magnetostatic case, since $\nabla \cdot B = 0$, there exist a vector potential A such that

$$B = \nabla \times A \tag{13.60}$$

and, thus,

$$H = \mu^{-1}\nabla \times A \tag{13.61}$$

Substituting this into Ampère's law $\nabla \times H = j$ yields

$$\nabla \times \mu^{-1}\nabla \times A = j \tag{13.62}$$

which is called the magnetostatic potential equation and is the fundamental equation of magnetostatics. For simplicity we shall in the following assume that the magnetic permittivity $\mu > 0$ is constant.

We observe that depending on if the space dimension d is 2 or 3, A is either a scalar or a vector. This is a consequence of the definition of the curl operator. The two dimensional case also follows from the three dimensional one by assuming no dependency of any quantity on, say, x_3, and $A = (0, 0, A_3)$. In this case, we then have $B = (B_1, B_2, 0)$ and $j = (0, 0, j_3)$.

Expressing B as the curl of A automatically implies $\nabla \cdot B = 0$, since the divergence of a curl is always zero.

Although the potential A is enough to uniquely specify the magnetic field B, the potential equation (13.62) itself is not enough to uniquely specify the potential A, not even with appropriate boundary conditions. This is a consequence of the fact that both the curl and divergence must be specified in order to uniquely specify a vector field. Therefore, it is necessary to impose a constraint, a so-called gauge, on A. The most common is the Coulomb gauge

$$\nabla \cdot A = 0 \tag{13.63}$$

With the imposition of the gauge A is determined uniquely up to a constant, which should be handled by the boundary conditions.

Using the vector identity $\nabla \times \nabla \times A = -\Delta A + \nabla(\nabla \cdot A)$ it is possible to rewrite (13.62) as $-\mu^{-1}\Delta A = j$. The rationale is that it decouples the original vector equation into a set of scalar equations for the components A_i of A. This is particularly useful in two dimensions, since it reduces the magnetostatic potential equation to a Poisson equation for A_3. In three dimensions, however, this rewriting is not so advantageous, since the boundary conditions are awkward and still introduce couplings between the components A_i.

In the following we shall mainly study the magnetostatic potential equation in the three dimensional case, as the two dimensional case has already been dealt with considering Poisson's equation.

13.6.1 Boundary Conditions

There are two common types of boundary conditions for the magnetic potential equation (13.62), namely:

- Dirichlet type boundary conditions of the form $A \times n = A_D$.
- Neumann type boundary conditions of the form $\mu^{-1}(\nabla \times A) \times n = j_N$, which is equivalent to $B \times n = j_N$.

13.6.2 Weak Form

To analyze the magnetostatic potential equation (13.62) we write it in weak form. To this end, let us assume that the boundary conditions are given by $A \times n = 0$. For uniqueness, we require A to satisfy the Coulomb gauge $\nabla \cdot A = 0$.

Multiplying $j = \nabla \times \nabla \times A$ with a vector $v \in H_0(\text{curl}; \Omega)$ and integrating by parts yields

$$(j, v) = (\mu^{-1} \nabla \times A, \nabla \times v) \tag{13.64}$$

Similarly, multiplying $0 = \nabla \cdot A$ by a function q in, say, $L^2(\Omega)$ yields

$$0 = (\nabla \cdot A, q) \tag{13.65}$$

or, if we assume q to be more regular, say, in $H_0^1(\Omega)$,

$$0 = (\nabla \cdot A, q) = (n \cdot A, q)_{\partial\Omega} - (A, \nabla q) = -(A, \nabla q) \tag{13.66}$$

Contemplating this we have, at least, two ways to enforce $\nabla \cdot A = 0$. One way is to choose $q = \nabla \cdot v$ in (13.65) and add $\nu(\nabla \cdot A, \nabla \cdot v)$, with $\nu > 0$ a (big) constant, to (13.64). The larger the constant ν, the more penalization on any divergence of A. Moreover, the form $(\mu^{-1}\nabla\times A, \nabla\times v)+\nu(\nabla\cdot A, \nabla\cdot v)$ is coercive on $V = [H^1(\Omega)]^d$, and can therefore be discretized by standard (i.e., nodal) finite elements. Indeed, the resulting finite element method works well as long as the domain is convex. However, if the domain is non-convex with re-entrant corners, it turns out that the solution A is so singular that it can not be approximated by continuous piecewise polynomials. In fact, it can be shown that A does not even belong to V. Thus, the weak form is posed on a faulty space V, and the finite element approximation is unable to converge. This clearly limits the practical use of this particular finite element method.

Another way to obtain a divergence free magnetic potential A is to use (13.66) in combination with a dummy Lagrangian multiplier Λ. This leads to the following weak form of (13.64): find $(A, \Lambda) \in H_0(\text{curl}; \Omega) \times H_0^1(\Omega)$ such that

$$a(A, v) + b(v, \Lambda) = l(v), \quad \forall v \in V \tag{13.67a}$$

$$b(A, q) = 0, \quad \forall q \in Q \tag{13.67b}$$

where we have introduced the linear forms

$$a(u, v) = (\mu^{-1}\nabla \times A, \nabla \times v) \tag{13.68}$$

$$b(A, q) = -(A, \nabla q) \tag{13.69}$$

$$l(v) = (j, v) \tag{13.70}$$

We recognize (13.67) as a saddle-point problem, similar to the Stokes system. Indeed, the magnetostatic potential equation and the Stokes system share many features, which is reflected by the fact that the mathematical analysis of them is the same in many aspects. For example, as we shall see, existence and uniqueness of a solution can be shown similarly. However, since the involved spaces and linear forms are different, many details differ.

In the following, for ease of notation, let us denote $V = H_0(\text{curl}; \Omega)$, and $Q = H_0^1(\Omega)$.

In compact form (13.67) can be written: find $(A, \Lambda) \in V \times Q$ such that

$$B((A, \Lambda), (v, q)) = F((v, q)), \quad \forall (v, q) \in V \times Q \tag{13.71}$$

where we have introduced the big linear forms

$$B((A, \Lambda), (v, q)) = a(u, v) + b(v, \Lambda) + b(A, q) \tag{13.72}$$

$$F((v, q)) = l(v) \tag{13.73}$$

13.6.3 The Helmholtz Decomposition

When the magnetic potential A itself is not of prime interest, but rather its curl, that is, the magnetic flux $B = \nabla \times A$, yet another weak form than (13.67) is sometimes used. This alternative, or reduced, weak form relies on the following result from functional analysis called the Helmholtz decomposition.

Theorem 13.1 (Helmholtz Decomposition). *Any vector $v \in H_0(curl; \Omega)$ can be written as*

$$v = z + \nabla \phi \tag{13.74}$$

where $z \in Z = \{v \in H_0(curl; \Omega) : (v, \nabla q) = 0, \ \forall q \in H_0^1(\Omega)\}$ is divergence free, and $\phi \in H_0^1(\Omega)$.

We note that, by definition, z and $\nabla\phi$ are L^2 orthogonal. We also note that the space Z coincides with the null space of the bilinear form $b(\cdot,\cdot)$. It can be shown that Z is a Hilbert space under the norm $\|\cdot\|_{H(\text{curl};\Omega)}$.

From the abstract point of view, the Helmholtz decomposition allows for a simple characterization of the space $H_0(\text{curl};\Omega)$ in the sense that it can be written as the direct sum

$$H_0(\text{curl};\Omega) = Z \oplus \nabla H_0^1(\Omega) \tag{13.75}$$

We remark that there are many variants of the Helmholtz decomposition. For example, $H(\text{curl};\Omega)$ can be written as the direct sum $H(\text{curl};\Omega) = Z \oplus \nabla H^1(\Omega)$.

But let us return to the derivation of a reduced weak form of the magnetostatic potential equation. Using the Helmholtz decomposition we can write A as $A = A_0 + \nabla\phi$ with $A_0 \in Z$ and $\phi \in H_0^1(\Omega)$. In doing so, we make the key observation that, due to the vector identity $\nabla\times(\nabla\phi) = 0$, only A_0 is required to compute B from A. To see this note that $B = \nabla\times A = \nabla\times(A_0+\nabla\phi) = \nabla\times A_0$. Therefore, let us try to eliminate ϕ all together. To this end, we substitute the Helmholtz decomposition of A into the bilinear form $a(\cdot,\cdot)$, which yields

$$a(A,v) = a(A_0 + \nabla\phi, v_0 + \nabla\eta) \tag{13.76}$$

$$= (\mu^{-1}\nabla\times(A_0 + \nabla\phi), \nabla\times(v_0 + \nabla\eta)) \tag{13.77}$$

$$= (\mu^{-1}\nabla\times A_0, \nabla\times v_0) \tag{13.78}$$

Here, we have made a Helmholtz decomposition of also the test function $v = v_0 + \nabla\eta$, with $v_0 \in Z$ and $\eta \in H_0^1(\Omega)$. Continuing similarly with the linear form $l(\cdot)$ we have

$$l(v) = (j, v_0 + \nabla\eta) \tag{13.79}$$

$$= (j, v_0) - (\nabla\cdot j, \eta) + (n\cdot j, \eta)_{\partial\Omega} \tag{13.80}$$

$$= (j, v_0) - (\nabla\cdot j, \eta) \tag{13.81}$$

where we have used integration by parts and that η vanish on the boundary. Now, assuming tacitly that

$$\nabla\cdot j = 0 \tag{13.82}$$

we are left with

$$l(v) = (j, v_0) \tag{13.83}$$

Under the assumption of a divergence free current density j, we observe that neither the trial function ϕ nor its test function η lingers in any of the linear forms

$a(\cdot, \cdot)$ and $l(\cdot)$. Thus, we have managed to eliminate ϕ from the equation as we wanted. As a consequence, we end up with the reduced weak form: find $A_0 \in Z$ such that

$$a(A_0, v_0) = l(v_0), \quad \forall v_0 \in Z \tag{13.84}$$

An advantage of the reduced weak form over the saddle-point problem is that it does not include the Lagrangian multiplier Λ, but a drawback is that it is posed on the space Z of divergence free vectors, which is difficult to discretize numerically.

We remark that the requirement $\nabla \cdot j = 0$ is not necessary for the saddle-point problem (13.67) to be well posed. However, if j is divergence free, then the Lagrangian multiplier Λ vanish for the exact solution A.

13.6.4 Existence and Uniqueness of the Solution

Because the weak form (13.67) is a saddle-point problem, existence and uniqueness of the solution must be deduced from Brezzi's Theorem 12.2. To this end, we need to show that the bilinear form $a(\cdot, \cdot)$ is coercive and continuous on the null space Z of $b(\cdot, \cdot)$, that the linear form $l(\cdot)$ is continuous on Z, and that the bilinear form $b(\cdot, \cdot)$ is continuous on Z and satisfies the inf-sup condition

$$\sup_{v \in V} \frac{b(v, q)}{\|v\|_V} \geq C\|q\|_Q, \quad \forall q \in Q \tag{13.85}$$

For the Stokes system, a great deal of effort is needed to show the inf-sup condition. Here, it is easy. By taking $v = \nabla q$ with $q \in Q$, we have

$$\frac{b(\nabla q, q)}{\|\nabla q\|_V} = \frac{(\nabla q, \nabla q)}{\|\nabla \times (\nabla q)\| + \|\nabla q\|} = \frac{\|\nabla q\|^2}{\|\nabla q\|} = \|\nabla q\| = \|q\|_Q \tag{13.86}$$

This is an admissible choice of v, since we know from the Helmholtz decomposition that ∇q is a member of $H(\text{curl}; \Omega)$.

Unlike for the Stokes system, the coercivity of $a(\cdot, \cdot)$ on Z is hard to establish. To do so, we need the following variant of the Poincaré inequality on Z

$$\|v\| \leq C\|\nabla \times v\| \tag{13.87}$$

Taking this result for granted, we immediately have

$$a(A, A) = \mu^{-1}\|\nabla \times A\|^2 \geq C(\|\nabla \times A\|^2 + \|A\|^2) = C\|A\|_V^2 \tag{13.88}$$

which shows the coercivity property, since by assumption $\mu^{-1} > 0$.

The linearity of $a(\cdot,\cdot)$ is easy to deduce using the Cauchy-Schwarz inequality.

$$a(A, v) = \mu^{-1}(\nabla \times A, \nabla \times v) \tag{13.89}$$

$$\leq C \|\nabla \times A\| \, \|\nabla \times v\| \tag{13.90}$$

$$\leq C \|A\|_V \|v\|_V \tag{13.91}$$

The linearity of $b(\cdot,\cdot)$ also follows from the Cauchy-Schwarz inequality.

$$b(A, q) = -(A, \nabla q) \tag{13.92}$$

$$\leq \|A\| \, \|\nabla q\| \tag{13.93}$$

$$\leq C \|A\|_V \|q\|_Q \tag{13.94}$$

Thus, we conclude that the requirements for Brezzi's Theorem 12.2 are fulfilled. Hence, the existence and uniqueness of A and Λ is guaranteed.

We remark that the existence and uniqueness of a solution $A_0 \in Z$ to the reduced weak form (13.84) follows from the Lax-Milgram lemma, since, as we have seen, $a(\cdot,\cdot)$ is coercive and continuous on Z.

13.6.5 Finite Element Approximation

In order to formulate a numerical method, let $\mathcal{K}$ be a mesh of the domain Ω into tetrahedral elements K. On this mesh, let V_h be the Nédélec space of lowest order, and Q_h the space of continuous piecewise linears, defined by

$$V_h = \{v \in V : v|_K = a + b \times x, \forall K \in \mathcal{K}\} \tag{13.95}$$

$$Q_h = \{v \in Q : v|_K \in P_1(K), \ \forall K \in \mathcal{K}\} \tag{13.96}$$

Now, the finite element approximation of (13.67) takes the form: find $(A_h, \Lambda_h) \in V_h \times Q_h$ such that

$$a(A_h, v) + b(\Lambda_h, v) = l(v), \quad \forall v \in V_h \tag{13.97a}$$

$$b(A_h, q) = 0, \quad \forall q \in Q_h \tag{13.97b}$$

or, in compact form: find $(A_h, \Lambda_h) \in V_h \times Q_h$ such that

$$B((A_h, \Lambda_h), (v, q)) = F((v, q)), \quad \forall (v, q) \in V_h \times Q_h \tag{13.98}$$

Existence and uniqueness of the finite element solution (A_h, Λ_h) follows from the fact that the combination of Nédélec and Lagrange elements form discrete inf-sup stable pairs.

To assert that the current density j is divergence free, it is common to express j as $j = \nabla \times t$, where t is a potential. Moreover, this potential is often interpolated onto the Nédélec space for ease of implementation.

13.6.6 A Priori Error Estimate

The derivation of an a priori error estimate for the magnetostatic potential equation follows the same line of reasoning as for the Stokes system. In doing so, recall that a key ingredient is the B-stability property

$$\sup_{(v,q)\in V\times Q} \frac{B((A,\Lambda),(v,q))}{\|(v,q)\|_{V\times Q}} \geq C\,\|(A,\Lambda)\|_{V\times Q} \tag{13.99}$$

In principle, we can show B-stability in the same way as we did for the Stokes system together with the Helmholtz decomposition. However, as the calculations are a bit lengthy let us tacitly assume that (13.99) holds.

Once the B-stability property has been established the following best approximation result is a simple consequence of the Triangle inequality and Galerkin orthogonality

$$\|A - A_h\|_V + \|\Lambda - \Lambda_h\|_Q \leq C(\|A - v\|_V + \|\Lambda - q\|_Q), \quad \forall (v,q) \in V \times Q \tag{13.100}$$

Now, without going into the detains it is possible to construct a Nédélec interpolant $\pi v \in V_h$ to a vector $v \in V$, which satisfies the interpolation estimates

$$\|v - \pi v\| \leq Ch|v|_{H^1(\Omega)} \tag{13.101}$$

$$\|\nabla \times (v - \pi v)\| \leq Ch|\nabla \times v|_{H^1(\Omega)} \tag{13.102}$$

provided v and $\nabla \times v$ are sufficiently regular.

Choosing finally $v = \pi A \in V_h$ in the best approximation result and using the above and standard interpolation estimates we readily obtain the following a priori error estimate.

Theorem 13.2. *The finite element approximation (A_h, Λ_h) satisfies the estimate*

$$\|A - A_h\|_V + \|\Lambda - \Lambda_h\|_Q \leq Ch\left(|A|_{H^1(\Omega)} + |\nabla \times A|_{H^1(\Omega)} + |\nabla \Lambda|_{H^1(\Omega)}\right) \tag{13.103}$$

This a priori error estimate can be further simplified by observing that if the current density j is divergence free, then both the continuous and discrete Lagrangian multiplier Λ and Λ_h vanish.

13.6.7 A Posteriori Error Estimate

The derivation of an a posteriori error estimate for the magnetostatic potential equation again follows the same line of reasoning as for the Stokes system.

From Galerkin orthogonality we have

$$B((A - A_h, \Lambda - \Lambda_h), (v, q)) = B((A - A_h, \Lambda - \Lambda_h), (v - \pi v, q - \pi q)) \tag{13.104}$$

where $\pi v \in V_h$ and $\pi q \in Q_h$ are interpolants of v and q. Using the Helmholtz decomposition we write $v - \pi v = r + \nabla \varphi$, with $r \in Z$ and $\varphi \in H_0^1(\Omega)$. It has been shown by Schöberl [59] that r and φ can essentially be chosen such that

$$h_K^{-1}\|r\|_K + \|\nabla \times r\|_K \leq \|\nabla \times v\|_K, \qquad h_K^{-1}\|\varphi\|_K + \|\nabla\varphi\|_K \leq \|v\|_K \tag{13.105}$$

on each element K. Substituting this into (13.104), yields

$$B((A - A_h, \Lambda - \Lambda_h), (v - \pi v, q - \pi q)) = F((v, q)) - B((A_h, \Lambda_h), (v - \pi v, q - \pi q)) \tag{13.106}$$

$$= l(v - \pi v) - a(A_h, v - \pi v) - b(v - \pi v, \Lambda_h) \tag{13.107}$$

$$- b(A_h, q - \pi q)$$

$$= l(r + \nabla\varphi) - a(A_h, r + \nabla\varphi) - b(r + \nabla\varphi, \Lambda_h) \tag{13.108}$$

$$- b(A_h, q - \pi q)$$

$$= l(r) - a(A_h, r) - b(\nabla\varphi, \Lambda_h) \tag{13.109}$$

$$- b(A_h, q - \pi q)$$

where we have used the vector identity $\nabla \times (\nabla\varphi) = 0$, and that r is orthogonal to $\nabla \Lambda_h$. Breaking the integrals into a sum over the elements as usual, and integrating by parts on each term, and estimating using the Cauchy-Schwarz inequality, further yields

$$l(r) - a(A_h, r) - b(\nabla\varphi, \Lambda_h) - b(A_h, q - \pi q) = (j, r) - (\mu^{-1}\nabla \times A_h, \nabla \times r)$$

$$+ (\nabla\varphi, \nabla\Lambda_h) \tag{13.110}$$

$$+ (A_h, \nabla(q - \pi q))$$

$$= \sum_{K \in \mathcal{K}} (j - \nabla \times \mu^{-1}\nabla \times A_h, r)_K$$

$$- \tfrac{1}{2}([\mu^{-1}(\nabla \times A_h) \times n], r)_{\partial K}$$

$$
\begin{aligned}
&+ (\nabla\varphi, \nabla \Lambda_h)_K \\
&- (\nabla \cdot A_h, q - \pi q)_K \\
&+ \tfrac{1}{2}([n \cdot A_h], q - \pi q)_{\partial K} \\
\leq \sum_{K \in \mathcal{K}} &\|j - \nabla \times \mu^{-1} \nabla \times A_h\|_K \|r\|_K \qquad (13.111) \\
&+ \tfrac{1}{2}\|[\mu^{-1}(\nabla \times A_h) \times n]\|_K \|r\|_{\partial K} \\
&+ \|\nabla\varphi\|_K \|\nabla \Lambda_h\|_K \\
&+ \|\nabla \cdot A_h\|_K \|q - \pi q\|_K \\
&+ \tfrac{1}{2}\|[n \cdot A_h]\|_{\partial K} \|q - \pi q\|_{\partial K}
\end{aligned}
$$

Now, because r is divergence free its normal component $r \cdot n$ is zero, and, thus, r has only tangent component $r \times n$ on the element boundary ∂K. On this boundary there holds the Trace inequality $\|v \times n\|_{\partial K} \leq C(h_K^{-1/2}\|v\|_K + h_K^{1/2}\|\nabla \times v\|_K)$. Thus, in view of the inequality for r in (13.105) we have $\|r\|_{\partial K} = \|r \times n\|_{\partial K} \leq Ch^{1/2}(\|v\|_K + \|\nabla \times v\|_K)$. Using this and standard interpolation estimates we can estimate the sum (13.111), which we denote by S, by

$$
\begin{aligned}
S \leq &C \sum_{K \in \mathcal{K}} h_K \|j - \nabla \times \mu^{-1} \nabla \times A_h\|_K \|\nabla \times v\|_K \qquad (13.112) \\
&+ \tfrac{1}{2} h_K^{1/2} \|[\mu^{-1}(\nabla \times A_h) \times n]\|_{\partial K \setminus \partial \Omega} (\|v\|_K + \|\nabla \times v\|_K) \\
&+ \|\nabla \Lambda_h\|_K \|v\|_K \\
&+ h_K \|\nabla \cdot A_h\|_K \|\nabla q\|_K \\
&+ \tfrac{1}{2} h_K^{1/2} \|[n \cdot A_h]\|_{\partial K \setminus \partial \Omega} \|\nabla q\|_K \\
\leq &\left(C \sum_{K \in \mathcal{K}} \eta_K \right) (\|v\| + \|\nabla \times v\| + \|\nabla q\|) \qquad (13.113)
\end{aligned}
$$

where we have introduced the element residual η_K, defined by

$$
\begin{aligned}
\eta_K = h_K \|j - \nabla \times \mu^{-1} \nabla \times A_h\|_K + \tfrac{1}{2} h_K^{1/2} \|[\mu^{-1}(\nabla \times A_h) \times n]\|_{\partial K \setminus \partial \Omega} \qquad (13.114) \\
+ \|\nabla \Lambda_h\|_K + h_K \|\nabla \cdot A_h\|_K + \tfrac{1}{2} h_K^{1/2} \|[n \cdot A_h]\|_{\partial K \setminus \partial \Omega}
\end{aligned}
$$

Summarizing, we thus have

$$
B((A - A_h, \Lambda - \Lambda_h), (v - \pi v, q - \pi q)) \leq C \left(\sum_{K \in \mathcal{K}} \eta_K \right) \|(v, q)\|_{V \times Q} \qquad (13.115)
$$

Using this the following a posteriori error estimate is a simple consequence of B-stability.

Theorem 13.3. *The finite element solution* (A_h, Λ_h) *satisfies the estimate*

$$\|(A - A_h, \Lambda - \Lambda_h)\|_{V \times Q} \leq C \sum_{K \in \mathcal{K}} \eta_K \tag{13.116}$$

where the element residual η_K *is the sum of the cell residual* $R_K = h_K(\|j - \nabla \times \nabla \times A_h\|_K + \|\nabla \cdot A_h\|_K) + \|\nabla \Lambda_h\|_K$ *and the edge residual* $r_K = \frac{1}{2} h_K^{1/2}(\|[\mu^{-1}(\nabla \times A_h) \times n]\|_{\partial K \setminus \partial \Omega} + \|[n \cdot A_h]\|_{\partial K \setminus \partial \Omega})$.

13.7 Further Reading

Introductory books on computational electromagnetics include the ones by Chari and Salon [16] and Bondesson et al. [13]. Both these books look at computational electromagnetics from a broad perspective and contain material on a wide range of numerical methods, such as FD, FDTD, FEM, BEM, and MoM, for instance. More advanced books include those of Jin [44], Solin [66] and Monk [49].

13.8 Problems

Exercise 13.1. In electrostatics, due to (13.27a), there exists a scalar potential ϕ, such that $E = -\nabla \phi$. Show, using (13.27b), that ϕ satisfies Poisson's equation $-\Delta \phi = q/\epsilon_0$ in free space. Can you interpret the Dirichlet and Neumann boundary conditions $\phi = 0$ and $n \cdot \nabla \phi = 0$ from the physical point of view?

Exercise 13.2. Derive the weak form of the electric wave equation $\nabla \times \mu^{-1} \nabla \times E - \epsilon \omega^2 E = 0$, subject to the boundary conditions $E \times n = g_D$ on Γ_D and $(\mu^{-1} \nabla \times E) \times n - i \epsilon \omega^2 = g_N$ on Γ_N.

Exercise 13.3. Show (13.55).

Exercise 13.4. Write down three benefits with using Nédélec edge elements instead of nodal Lagrange elements to discretize a problem involving the curl-curl operator $\nabla \times \nabla \times$.

Chapter 14
Discontinuous Galerkin Methods

Abstract In this final chapter we present the discontinuous Galerkin (dG) method. This method is based on finite element spaces that consist of discontinuous piecewise polynomials defined on a partition of the computational domain. Such methods are very flexible, for example, since they allow construction of more general methods and since they allow for simple adaptation. Discontinuous Galerkin methods were originally developed for first order problems and were later extended to second order problems. We cover both categories, and derive basic stability and error estimates. Due to the discontinuous nature of the finite element space additional terms in the weak form are necessary to enforce the proper continuity conditions between adjacent elements. We also consider how to handle these additional terms in the implementation of the method.

14.1 A First Order Problem

14.1.1 Model Problem

Let Ω be a domain in $\mathbb{R}^d$, $d = 1, 2$, or 3 with boundary $\partial\Omega$. Let $b = [b_i]_{i=1}^d$ be a given vector field and c a given scalar function. We consider the following first order problem modeling convection and reaction: find u such that

$$cu + b \cdot \nabla u = f, \quad \text{in } \Omega \tag{14.1a}$$

$$u = g, \quad \text{on } \partial\Omega_- \tag{14.1b}$$

where

$$\partial\Omega_- = \{x \in \partial\Omega : n(x) \cdot b(x) < 0\} \tag{14.2}$$

is the so-called inflow part of the boundary. For simplicity, we assume that b is a constant vector and c a constant function.

M.G. Larson and F. Bengzon, *The Finite Element Method: Theory, Implementation, and Applications*, Texts in Computational Science and Engineering 10,
DOI 10.1007/978-3-642-33287-6_14,

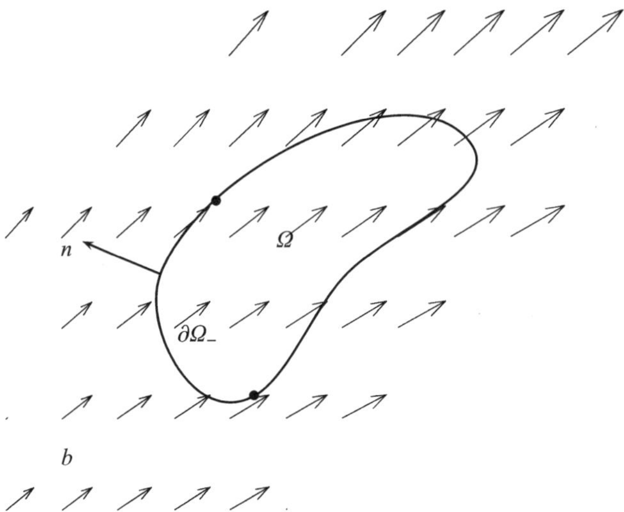

Fig. 14.1 The inflow boundary $\partial\Omega_-$, where $n \cdot b < 0$

The operators c and $b \cdot \nabla$ can be simply interpreted. The first scales u so that it is proportional to f. The second, transports u along the direction of b. To describe this we speak about reaction and convection. Note that there is no diffusion term like $-\epsilon \Delta u$, which can smooth u and make it adhere to a boundary condition. Therefore, we can only have boundary conditions on the inflow part of the boundary. That is, the parts of $\partial\Omega$ where the vectors of b point into Ω. See Fig. 14.1.

14.1.2 Discontinuous Finite Element Spaces

Let $\mathcal{K} = \{K\}$ be a mesh of Ω and define the space of discontinuous piecewise linear functions

$$V_h = \{v : v|_K \in P_1(K), \ \forall K \in \mathcal{K}\} \tag{14.3}$$

where $P_1(K)$ is the space of linear polynomials on element K.

Thus, the members of V_h are linear on each element K, but generally discontinuous across the element boundaries ∂K.

As before, we let $\mathcal{E}_I$ denote the set of interior edges and with each interior edge E we associate a fixed unit normal n. We denote by K^+ the element for which n is the exterior normal, and K^- the element for which $-n$ is the exterior normal. For edges on the boundary $\partial\Omega$ we let n be the exterior unit normal to Ω. Further, we define the jump and the average of a function $v \in V_h$ at the edge E by

$$[v] = v^+ - v^-, \qquad \langle v \rangle = \frac{u^+ + u^-}{2} \tag{14.4}$$

14.1.3 The Discontinuous Galerkin Method

To derive a discontinuous Galerkin method we multiply the transport equation by $v \in V_h$ and integrate over Ω. Integrating by parts on each element gives

$$(f, v) = \sum_{K \in \mathcal{K}} (cu + b \cdot \nabla u, v)_K \tag{14.5}$$

$$= \sum_{K \in \mathcal{K}} (cu, v)_K - (u, b \cdot \nabla v)_K + (n \cdot bu, v)_{\partial K} \tag{14.6}$$

$$= \sum_{K \in \mathcal{K}} (cu, v)_K - (u, b \cdot \nabla v)_K \tag{14.7}$$

$$+ \sum_{E \in \mathcal{E}_I} (n \cdot bu, [v])_E + (n \cdot bg, v)_{\partial\Omega_-} + (n \cdot bu, v)_{\partial\Omega \setminus \partial\Omega_-}$$

where we have used the boundary condition $u = g$ on $\partial\Omega_-$, the fact that u is continuous along the characteristics, and that $\nabla \cdot b = 0$ so that $\nabla \cdot (bu) = b \cdot \nabla u$. In order to make sense of this form also for $u \in V_h$ we replace u by the average $\langle u \rangle + \gamma[u]$, where γ is a parameter on interior edges to be determined, which gives

$$(f, v) - (n \cdot bg, v)_{\partial\Omega_-} = \sum_{K\mathcal{K}} (cu, v)_K - (u, b \cdot \nabla v)_K \tag{14.8}$$

$$+ \sum_{E \in \mathcal{E}_I} (n \cdot b\langle u \rangle, [v])_E + (\gamma n \cdot b[u], [v])_E + (n \cdot bu, v)_{\partial\Omega \setminus \partial\Omega_-}$$

$$= \sum_{K \in \mathcal{K}} (cu, v)_K + (b \cdot \nabla u, v)_K \tag{14.9}$$

$$- \sum_{E \in \mathcal{E}_I} (n \cdot b[u], \langle v \rangle)_E + (\gamma n \cdot b[u], [v])_E - (n \cdot bu, v)_{\partial\Omega_-}$$

Here, we finally have used integration by parts on each element together with the identity

$$[uv] = [u]\langle v \rangle + \langle u \rangle [v] \tag{14.10}$$

We may thus formulate the following discontinuous Galerkin method: find $u_h \in V_h$ such that

$$a_h(u_h, v) = l(v), \quad \forall v \in V_h \tag{14.11}$$

where the bilinear and linear form $a_h(\cdot, \cdot)$ and $l_h(\cdot)$ are defined by

$$a_h(u,v) = \sum_{K\in\mathcal{K}} (cu,v)_K + (b\cdot\nabla u, v)_K \tag{14.12}$$
$$- \sum_{E\in\mathcal{E}_I} (n\cdot b[u], \langle v\rangle)_E + (\gamma n\cdot b[u],[v])_E - (n\cdot bu, v)_{\partial\Omega_-}$$
$$l(v) = (f,v) - (n\cdot bg, v)_{\partial\Omega_-} \tag{14.13}$$

respectively.

We note that the method satisfies the Galerkin orthogonality

$$a_h(u-u_h,v) = 0, \quad \forall v\in V_h \tag{14.14}$$

Of course, this is due to the fact that u satisfies (14.11). We say that the method is consistent.

14.1.4 Stability Estimates

For the stability analysis we introduce the weighted norms

$$\|v\|^2_{b,\mathcal{E}_I} = \sum_{E\in\mathcal{E}_I} (|n\cdot b|v,v)_E, \quad \|v\|^2_{\partial\Omega} = (|n\cdot b|v,v)_{\partial\Omega} \tag{14.15}$$

Assuming that there is a constant $C>0$ such that

$$C|n\cdot b| \leq \gamma n\cdot b, \quad \forall E\in\mathcal{E}_I \tag{14.16}$$

the following stability estimate holds

$$C(\|u\|^2_\Omega + \|[u]\|^2_{b,\mathcal{E}_I} + \|u\|^2_{b,\partial\Omega}) \leq a_h(u,u) \tag{14.17}$$

where $\|u\|_\Omega$ is to be interpreted as $\sum_{K\in\mathcal{K}} \|u\|_K$. Before we turn to the proof of this estimate, let us investigate the meaning of condition (14.16) on the parameter γ. We first note that the average is a convex combination

$$\langle u\rangle + \gamma[u] = \left(\frac{1}{2}+\gamma\right)u^+ + \left(\frac{1}{2}-\gamma\right)u^- \tag{14.18}$$

Next we note that if $n\cdot b>0$ then γ is positive and we get larger weight on K^+, the element located upstream of the edge, while if $n\cdot b<0$ we get a negative γ and thus a larger weight on K^- which is the upstream element in this case. Thus we conclude that upwinding, in the sense that a larger weight is used on the upstream side of the edge, yields stability of the method. In particular, choosing $\gamma=\text{sign}(n\cdot b)/2$ we obtain the traditional discontinuous Galerkin method for first order problems where

$$[u] + \gamma\langle u\rangle = \begin{cases} u^+, & \text{if } n\cdot b > 0 \\ u^-, & \text{if } n\cdot b < 0 \end{cases} \tag{14.19}$$

is precisely the upstream value at the face.

We now turn to the proof of (14.17). Setting $v = u$ we get

$$a_h(u,u) = \sum_{K\in\mathcal{K}} (cu,u)_K + (b\cdot\nabla u, u)_K - \sum_{E\in\mathcal{E}_I} (n\cdot b[u], \langle u\rangle)_E - (n\cdot bu, u)_{\partial\Omega_-} \tag{14.20}$$

Focusing on the second term and using integration by parts we obtain

$$\sum_{K\in\mathcal{K}} (b\cdot\nabla u, u)_K = -\sum_{K\in\mathcal{K}} (u, b\cdot\nabla u)_K + (n\cdot bu, u)_{\partial K} \tag{14.21}$$

$$= -\sum_{K\in\mathcal{K}} (b\cdot\nabla u, u)_K + 2\sum_{E\in\mathcal{E}_I} (n\cdot b[u], \langle u\rangle)_E + (n\cdot bu, u)_{\partial\Omega} \tag{14.22}$$

where we have used the identity

$$n^+\cdot b(u^+)^2 + n^-\cdot b(u^-)^2 = n\cdot b((u^+)^2 - (u^-)^2) = 2n\cdot b[u]\langle u\rangle \tag{14.23}$$

Thus, we conclude that the following identity holds for the second term

$$\sum_{K\in\mathcal{K}} (b\cdot\nabla u, u)_K = \sum_{E\in\mathcal{E}_I} (n\cdot b[u], \langle u\rangle)_E + \frac{1}{2}(n\cdot bu, u)_{\partial\Omega} \tag{14.24}$$

Inserting this identity into (14.20) we get the desired estimate.

Next we seek to improve the control of the streamline derivative $b\cdot\nabla v$. Since we are using discontinuous functions we may choose $v = hb\cdot\nabla u$, which gives

$$a_h(u, hb\cdot\nabla u) = \sum_{K\in\mathcal{K}} h(cu, b\cdot\nabla u)_K + h(b\cdot\nabla u, b\cdot\nabla u)_K \tag{14.25}$$

$$- \sum_{E\in\mathcal{E}_I} h(n\cdot b[u], \langle b\cdot\nabla u\rangle)_E - h(n\cdot bu, b\cdot\nabla u)_{\partial\Omega_-}$$

$$\geq h\|b\cdot\nabla u\|^2_\Omega - Ch\|u\|_\Omega\|b\cdot\nabla u\|_\Omega \tag{14.26}$$

$$- C\,\|[u]\|_{b,\mathcal{E}_I} h^{1/2}\|b\cdot\nabla u\|_\Omega - C\,\|u\|_{b,\partial\Omega}\|b\cdot\nabla u\|_{\partial\Omega}$$

$$\geq C_1\sum_{K\in\mathcal{K}} h\|b\cdot\nabla u\|^2_K - C_2(\|u\|^2_\Omega + \|[u]\|^2_{b,\mathcal{E}_I} + \|u\|^2_{b,\partial\Omega}) \tag{14.27}$$

Thus, we conclude that there are constants C_1 and C_2 such that

$$a_h(u, hb\cdot\nabla u) \geq C_1 \sum_{K\in\mathcal{K}} h\|b\cdot\nabla u\|_K^2 - C_2(\|u\|_\Omega^2 + \|[u]\|_{b,\mathcal{E}_I}^2 + \|u\|_{b,\partial\Omega}^2) \quad (14.28)$$

In order to state our final stability result, we introduce the norm

$$|||u|||^2 = \|u\|_\Omega^2 + \sum_{K\in\mathcal{K}} h\|b\cdot\nabla u\|_K^2 + \|[u]\|_{b,\mathcal{E}_I}^2 + \|u\|_{b,\partial\Omega}^2 \quad (14.29)$$

Theorem 14.1. *The following inf-sup condition holds*

$$C\,|||u||| \leq \sup_{v\in V_h} \frac{a_h(u,v)}{|||v|||} \quad (14.30)$$

Proof. In order to prove this result, we first set $v = u + \delta hb\cdot\nabla u$, where δ is a positive parameter. Then, we have

$$\begin{aligned}
a_h(u, u+\delta hb\cdot\nabla u) &= a_h(u,u) + \delta a_h(u, hb\cdot\nabla u) && (14.31)\\
&\geq C(\|u\|_\Omega^2 + \|[u]\|_{b,\mathcal{E}_I}^2 + \|u\|_{b,\partial\Omega}^2) && (14.32)\\
&\quad + \delta(C_1 \sum_{K\in\mathcal{K}} h\|b\cdot\nabla u\|_K^2 - C_2(\|u\|_\Omega^2 + \|[u]\|_{b,\mathcal{E}_I}^2 + \|u\|_{b,\partial\Omega}^2))\\
&\geq C\,|||u|||^2 && (14.33)
\end{aligned}$$

for δ small enough. Next, we have the estimate

$$|||v||| \leq |||u||| + \delta|||hb\cdot\nabla u||| \leq C\,|||u||| \quad (14.34)$$

and the inf-sup condition follows. □

Finally, in order to verify (14.34) we note that

$$\begin{aligned}
|||hb\cdot\nabla u|||^2 &= h^2\|b\cdot\nabla u\|_\Omega^2 + \sum_{K\in\mathcal{K}} h^3\|b\cdot\nabla(b\cdot\nabla u)\|_K^2 && (14.35)\\
&\quad + h^2\|[b\cdot\nabla u]\|_{b,\mathcal{E}_I}^2 + h^2\|b\cdot\nabla u\|_{b,\partial\Omega}^2\\
&\leq h^2\|b\cdot\nabla u\|_\Omega^2 + C\sum_{K\in\mathcal{K}} h^2\|b\cdot\nabla u\|_{\partial K}^2 && (14.36)\\
&\leq h^2\|b\cdot\nabla u\|_\Omega^2 + C\sum_{K\in\mathcal{K}} h\|b\cdot\nabla u\|_K^2 && (14.37)\\
&\leq C\,|||u|||^2 && (14.38)
\end{aligned}$$

where we have used the fact that $b \cdot \nabla u$ is constant, since we assumed that b is constant and u is piecewise linear, so that all second order derivatives of $b \cdot \nabla u$ vanish together with the Trace inequality.

14.1.5 A Priori Error Estimates

The following a priori error estimate holds.

Theorem 14.2. *The dG solution u_h, defined by* (14.11), *satisfies the estimate*

$$|||u - u_h||| \leq C h^{3/2} |u|_{H^2(\Omega)} \tag{14.39}$$

Proof. To prove the estimate, we use the Triangle inequality to divide the error into an interpolation error and a discrete error

$$|||u - u_h||| \leq |||u - \pi u||| + |||\pi u - u_h||| \tag{14.40}$$

For the second term we first use the inf-sup condition (14.30) and then Galerkin orthogonality (14.14) to get

$$C\,|||\pi u - u_h||| \leq \sup_{v \in V_h} \frac{a_h(\pi u - u_h, v)}{|||v|||} = \sup_{v \in V_h} \frac{a_h(\pi u - u, v)}{|||v|||} \tag{14.41}$$

Setting $\pi u - u = \eta$ we have using integration by parts on each element

$$a_h(\eta, v) = (c\eta, v) + \sum_{K \in \mathcal{K}} -(\eta, b \cdot \nabla v)_K \tag{14.42}$$

$$+ \sum_{E \in \mathcal{E}_I} (n \cdot b\langle\eta\rangle, [v])_E + (n \cdot b\eta, u)_{\partial\Omega \setminus \partial\Omega_-}$$

$$\leq C\,\|\eta\|_\Omega \|v\|_\Omega + C \sum_{K \in \mathcal{K}} h^{-1/2} \|\eta\|_K h^{1/2} \|b \cdot \nabla v\|_K \tag{14.43}$$

$$+ \|\langle\eta\rangle\|_{b,\mathcal{E}_I} \|[v]\|_{b,\mathcal{E}_I} + \|\eta\|_{b,\partial\Omega} \|v\|_{b,\partial\Omega}$$

$$\leq C \left(\sum_{K \in \mathcal{K}} h^{-1} \|\eta\|_K^2 + h \|\nabla\eta\|_K^2 \right)^{1/2} |||v||| \tag{14.44}$$

$$\leq C h^{3/2} |u|_{H^2(\Omega)} \tag{14.45}$$

where we used the following estimate

$$\|\langle\eta\rangle\|_{b,\mathcal{E}_I}^2 \leq C \sum_{K \in \mathcal{K}} \|\eta\|_{\partial K}^2 \leq C \sum_{K \in \mathcal{K}} h^{-1} \|\eta\|_K^2 + h \|\nabla\eta\|_K^2 \tag{14.46}$$

and finally standard interpolation error estimates. □

14.2 A Second Order Elliptic Problem

14.2.1 Model Problem

During the last decade, there has been a revived interest in dG methods. This is partially due to the fact that efficient discretizations of second order terms have been derived. Therefore, let us revisit the familiar Poisson equation: find u such that

$$-\Delta u = f, \quad \text{in } \Omega \tag{14.47a}$$

$$u = 0, \quad \text{on } \partial\Omega \tag{14.47b}$$

where Δ is the second order Laplacian operator.

14.2.2 The Symmetric Interior Penalty Method

To derive a discontinuous Galerkin method we multiply the equation with a test function $v \in V_h$. Integration by parts on each element gives us

$$(f, v) = \sum_{K\in\mathcal{K}} (\nabla u, \nabla v)_K - (n \cdot \nabla u, v)_{\partial K} \tag{14.48}$$

$$= \sum_{K\in\mathcal{K}} (\nabla u, \nabla v)_K - \sum_{E\in\mathcal{E}_I} (n \cdot \nabla u, [v])_E - (n \cdot \nabla u, v)_{\partial\Omega} \tag{14.49}$$

where we used that fact that $n \cdot \nabla u$ is continuous across the element faces, or in other words $[n \cdot \nabla u] = 0$. To make sense of this expression also for $u \in V_h$ we replace the normal flux $n \cdot \nabla u$ by the discrete flux $\langle n \cdot \nabla u\rangle - \beta h^{-1}[u]$, where β is a positive parameter. Note that the discrete flux may be viewed as a certain approximation of the normal derivative across the face that take the average slope $\langle n \cdot \nabla u\rangle$ as well as the negative jump divided by the meshsize $-h^{-1}[u] = h^{-1}(u^- - u^+)$, which may be interpreted as a finite difference approximation of the contribution of the jump to the normal flux, into account. Inserting the discrete flux we immediately arrive at

$$(f, v) = \sum_{K\in\mathcal{K}} (\nabla u, \nabla v)_K - \sum_{E\in\mathcal{E}_I} (\langle n \cdot \nabla u\rangle, [v])_E - (n \cdot \nabla u, v)_{\partial\Omega} \tag{14.50}$$

$$+ \sum_{E\in\mathcal{E}_I} (\beta h^{-1}[u], [v])_E + (\beta h^{-1} u, v)_{\partial\Omega}$$

Finally, we note that the following term is zero when u is the exact solution

$$\sum_{E\in\mathcal{E}_I}([u],\langle n\cdot\nabla v\rangle)_E+(u,n\cdot\nabla v)_{\partial\Omega} \tag{14.51}$$

and therefore we can subtract it on the right hand side to obtain a symmetric form without losing consistency.

We thus define the bilinear and linear form

$$\begin{aligned} a_h(u,v) &= \sum_{K\in\mathcal{K}}(\nabla u,\nabla v)_K-\sum_{E\in\mathcal{E}_I}(\langle n\cdot\nabla u\rangle,[v])_E-(n\cdot\nabla u,v)_{\partial\Omega} && (14.52)\\ &\quad-\sum_{E\in\mathcal{E}_I}([u],\langle n\cdot\nabla v\rangle)_E-(u,n\cdot\nabla v)_{\partial\Omega}\\ &\quad+\sum_{E\in\mathcal{E}_I}\beta h^{-1}([u],[v])_E+(\beta h^{-1}[u],[v])_{\partial\Omega}\\ l_h(v) &= (f,v) && (14.53)\end{aligned}$$

where we note that the first edge and boundary term stems from elementwise partial integration, the second edge and boundary term is added for symmetry, and the third edge term penalizes the jump of the solution u between adjacent elements and the third boundary term enforces the Dirichlet boundary condition. The parameter $\beta>0$ controls the amount of penalization.

With the above definitions of $a_h(\cdot,\cdot)$ and $l_h(\cdot)$, the finite element method reads: find $u_h\in V_h$ such that

$$a_h(u_h,v)=l_h(v),\quad \forall v\in V_h \tag{14.54}$$

This dG method is called the Nitsche's method or the Symmetric Interior Penalty Galerkin method (SIPG).

We note that the SIPG method is consistent, and satisfies the Galerkin orthogonality condition

$$a_h(u-u_h,v)=0,\quad \forall v\in V_h \tag{14.55}$$

14.2.3 Approximation Properties

For the analysis of the method we define the following energy type norm

$$|||v|||^2 = \sum_{K\in\mathcal{K}} \|\nabla v\|_K^2 + \sum_{E\in\mathcal{E}_I} h\|\langle n\cdot\nabla v\rangle\|_E^2 + h\|n\cdot\nabla v\|_{\partial\Omega}^2 \tag{14.56}$$
$$+ \sum_{E\in\mathcal{E}_I} h^{-1}\|[v]\|_E^2 + h^{-1}\|v\|_{\partial\Omega}^2$$

We also define the interpolation operator π, so that $(\pi u)|_K = \pi_K u$, and prove the interpolation error estimate

$$|||u-\pi u||| \le Ch|u|_{H^2(\Omega)} \tag{14.57}$$

Setting $\eta = u - \pi u$ and using the Trace inequality

$$\|\eta\|_{\partial K}^2 \le C(h^{-1}\|\eta\|_K^2 + h\|\nabla\eta\|_K^2) \tag{14.58}$$

we get the estimate

$$|||\eta|||^2 = \sum_{K\in\mathcal{K}} \|\nabla\eta\|_K^2 + \sum_{E\in\mathcal{E}_I} h\|\langle n\cdot\nabla\eta\rangle\|_E^2 + h\|n\cdot\nabla\eta\|_{\partial\Omega}^2 \tag{14.59}$$
$$+ \sum_{E\in\mathcal{E}_I} h^{-1}\|[\eta]\|_E^2 + h^{-1}\|\eta\|_{\partial\Omega}^2$$
$$\le \sum_{K\in\mathcal{K}} \|\nabla\eta\|_K^2 + Ch(h^{-1}\|\nabla\eta\|_K^2 + h\|\nabla(\nabla\eta)\|_K^2) \tag{14.60}$$
$$+ Ch^{-1}(h^{-1}\|\eta\|_K^2 + \|\nabla\eta\|_K^2)$$
$$\le Ch^2\sum_{K\in\mathcal{K}} |u|_{H^2(K)}^2 \tag{14.61}$$

14.2.4 A Priori Error Estimates

In order to prove a priori error estimates we first establish coercivity and continuity of the bilinear form. In this case the bilinear form is coercive only on the discrete space V_h and it also requires the penalty parameter β to be large enough. Continuity holds for sufficiently elementwise regular functions.

The following coercivity holds on V_h.

$$C|||v|||^2 \le a_h(v,v), \quad \forall v\in V_h \tag{14.62}$$

In order to prove this result we first note that the following inverse inequality holds

$$h^{1/2}\|n\cdot\nabla v\|_{\partial K} \le C\|\nabla v\|_K, \quad \forall v\in P_1(K) \tag{14.63}$$

For piecewise linear elements this inequality is a consequence of the Trace inequality, since

$$\|n \cdot \nabla v\|_{\partial K}^2 \leq \|\nabla v\|_{\partial K}^2 \leq C(h^{-1}\|\nabla v\|_K^2 + h\|\nabla(\nabla v)\|_K^2) \leq Ch^{-1}\|\nabla v\|_K^2 \quad (14.64)$$

where we finally used that $\nabla(\nabla v) = 0$ for any $v \in P_1(K)$. We then have

$$\begin{aligned}
a_h(u,u) &\geq \sum_{K\in\mathcal{K}} \|\nabla u\|_K^2 && (14.65)\\
&\quad - \sum_{E\in\mathcal{E}_I} 2h^{1/2}\|\langle n\cdot\nabla u\rangle\|_E h^{-1/2}\|[u]\|_E - 2h^{1/2}\|n\cdot\nabla u\|_{\partial\Omega} h^{-1/2}\|u\|_{\partial\Omega}\\
&\quad + \sum_{E\in\mathcal{E}_I} \beta h^{-1}\|[u]\|_E^2 + \beta h^{-1}\|u\|_{\partial\Omega}^2\\
&\geq \sum_{K\in\mathcal{K}} \|\nabla u\|_K^2 && (14.66)\\
&\quad - \sum_{E\in\mathcal{E}_I} (\delta h\|\langle n\cdot\nabla u\rangle\|_E^2 + \delta^{-1}h^{-1}\|[u]\|_E^2) - (\delta h\|n\cdot\nabla u\|_{\partial\Omega}^2 + \delta^{-1}h^{-1}\|u\|_{\partial\Omega}^2)\\
&\quad + \sum_{E\in\mathcal{E}_I} \beta h^{-1}\|[u]\|_E^2 + \beta h^{-1}\|u\|_{\partial\Omega}^2\\
&\geq \sum_{K\in\mathcal{K}} \|\nabla u\|_K^2 - \sum_{K\in\mathcal{K}} \delta C\|\nabla u\|_K^2 && (14.67)\\
&\quad + \sum_{E\in\mathcal{E}_I} (\beta-\delta^{-1})h^{-1}\|[u]\|_E^2 + (\beta-\delta^{-1})h^{-1}\|u\|_{\partial\Omega}^2
\end{aligned}$$

where we have used the scaled arithmetic-geometric mean inequality $2ab \leq \delta a^2 + \delta^{-1}b^2$ and the inverse inequality (14.63). Choosing $\delta > 0$ small enough and β large enough the estimate follows. Next, we have the continuity

$$a_h(u,v) \leq |||u|||\,|||v|||, \qquad \forall v \in H^2 \cup V_h \quad (14.68)$$

which follows directly from the Cauchy-Schwarz inequality

$$\begin{aligned}
a_h(u,v) &\leq \sum_{K\in\mathcal{K}} \|\nabla u\|_K\|\nabla v\|_K && (14.69)\\
&\quad + \sum_{E\in\mathcal{E}_I} \|\langle n\cdot\nabla u\rangle\|_E\|[v]\|_E + \|[u]\|_E\|\langle n\cdot\nabla v\rangle\|_E\\
&\quad + \|n\cdot u\|_{\partial\Omega}\|v\|_{\partial\Omega} + \|n\cdot v\|_{\partial\Omega}\|u\|_{\partial\Omega}\\
&\quad + \beta h^{-1}\|[u]\|_E\|[v]\|_E + \beta h^{-1}\|u\|_{\partial\Omega}\|v\|_{\partial\Omega}
\end{aligned}$$

$$\leq C|||u||| \, |||v||| \tag{14.70}$$

Using these estimates we may prove the following a priori error estimate.

Theorem 14.3. *The dG solution u_h, defined by* (14.54)*, satisfies the estimate*

$$|||u - u_h||| \leq Ch|u|_{H^2(\Omega)} \tag{14.71}$$

Proof. To prove this estimate, we use the triangle inequality to divide the error into an interpolation error and a discrete error

$$|||u - u_h||| \leq |||u - \pi u||| + |||\pi u - u_h||| \tag{14.72}$$

The first contribution can be estimated directly using the interpolation error estimate. For the second we use the coercivity followed by Galerkin orthogonality

$$C|||\pi u - u_h|||^2 \leq a_h(\pi u - u_h, \pi u - u_h) \tag{14.73}$$

$$\leq a_h(u - \pi u, \pi u - u_h) \tag{14.74}$$

$$\leq |||u - \pi u||| \, |||\pi u - u_h||| \tag{14.75}$$

$$\leq Ch|u|_{H^2(\Omega)}|||\pi u - u_h||| \tag{14.76}$$

□

We may also use a duality argument to prove an estimate for the L^2-norm of the error of the form

$$\|u - u_h\| \leq Ch^2|u|_{H^2(\Omega)} \tag{14.77}$$

14.2.5 Non-symmetric Versions of the Method

For Poisson's equation it is also possible to use the following non-symmetric form

$$a_h(u, v) = \sum_{K \in \mathcal{K}} (\nabla u, \nabla v)_K \tag{14.78}$$

$$- \sum_{E \in \mathcal{E}_I} (\langle n \cdot \nabla u \rangle, [v])_E + \alpha \sum_{E \in \mathcal{E}_I} ([u], \langle n \cdot \nabla v \rangle)_E + \sum_{E \in \mathcal{E}_I} \beta h^{-1}([u], [v])_E$$

where α is a parameter. Common choices include $\alpha = 0$ and $\alpha = 1$. The latter choice leads to a very simple coercivity proof which does not require the inverse inequality. However, the standard proof for the L^2 error estimate is not applicable and indeed the behavior of the method in L^2 is complicated. The former choice $\alpha = 0$ necessitates a sufficiently large value of β in order for coercivity to hold.

14.2.6 Computer Implementation

To implement the SIPG method we note that a basis for the polynomial space $P_1(K)$ on each element K is given by the usual hat basis functions $\varphi_i = a_i + b_i x_1 + c_i x_2$, $i = 1, 2, 3$. However, unlike the usual case these are not continuous across element boundary ∂K. Indeed, each node or triangle vertex has a different number and is a different degree of freedom on each element. More precisely, we let K_i, $i = 1, 2, \dots, n_t$ contain the degrees of freedom $3i$, $3i-1$ and $3i-2$, with n_t the number of triangles.

Let us write a routine to evaluate the hat functions at a given point within a triangle element.

```
function [v,b,c] = HatCoefficients(xc,yc,ex,ey)
V=[ones(3,1) xc yc];
A=V\eye(3,3);
a=A(1,:); b=A(2,:); c=A(3,:);
v=a+b*ex+c*ey;
```

Input is the triangle vertex coordinates `xv` and `yv` and the evaluation point coordinates `ex` and `ey`. Output is the hat function values `v`, and the partial derivatives `b` and `c`. These can also be computed with `HatGradients`.

Now, the 3×3 element stiffness matrix $A_{ij}^K = (\nabla\varphi_j, \nabla\varphi_i)_K$, and the 3×1 element load vector $b_i^K = (f, \varphi_i)_K$ are easy to compute and assemble by looping over the elements and using a simple one point quadrature rule on each element. In doing so, we obtain the following routine.

```
function [A,B] = dG1CellAssembler2D(p,t,force)
nt=size(t,2);
A=sparse(3*nt,3*nt); % stiffness matrix
B=zeros(3*nt,1); % load vector
for i=1:nt
  nodes=t(1:3,i);
  xv=p(1,nodes)';
  yv=p(2,nodes)';
  [area,b,c]=HatGradients(xv,yv);
  dofs=[1:3]+3*(i-1); % element degrees of freedom
  AK=(b*b'+c*c')*area;
  BK=force(mean(xv),mean(yv))*ones(3,1)/3*area;
  A(dofs,dofs)=A(dofs,dofs)+AK;
  B(dofs)=B(dofs)+BK;
end
```

Here, input is the usual point and connectivity matrices `p` and `t`, and a function handle to a subroutine `force` describing the force f. Note that the degrees of freedom `dofs=[1:3]+3*(i-1)` on each element are based on the triangle number `i`.

The edge flux matrix $S^K = (\langle n \cdot \nabla\varphi_j\rangle, [\varphi_i])_E$, and the edge penalty matrix $P^E = ([\varphi_j], [\varphi_i])_E$ are more difficult to compute and assemble. To this end, we loop over the elements and consider the edges $\{E\}$. For each E, we consider the pair of adjacent elements K^+ and K^- that share E. As these two element involves three basis functions the size of the matrices S^E and P^E are 6×6. To compute these, we put these basis functions into a 6×1 vector, viz.,

$$\varphi^E = \begin{bmatrix} \varphi_1^+ \\ \varphi_2^+ \\ \varphi_3^+ \\ \varphi_1^- \\ \varphi_2^- \\ \varphi_3^- \end{bmatrix} \tag{14.79}$$

Assuming that the current element is K^-, say, with neighbour K^-, we than have

$$[\varphi^E] = \begin{bmatrix} \varphi_1^+ \\ \varphi_2^+ \\ \varphi_3^+ \\ -\varphi_1^- \\ -\varphi_2^- \\ -\varphi_3^- \end{bmatrix} \tag{14.80}$$

since all hats $\varphi_i^\pm$, $i = 1, 2, 3$, are zero outside their corresponding element $K^\pm$. Further, since $n_E = n^+$, we have

$$\langle n \cdot \nabla\varphi^E\rangle = \frac{1}{2} n_E \cdot \nabla\varphi^E \tag{14.81}$$

The outer products $[\varphi^E]\langle n \cdot \nabla\varphi^E\rangle^T$ and $[\varphi^E][\varphi^E]^T$ are the integrands of S^E and P^E, respectively. For the actual integration we use Simpson's formula mapped from $[0, 1]$ onto E, which is sufficient for integrating these products of hat functions. Using this approach we obtain the following routine.

```
function [P,S] = dG1EdgeAssembler2D(p,t,neighbours)
nt=size(t,2);
S=sparse(3*nt,3*nt); % flux matrix
P=sparse(3*nt,3*nt); % penalty matrix
edge2node=[2 3; 1 3; 1 2]; % edge-to-node lookup table
cmat=[2 0; 1 1; 0 2]/2; wvec=[1 4 1]/6; % Simpson's formula
for i=1:nt
  pnodes=t(1:3,i); % nodes on "plus" element
  xp=p(1,pnodes)';
  yp=p(2,pnodes)';
  [area,ds,nx,ny]=Triutils(xp,yp);
  for j=1:3 % loop over edges
    n=neighbours(i,j); % element neighbour
```

```
    if n>i, continue; end % only assemble once on each edge
    if n<0, n=i; end % boundary?
    e2n=edge2node(j,:); % nodes on edge
    ex=cmat*xp(e2n); % x-coordinates of quadrature points
    ey=cmat*yp(e2n); % y-
    SE=zeros(6,6);
    PE=zeros(6,6);
    mnodes=t(1:3,n); % nodes on "minus" element
    xm=p(1,mnodes)';
    ym=p(2,mnodes)';
    for q=1:length(wvec) % quadrature loop on edge
      wxlen=wvec(q)*ds(j); % quadrature weight times edge length
      [vp,bp,cp]=HatCoefficients(xp,yp,ex(q),ey(q));
      [vm,bm,cm]=HatCoefficients(xm,ym,ex(q),ey(q));
      jump=[vp -vm]; % jump
      avgdn=[nx(j)*bp+ny(j)*cp nx(j)*bm+ny(j)*cm]/2; % average
      PE=PE+jump'*jump*wxlen;
      SE=SE+jump'*avgdn*wxlen;
    end
    dofs=[[1:3]+3*(i-1) [1:3]+3*(n-1)];
    if n==i % boundary
      PE=PE(1:3,1:3);
      SE=SE(1:3,1:3)*2; % no average on boundary
      dofs=dofs(1:3);
    end
    P(dofs,dofs)=P(dofs,dofs)+PE;
    S(dofs,dofs)=S(dofs,dofs)+SE;
  end
end
```

Here, we reuse the routine `Triutils` for computing edge lengths and normals.

Note that S^K and P^K must be modified on the boundary of Ω, since the average $\langle n \cdot \nabla\varphi^E\rangle$ and jump $[\varphi^E]$ is only one-sided on $\partial\Omega$ (i.e., there is no K^- element).

Putting the pieces together we obtain the following main routine.

```
function dG1PoissonSolver2D()
clear all, close all
g=Rectg(0,0,1,1); % unit square
beta=9; % penalty parameter
alpha=-1; % SIPG parameter
h=0.0625; % mesh size
[p,e,t]=initmesh(g,'hmax',h);
neighbours=Tri2Tri(p,t); % element neighbours
force = inline('2*pi^2*sin(pi*x)*sin(pi*y)','x','y');
[A,B]=dG1CellAssembler2D(p,t,force);
[P,S]=dG1EdgeAssembler2D(p,t,neighbours);
U=(A-S+alpha*S'+beta/h*P)\B;
% - visualization ---
nt=size(t,2);
X=zeros(3*nt,1); Y=zeros(3*nt,1);
i=t(1,:); j=t(2,:); k=t(3,:);
```

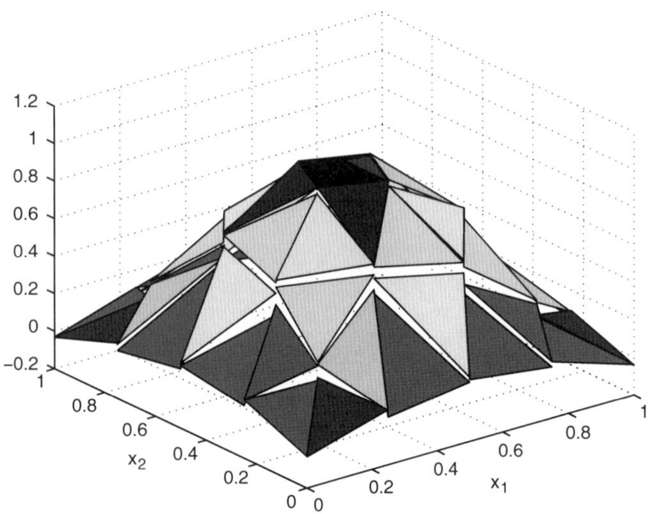

Fig. 14.2 Plot of u_h with $\beta = 3$ and $h = 0.25$

```
X(1:3:end)=p(1,i); X(2:3:end)=p(1,j); X(3:3:end)=p(1,k);
Y(1:3:end)=p(2,i); Y(2:3:end)=p(2,j); Y(3:3:end)=p(2,k);
trisurf(reshape([1:3*nt],3,nt)',X,Y,U)
xlabel('x_1'), ylabel('x_2')
```

Running this code with $f = 2\pi^2 \sin(\pi x_1) \sin(\pi x_2)$, which corresponding to $u = \sin(\pi x_1) \sin(\pi x_2)$ on the unit square $\Omega = [0, 1]^2$, and $\beta = 3$ and $\beta = 36$ on a coarse mesh $h = 0.25$ we obtain the dG solution u_h of Figs. 14.2 and 14.3. From this we see that β controls the amount of discontinuity of u_h. As β is increased u_h becomes more continuous. A value of β in the range 1–10 is common. Indeed, using $\beta = 9$ and a finer mesh with $h = 0.125$ we obtain the u_h shown in Fig. 14.4.

14.3 The Transport Equation with Diffusion

A natural extension of the dG methods for the first and second order equations presented above is to combine them to get a numerical method for the transport equation: find u such that

$$-\epsilon \Delta u + b \cdot \nabla u + cu = f, \quad \text{in } \Omega \tag{14.82a}$$

$$u = g, \quad \text{on } \partial\Omega \tag{14.82b}$$

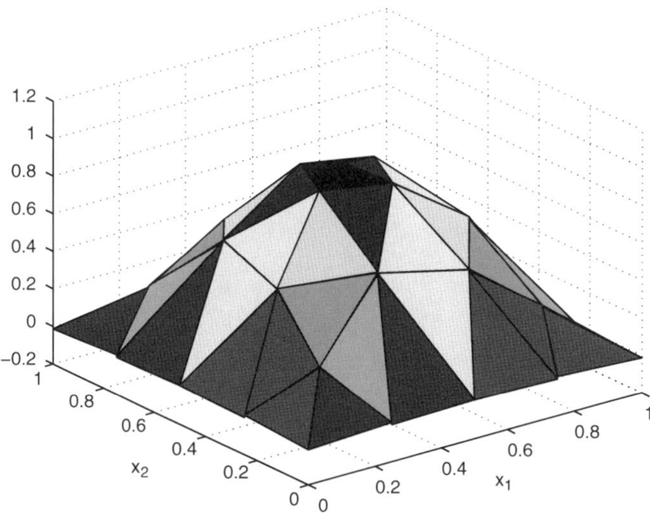

Fig. 14.3 Plot of u_h with $\beta = 36$ and $h = 0.25$

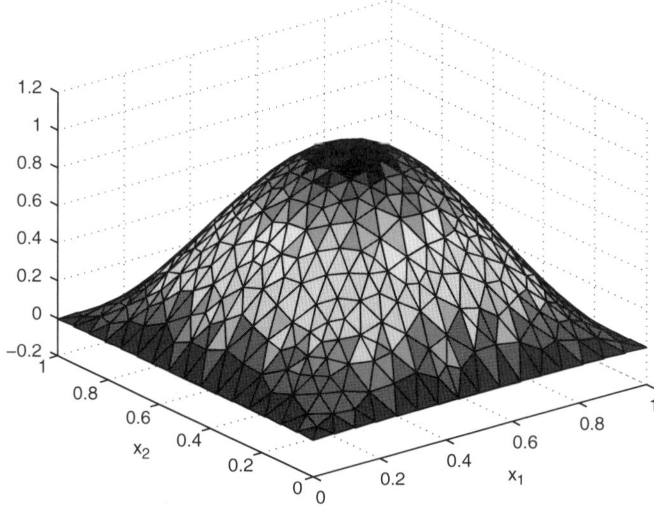

Fig. 14.4 Plot of u_h with $\beta = 9$ and $h = 0.0625$

The dG method takes the form: find $u_h \in V_h$ such that $a(u_h, v) = l_h(v)$ for all $v \in V_h$, where

$$a_h(u, v) = \sum_{K \in \mathcal{K}} \epsilon(\nabla u, \nabla v)_K + (b \cdot \nabla u, v)_K + (cu, v)_K \tag{14.83}$$

$$
\begin{aligned}
&- \sum_{E \in \mathcal{E}_I} \epsilon(\langle n \cdot \nabla u \rangle, [v])_E + \alpha \sum_{E \in \mathcal{E}_I} \epsilon([u], \langle n \cdot \nabla v \rangle)_E + \sum_{E \in \mathcal{E}_I} \beta h^{-1} \epsilon([u], [v])_E \\
&- \epsilon(n \cdot \nabla u, v)_{\partial\Omega} + \alpha\epsilon(u, n \cdot \nabla v)_{\partial\Omega} + \beta h^{-1}\epsilon(u, v)_{\partial\Omega} \\
&- \sum_{E \in \mathcal{E}_I} (n \cdot b[u], \langle v \rangle)_E + (\gamma n \cdot b[u], [v])_E - (n \cdot bu, v)_{\partial\Omega_-}
\end{aligned}
$$

$$
l(v) = (f, v) + \alpha\epsilon(g, n \cdot \nabla v)_{\partial\Omega} + \beta h^{-1}\epsilon(g, v)_{\partial\Omega} - (n \cdot bg, v)_{\partial\Omega_-} \tag{14.84}
$$

Since this dG method includes upwinding, which is stabilizing in case of high convection and low diffusion, it can be used as an alternative to the SD or GLS method.

Note that the smaller ϵ is the weaker the boundary condition $u = g$ is enforced.

14.4 Further Reading

We refer to the books by Di Pietro and Ern [24], Riviere [54], Cockburn, Karniadakis, Shu [21], and Hesthaven [42].

14.5 Problems

Exercise 14.1. Use `dG1PoissonSolver2D` to test the so-called NIPG method by changing the parameter α to $+1$.

Appendix A
Some Additional Matlab Code

A.1 Tri2Edge.m

The following routine numbers the edges of a triangle mesh.

```
function edges = Tri2Edge(p,t)
np=size(p,2); % number of vertices
nt=size(t,2); % number of triangles
i=t(1,:); % i=1st vertex within all elements
j=t(2,:); % j=2nd
k=t(3,:); % k=3rd
A=sparse(j,k,-1,np,np); % 1st edge is between (j,k)
A=A+sparse(i,k,-1,np,np); % 2nd               (i,k)
A=A+sparse(i,j,-1,np,np); % 3rd               (i,j)
A=-((A+A.')<0);
A=triu(A); % extract upper triangle of A
[r,c,v]=find(A); % rows, columns, and values(=-1)
v=[1:length(v)]; % renumber values (ie. edges)
A=sparse(r,c,v,np,np); % reassemble A
A=A+A'; % expand A to a symmetric matrix
edges=zeros(nt,3);
for k=1:nt
  edges(k,:)=[A(t(2,k),t(3,k))
              A(t(1,k),t(3,k))
              A(t(1,k),t(2,k))]';
end
```

Input is the standard point and triangle matrix `p` and `t`. Output is a $n_t \times 3$ matrix, with n_t the number of triangles, `edges` containing the edge numbers. In element i the global edge number of local edge j is given by `edges(i,j)`. In triangle i local edge j lies opposite local vertex j.

M.G. Larson and F. Bengzon, *The Finite Element Method: Theory, Implementation, and Applications*, Texts in Computational Science and Engineering 10,
DOI 10.1007/978-3-642-33287-6, © Springer-Verlag Berlin Heidelberg 2013

A.2 EdgeMidPoints.m

The following routine computes the coordinates of edge midpoints.

```
function [xmid,ymid,e] = EdgeMidPoints(p,t2e,t)
i=t(1,:); j=t(2,:); k=t(3,:); % triangle vertices
t2e=t2e(:); % all edges in a long row
start=[j i i]; % start vertices of all edges
stop =[k k j]; % stop
xmid=(p(1,start)+p(1,stop))/2; % mid point x-coordinates
ymid=(p(2,start)+p(2,stop))/2; %            y-
[e,idx]=unique(t2e); % remove duplicate edges
xmid=xmid(idx); % unique edge x-coordinates
ymid=ymid(idx); %             y-
```

Input is the standard point and triangle matrix `p` and `t`, and the edges `t2e` of every triangle, as defined by the output from `Tri2Edge.m`. Output, are the three vectors `e`, `xmid`, and `ymid`, which are the same length n_e, with `ne=max(t2n(:))` the number of unique edges, so that `xmid(i)` and `ymid(i)` are the midpoint of edge `e(i)`, $i = 1, \dots, n_e$.

A.3 Tri2Tri.m

The following routine finds neighbouring elements in a triangle mesh.

```
function neighbors = Tri2Tri(p,t)
np=size(p,2); % number of vertices
nt=size(t,2); % number of triangles
n2e=sparse(np,nt); % node-to-element adjacency matrix
                   % n2e(i,j)=1 means node "i" is in element "j"
for i=1:nt
  n2e(t(1:3,i),i)=ones(3,1);
end
neighbors=-ones(nt,3); % -1 means no neighbor
for i=1:nt
    % 1st edge lies between nodes t(2,i) and t(3,i), so search
    % the adjacency matrix for elements sharing these two nodes
    nb=intersect(find(n2e(t(2,i),:)),find(n2e(t(3,i),:)));
    nb=setdiff(nb,i); % remove element "i" from neighbors "nb"
    if isscalar(nb), neighbors(i,1)=nb(1); end
    % 2nd edge
    nb=intersect(find(n2e(t(3,i),:)),find(n2e(t(1,i),:)));
    nb=setdiff(nb,i);
    if isscalar(nb), neighbors(i,2)=nb(1); end
    % 3rd edge
    nb=intersect(find(n2e(t(1,i),:)),find(n2e(t(2,i),:)));
    nb=setdiff(nb,i);
    if isscalar(nb), neighbors(i,3)=nb(1); end
end
```

Input is the standard point and triangle matrix `p` and `t`. Output `neighbors` is a $n_t \times 3$ matrix, with n_t the number of triangles, in which row i contains the three element neighbours to element i. No neighbour is indicated by -1. Each row is ordered in the sense that the first neighbour shares edge 1 with the element, the second neighbour shares edge 2, and so on.

A.4 Dslitg.m

Geometry matrix for the double slit geometry.

```
function g = Dslitg()
g = [2          0     1.0000          0          0     1     0
     2     1.0000     1.0000          0     1.0000     1     0
     2     1.0000          0     1.0000     1.0000     1     0
     2    -0.2500          0     0.3333     0.3333     1     0
     2          0    -0.2500     0.4167     0.4167     1     0
     2    -0.2500          0     0.5833     0.5833     1     0
     2          0    -0.2500     0.6667     0.6667     1     0
     2          0          0          0     0.3333     0     1
     2          0          0     0.4167     0.5833     0     1
     2          0          0     0.6667     1.0000     0     1
     2    -0.2500    -0.2500     0.3333     0.4167     0     1
     2    -0.2500    -0.2500     0.5833     0.6667     0     1
    ]';
```

A.5 Airfoilg.m

Geometry matrix for a wing.

```
function g=Airfoilg()
g=[2    17.7218    16.0116     1.5737     1.6675     1     0
   2    16.0116     9.0610     1.6675     1.3668     1     0
   2     9.0610    -0.5759     1.3668    -0.1102     1     0
   2    -0.5759    -9.5198    -0.1102    -1.8942     1     0
   2    -9.5198   -15.6511    -1.8942    -2.5938     1     0
   2   -15.6511   -18.1571    -2.5938    -1.7234     1     0
   2   -18.1571   -16.9459    -1.7234     0.2051     1     0
   2   -16.9459   -12.4137     0.2051     2.2238     1     0
   2   -12.4137    -5.4090     2.2238     3.4543     1     0
   2    -5.4090     2.8155     3.4543     3.5046     1     0
   2     2.8155    10.6777     3.5046     2.6664     1     0
   2    10.6777    16.3037     2.6664     1.7834     1     0
```

```
   2    16.3037    17.7218     1.7834     1.5737     1     0
   2   -30.0000    30.0000   -15.0000   -15.0000     1     0
   2    30.0000    30.0000   -15.0000    15.0000     1     0
   2    30.0000   -30.0000    15.0000    15.0000     1     0
   2   -30.0000   -30.0000    15.0000   -15.0000     1     0]';
```

A.6 RectCircg.m

Geometry matrix for a rectangle with a cut out circle.

```
function g = RectCircg()
g=[ 2    2    2    2    1    1   1    1
    6    6   -2   -2   -1    0   1    0
    6   -2   -2    6    0    1   0   -1
   -2    2   -2   -2   -0   -1   0    1
    2    2    2   -2   -1    0   1    0
    1    1    0    1    0    0   0    0
    0    0    1    0    1    1   1    1
    0    0    0    0    0    0   0    0
    0    0    0    0    0    0   0    0
    0    0    0    0    1    1   1    1];
```

A.7 DFGg.m

Geometry matrix for the DFG benchmark.

```
function g = DFGg()
g=[2      2      2      2      1      1      1      1
   2.20   2.20   0      0      0.15   0.20   0.25   0.20
   2.20   0      0      2.20   0.20   0.25   0.20   0.15
   0      0.41   0      0      0.20   0.15   0.20   0.25
   0.41   0.41   0.41   0      0.15   0.20   0.25   0.20
   1      1      0      1      0      0      0      0
   0      0      1      0      1      1      1      1
   0      0      0      0      0.20   0.20   0.20   0.20
   0      0      0      0      0.20   0.20   0.20   0.20
   0      0      0      0      0.05   0.05   0.05   0.05];
```

A.8 Scatterg.m

Geometry matrix for the scattering cylinder.

```
function g = Scatterg()
g=[2     5     5    -5    -3     5     0     0     0     0
   2     5     5    -3     3     8     0     0     0     0
   2     5     5     3     5     3     0     0     0     0
   2    -5    -3     5     5     0     1     0     0     0
   2    -3     3     5     5     0     2     0     0     0
   2     3     5     5     5     0     3     0     0     0
   2    -5    -3    -3    -3     6     4     0     0     0
   2     3     5    -3    -3     8     5     0     0     0
   2    -5    -3     3     3     1     6     0     0     0
   2    -3     3     3     3     2     7     0     0     0
   2     3     5     3     3     3     8     0     0     0
   2     3     3    -5    -3     9     5     0     0     0
   2     3     3    -3     3     7     8     0     0     0
   2     3     3     3     5     2     3     0     0     0
   2    -3    -3    -5    -3     4     9     0     0     0
   2    -3    -3     3     5     1     2     0     0     0
   2    -5    -5    -5    -3     0     4     0     0     0
   2    -5    -5    -3     3     0     6     0     0     0
   2    -5    -5     3     5     0     1     0     0     0
   2    -5    -3    -5    -5     4     0     0     0     0
   2    -3     3    -5    -5     9     0     0     0     0
   2     3     5    -5    -5     5     0     0     0     0
   2    -3     3    -3    -3     7     9     0     0     0
   2    -3    -3    -3     3     6     7     0     0     0
   1    -1     0    -0    -1     0     7     0     0     1
   1     0     1    -1     0     0     7     0     0     1
   1     1     0     0     1     0     7     0     0     1
   1     0    -1     1    -0     0     7     0     0     1
]';
```

References

1. R. Adams and J. Fournier. *Sobolev Spaces*. Pure and Applied Mathematics. Academic Press, 2003.
2. M. Ainsworth and J. T. Oden. *A Posteriori Error Estimation in Finite Element Analysis*. John Wiley & Sons, 2000.
3. O. Axelsson and V. Barker. *Finite Element Solution of Boundary Value Problems*. Academic Press, 1984.
4. H. Bang and Y. W. Kwon. *The Finite Element Method Using MATLAB*. Mechanical Engineering Series. CRC Press, 2000.
5. W. Bangerth and R. Rannacher. *Adaptive Finite Element Methods for Differential Equations*. Birkhäuser, 2003.
6. R. Bank, A. Sherman, and A. Weiser. Refinement algorithms and data structures for regular local mesh refinement. In R. Stepleman, editor, *Scientific Computing*, pages 3–17, 1983.
7. R. Barrett, M. Berry, T. F. Chan, J. Demmel, J. Donato, J. Dongarra, V. Eijkhout, R. Pozo, C. Romine, and H. Van der Vorst. *Templates for the Solution of Linear Systems: Building Blocks for Iterative Methods, 2nd Edition*. SIAM, 1994.
8. K.-J. Bathe. *Finite Element Procedures*. Prentice-Hall, 1996.
9. T. Belytschko, W. Liu, and B. Moran. *Nonlinear Finite Elements for Continua and Structures*. J. Wiley & Sons, Chichester, 2000.
10. M. Benzi. Preconditioning techniques for large linear systems: A survey. *Journal of Computational Physics*, 182(2):418–477, 2002.
11. J. Bey. Tetrahedral grid refinement. *Computing*, 55:355–378, October 1995.
12. D. Boffi, F. Brezzi, and M. Fortin. Finite elements for the stokes problem. In *Mixed Finite Elements, Compatibility Conditions, and Applications*, volume 1939 of *Lecture Notes in Mathematics*, pages 45–100. Springer, 2008.
13. A. Bondesson, T. Rylander, and P. Ingelström. *Computational Electromagnetics*. Springer, 2000.
14. J. Bonet and R. Wood. *Nonlinear Continuum Mechanics for Finite Element Analysis*. Cambridge University Press, 2000.
15. D. Braess. *Finite Elements*. Cambridge University Press, 2007.
16. M. Chari and S. Salon. *Numerical Methods in Electromagnetism*. Academic Press, 2000.
17. A. Chorin. Numerical solution of the navier-stokes equations. *Mathematics of Computation*, 22:745–762, 1968.
18. A. Chorin and J. Marsden. *A Mathematical Introduction to Fluid Mechanics*. Springer, 1993.
19. P. Ciarlet. *The Finite Element Method for Elliptic Problems*. Classics in Applied Mathematics, 40. SIAM, 2002.

M.G. Larson and F. Bengzon, *The Finite Element Method: Theory, Implementation, and Applications*, Texts in Computational Science and Engineering 10,
DOI 10.1007/978-3-642-33287-6, © Springer-Verlag Berlin Heidelberg 2013

20. P. Clément. Approximation by finite element functions using local regularization. *Rev. Française Automat. Informat. Recherche Opérationnelle Sér. Analyse Numérique*, 9:77–84, 1975.
21. B. Cockburn, G. Karniadakis, and C.-W. Shu. *Discontinuous Galerkin Methods: Theory, Computation, and Applications*. Springer, 2000.
22. M. Crisfield, R. De Borst, J. Remmers, and C. Verhoosel. *Nonlinear Finite Element Analysis of Solids and Structures*. J. Wiley & Sons, 2012.
23. J. Demmel:. *Applied Numerical Linear Algebra*. SIAM, 1997.
24. D.A. Di Pietro and A. Ern. *Mathematical aspects of discontinuous Galerkin methods*, volume 69. Springer Verlag, 2011.
25. I. Duff, M. Erisman, and J. Reid. *Direct Methods for Sparse Matrices*. Oxford Science Publications, 1986.
26. H. Elman, D. Silvester, and A. Wathen. *Finite Elements and Fast Iterative Solvers*. Oxford University Press, 2005.
27. K. Eriksson, D. Estep, P. Hansbo, and C. Johnson. *Computational Differential Equations*. Studentliteratur, 1996.
28. G. Evans. *Practical Numerical Integration*. John Wiley & Sons, 1993.
29. L. Evans. *Partial Differential Equations*, volume 19 of *Graduate Studies in Mathematics*. American Mathematical Society, 1998.
30. G. Folland. *Introduction to Partial Differential Equations*. Princeton University Press, 1995.
31. M. Fortin and F. Brezzi. *Mixed and Hybrid Finite Element Methods*. Springer Series in Computational Mathematics. Springer, 1991.
32. A. George and J. Liu. *Computer Solution of Large Sparse Positive Definite Systems*. Prentice Hall, 1981.
33. P.-L. George and S. Frey. *Mesh Generation: Application to Finite Elements*. John Wiley & Sons, 2008.
34. V. Girault and P.-A. Raviart. *Finite Element Methods for Navier-Stokes Equations: Theory and Algorithms*. Springer-Verlag, 1986.
35. M. Gockenbach. *Partial Differential Equations: Analytical and Numerical Methods*. SIAM, 2002.
36. M. Gockenbach. *Understanding and Implementing the Finite Element Method*. SIAM, 2006.
37. J.-L. Guermond and A. Ern. *Theory and Practice of Finite Elements*. Applies Mathematical Sciences. Springer-Verlag, 2004.
38. M. Gunzburger and R. Nicolaides. *Incompressible Computational Fluid Dynamics*. Cambridge University Press, 1993.
39. W. Hackbusch. *Multi-Grid Methods and Applications*. Springer, 2003.
40. E. Hairer, S.P. Nørsett, and G. Wanner. *Solving Ordinary Differential Equations: Stiff and differential-algebraic problems*. Springer series in computational mathematics. Springer-Verlag, 1993.
41. M. Heath. *Scientific Computing*. McGraw Hill, 1996.
42. J. S. Hesthaven and T. Warburton. *Nodal Discontinuous Galerkin Methods: Algorithms, Analysis, and Applications*. Springer, 2008.
43. T. Hughes. *The Finite Element Method, Linear Static and Dynamic Finite Element Analysis*. Dover Publications, 2000.
44. J.-M. Jin. *The Finite Element Method in Electromagnetism*. IEEE Computer Society Press, 2002.
45. C. Johnson. *Numerical Solution of Partial Differential Equations by the Finite Element Method*. Studentlitteratur, 1987.
46. G. Karypis and V. Kumar. Metis – unstructured graph partitioning and sparse matrix ordering system, version 2.0. Technical report, 1995.
47. P. Knabner and L. Angermann. *Numerical Methods for Elliptic and Parabolic Partial Differential Equations*. Springer-Verlag, 2000.
48. E. Kreyszig. *Introductory Functional Analysis with Application*. Wiley, 1989.
49. P. Monk. *Finite Element Methods for Maxwell's Equations*. Clarendon Press, 2003.

50. J. Nitsche. Ein kriterium für die quasi-optimalität des ritzschen verfahrens. *Numerische Mathematik*, 11:346–348, 1968.
51. N. Ottosen and H. Petersson. *Introduction to the Finite Element Method*. Prentice-Hall, 1992.
52. J. Reddy. *An Introduction To Nonlinear Finite Element Analysis*. Oxford University Press, 2004.
53. M.-C. Rivara, C. Calderon, A. Fedorov, and N. Chrisochoides. Parallel decoupled terminal-edge bisection method for 3d mesh generation. *Engineering with Computers*, 22:111–119, August 2006.
54. B. Riviere. *Discontinuous Galerkin methods for solving elliptic and parabolic equations: theory and implementation*, volume 35. Siam, 2008.
55. H. G. Roos, M. Stynes, and L. Tobiska. *Numerical Methods for Singularly Perturbed Differential Equations*. Springer, 1996.
56. W. Rudin. *Functional Analysis*. International Series in Pure and Applied Mathematics. McGraw-Hill, 1991.
57. Y. Saad. *Iterative Methods for Sparse Linear Systems*. SIAM, 2003.
58. R. L. Sani and P. M. Gresho. *Incompressible Flow and the Finite Element Method, Isothermal Laminar Flow*. Mechanical Engineering. John Wiley & Sons, 2000.
59. J. Schöeberl. A posteriori error estimates for Maxwell equations. *Mathematics of Computation*, 77:633–649, 2008.
60. R. Scott and S. Brenner. *The Mathematical Theory of Finite Element Methods*. Springer, 2008.
61. R. Scott and S. Zhang. Finite element interpolation of non-smooth functions satisfying boundary conditions. *Mathematics of Computation*, 54(190):483–493, 1990.
62. K. Segeth, I. Dolezel, and P. Solin. *Higher-order Finite Element Methods*. Chapman & Hall/CRC, 2003.
63. J. Shewchuk. An introduction to the conjugate gradient method without the agonizing pain. Technical report, Pittsburgh, PA, USA, 1994.
64. J. Shewchuk. *Tetrahedral Mesh Generation by Delaunay Refinement*, volume 4, pages 86–95. ACM Press, 1998.
65. J. Shewchuk. Delaunay refinement algorithms for triangular mesh generation. *Computational Geometry*, 22(1-3):21–74, 2002.
66. P. Solin. *Partial Differential Equations and the Finite Element Method*. Wiley-Interscience, 2006.
67. J. Strikwerda. *Finite Difference Schemes and Partial Differential Equations*. SIAM, 2006.
68. B. Szabo and I. Babuska. *Finite Element Analysis*. North Holland, 1991.
69. R. Temam. *Navier-Stokes Equations*. North Holland, 1977.
70. V. Thomée and S. Larsson. *Partial Differential Equations with Numerical Methods*. Texts in applies Mathematics. Springer, 2005.
71. S. Turek. *Efficient Solvers for Incompressible Flow problems: An Algorithmic and Computational Approach*, volume 6. Springer-Verlag, 1999.
72. A. Valli and A. Quarteroni. *Numerical Approximation of Partial Differential Equations*. Springer-Verlag, 1994.
73. P. Wriggers. *Nonlinear Finite Element Methods*. Springer, 2009.
74. O. Zienkiewicz and R. Taylor. *The Finite Element Method for Solid and Structural Mechanics*, volume 2. Elsevier Butterworth-Heinemann, 2005.
75. O. Zienkiewicz, R. Taylor, and P. Nithiarasu. *The Finite Element Method for Fluid Dynamics*, volume 3. Elsevier Butterworth-Heinemann, 2005.
76. O. Zienkiewicz, R. Taylor, and J. Zhu. *The Finite Element Method: Its Basis and Fundamentals*, volume 1. Elsevier Butterworth-Heinemann, 2005.

Index

M.G. Larson and F. Bengzon, *The Finite Element Method: Theory, Implementation, and Applications*, Texts in Computational Science and Engineering 10,
DOI 10.1007/978-3-642-33287-6,

Editorial Policy

1. Textbooks on topics in the field of computational science and engineering will be considered. They should be written for courses in CSE education. Both graduate and undergraduate textbooks will be published in TCSE. Multidisciplinary topics and multidisciplinary teams of authors are especially welcome.
2. Format: Only works in English will be considered. For evaluation purposes, manuscripts may be submitted in print or electronic form, in the latter case, preferably as pdf- or zipped ps-files. Authors are requested to use the LaTeX style files available from Springer at: http://www.springer.com/authors/book+authors/helpdesk?SGWID=0-1723113-12-971304-0 (Click on → Templates → LaTeX → monographs)
   Electronic material can be included if appropriate. Please contact the publisher.
3. Those considering a book which might be suitable for the series are strongly advised to contact the publisher or the series editors at an early stage.

General Remarks

Careful preparation of manuscripts will help keep production time short and ensure a satisfactory appearance of the finished book.

The following terms and conditions hold:

Regarding free copies and royalties, the standard terms for Springer mathematics textbooks hold. Please write to martin.peters@springer.com for details.

Authors are entitled to purchase further copies of their book and other Springer books for their personal use, at a discount of 33.3% directly from Springer-Verlag.

Series Editors

Texts in Computational Science and Engineering

1. H. P. Langtangen, *Computational Partial Differential Equations.* Numerical Methods and Diffpack Programming, 2nd Edition.
2. A. Quarteroni, F. Saleri, P. Gervasio, *Scientific Computing with MATLAB and Octave*, 3rd Edition.
3. H. P. Langtangen, *Python Scripting for Computational Science*, 3rd Edition.
4. H. Gardner, G. Manduchi, *Design Patterns for e-Science.*
5. M. Griebel, S. Knapek, G. Zumbusch, *Numerical Simulation in Molecular Dynamics.*
6. H. P. Langtangen, *A Primer on Scientific Programming with Python*, 3rd Edition.
7. A. Tveito, H. P. Langtangen, B. F. Nielsen, X. Cai, *Elements of Scientific Computing.*
8. B. Gustafsson, *Fundamentals of Scientific Computing.*
9. M. Bader, *Space-Filling Curves.*
10. M.G. Larson, F. Bengzon, *The Finite Element Method: Theory, Implementation, and Applications.*

For further information on these books please have a look at our mathematics catalogue at the following URL: www.springer.com/series/5151

Monographs in Computational Science and Engineering

1. J. Sundnes, G.T. Lines, X. Cai, B.F. Nielsen, K.-A. Mardal, A. Tveito, *Computing the Electrical Activity in the Heart.*

For further information on this book, please have a look at our mathematics catalogue at the following URL: www.springer.com/series/7417

Lecture Notes in Computational Science and Engineering

1. D. Funaro, *Spectral Elements for Transport-Dominated Equations.*
2. H.P. Langtangen, *Computational Partial Differential Equations.* Numerical Methods and Diffpack Programming.
3. W. Hackbusch, G. Wittum (eds.), *Multigrid Methods V.*
4. P. Deuflhard, J. Hermans, B. Leimkuhler, A.E. Mark, S. Reich, R.D. Skeel (eds.), *Computational Molecular Dynamics: Challenges, Methods, Ideas.*
5. D. Kröner, M. Ohlberger, C. Rohde (eds.), *An Introduction to Recent Developments in Theory and Numerics for Conservation Laws.*

6. S. Turek, *Efficient Solvers for Incompressible Flow Problems.* An Algorithmic and Computational Approach.
7. R. von Schwerin, ***M**ulti **B**ody **S**ystem **SIM**ulation.* Numerical Methods, Algorithms, and Software.
8. H.-J. Bungartz, F. Durst, C. Zenger (eds.), *High Performance Scientific and Engineering Computing.*
9. T.J. Barth, H. Deconinck (eds.), *High-Order Methods for Computational Physics.*
10. H.P. Langtangen, A.M. Bruaset, E. Quak (eds.), *Advances in Software Tools for Scientific Computing.*
11. B. Cockburn, G.E. Karniadakis, C.-W. Shu (eds.), *Discontinuous Galerkin Methods.* Theory, Computation and Applications.
12. U. van Rienen, *Numerical Methods in Computational Electrodynamics.* Linear Systems in Practical Applications.
13. B. Engquist, L. Johnsson, M. Hammill, F. Short (eds.), *Simulation and Visualization on the Grid.*
14. E. Dick, K. Riemslagh, J. Vierendeels (eds.), *Multigrid Methods VI.*
15. A. Frommer, T. Lippert, B. Medeke, K. Schilling (eds.), *Numerical Challenges in Lattice Quantum Chromodynamics.*
16. J. Lang, *Adaptive Multilevel Solution of Nonlinear Parabolic PDE Systems.* Theory, Algorithm, and Applications.
17. B.I. Wohlmuth, *Discretization Methods and Iterative Solvers Based on Domain Decomposition.*
18. U. van Rienen, M. Günther, D. Hecht (eds.), *Scientific Computing in Electrical Engineering.*
19. I. Babuška, P.G. Ciarlet, T. Miyoshi (eds.), *Mathematical Modeling and Numerical Simulation in Continuum Mechanics.*
20. T.J. Barth, T. Chan, R. Haimes (eds.), *Multiscale and Multiresolution Methods.* Theory and Applications.
21. M. Breuer, F. Durst, C. Zenger (eds.), *High Performance Scientific and Engineering Computing.*
22. K. Urban, *Wavelets in Numerical Simulation.* Problem Adapted Construction and Applications.
23. L.F. Pavarino, A. Toselli (eds.), *Recent Developments in Domain Decomposition Methods.*
24. T. Schlick, H.H. Gan (eds.), *Computational Methods for Macromolecules: Challenges and Applications.*
25. T.J. Barth, H. Deconinck (eds.), *Error Estimation and Adaptive Discretization Methods in Computational Fluid Dynamics.*
26. M. Griebel, M.A. Schweitzer (eds.), *Meshfree Methods for Partial Differential Equations.*

27. S. Müller, *Adaptive Multiscale Schemes for Conservation Laws.*
28. C. Carstensen, S. Funken, W. Hackbusch, R.H.W. Hoppe, P. Monk (eds.), *Computational Electromagnetics.*
29. M.A. Schweitzer, *A Parallel Multilevel Partition of Unity Method for Elliptic Partial Differential Equations.*
30. T. Biegler, O. Ghattas, M. Heinkenschloss, B. van Bloemen Waanders (eds.), *Large-Scale PDE-Constrained Optimization.*
31. M. Ainsworth, P. Davies, D. Duncan, P. Martin, B. Rynne (eds.), *Topics in Computational Wave Propagation.* Direct and Inverse Problems.
32. H. Emmerich, B. Nestler, M. Schreckenberg (eds.), *Interface and Transport Dynamics.* Computa- tional Modelling.
33. H.P. Langtangen, A. Tveito (eds.), *Advanced Topics in Computational Partial Differential Equations.* Numerical Methods and Diffpack Programming.
34. V. John, *Large Eddy Simulation of Turbulent Incompressible Flows.* Analytical and Numerical Results for a Class of LES Models.
35. E. Bänsch (ed.), *Challenges in Scientific Computing - CISC 2002.*
36. B.N. Khoromskij, G. Wittum, *Numerical Solution of Elliptic Differential Equations by Reduction to the Interface.*
37. A. Iske, *Multiresolution Methods in Scattered Data Modelling.*
38. S.-I. Niculescu, K. Gu (eds.), *Advances in Time-Delay Systems.*
39. S. Attinger, P. Koumoutsakos (eds.), *Multiscale Modelling and Simulation.*
40. R. Kornhuber, R. Hoppe, J. Périaux, O. Pironneau, O. Wildlund, J. Xu (eds.), *Domain Decomposition Methods in Science and Engineering.*
41. T. Plewa, T. Linde, V.G. Weirs (eds.), *Adaptive Mesh Refinement – Theory and Applications.*
42. A. Schmidt, K.G. Siebert, *Design of Adaptive Finite Element Software.* The Finite Element Toolbox ALBERTA.
43. M. Griebel, M.A. Schweitzer (eds.), *Meshfree Methods for Partial Differential Equations II.*
44. B. Engquist, P. Lötstedt, O. Runborg (eds.), *Multiscale Methods in Science and Engineering.*
45. P. Benner, V. Mehrmann, D.C. Sorensen (eds.), *Dimension Reduction of Large-Scale Systems.*
46. D. Kressner, *Numerical Methods for General and Structured Eigenvalue Problems.*
47. A. Boriçi, A. Frommer, B. Joó, A. Kennedy, B. Pendleton (eds.), *QCD and Numerical Analysis III.*
48. F. Graziani (ed.), *Computational Methods in Transport.*
49. B. Leimkuhler, C. Chipot, R. Elber, A. Laaksonen, A. Mark, T. Schlick, C. Schütte, R. Skeel (eds.), *New Algorithms for Macromolecular Simulation.*

50. M. Bücker, G. Corliss, P. Hovland, U. Naumann, B. Norris (eds.), *Automatic Differentiation: Applications, Theory, and Implementations.*
51. A.M. Bruaset, A. Tveito (eds.), *Numerical Solution of Partial Differential Equations on Parallel Computers.*
52. K.H. Hoffmann, A. Meyer (eds.), *Parallel Algorithms and Cluster Computing.*
53. H.-J. Bungartz, M. Schäfer (eds.), *Fluid-Structure Interaction.*
54. J. Behrens, *Adaptive Atmospheric Modeling.*
55. O. Widlund, D. Keyes (eds.), *Domain Decomposition Methods in Science and Engineering XVI.*
56. S. Kassinos, C. Langer, G. Iaccarino, P. Moin (eds.), *Complex Effects in Large Eddy Simulations.*
57. M. Griebel, M.A Schweitzer (eds.), *Meshfree Methods for Partial Differential Equations III.*
58. A.N. Gorban, B. Kégl, D.C. Wunsch, A. Zinovyev (eds.), *Principal Manifolds for Data Visualization and Dimension Reduction.*
59. H. Ammari (ed.), *Modeling and Computations in Electromagnetics: A Volume Dedicated to Jean-Claude Nédélec.*
60. U. Langer, M. Discacciati, D. Keyes, O. Widlund, W. Zulehner (eds.), *Domain Decomposition Methods in Science and Engineering XVII.*
61. T. Mathew, *Domain Decomposition Methods for the Numerical Solution of Partial Differential Equations.*
62. F. Graziani (ed.), *Computational Methods in Transport: Verification and Validation.*
63. M. Bebendorf, *Hierarchical Matrices.* A Means to Efficiently Solve Elliptic Boundary Value Problems.
64. C.H. Bischof, H.M. Bücker, P. Hovland, U. Naumann, J. Utke (eds.), *Advances in Automatic Differentiation.*
65. M. Griebel, M.A. Schweitzer (eds.), *Meshfree Methods for Partial Differential Equations IV.*
66. B. Engquist, P. Lötstedt, O. Runborg (eds.), *Multiscale Modeling and Simulation in Science.*
67. I.H. Tuncer, Ü. Gülcat, D.R. Emerson, K. Matsuno (eds.), *Parallel Computational Fluid Dynamics 2007.*
68. S. Yip, T. Diaz de la Rubia (eds.), *Scientific Modeling and Simulations.*
69. A. Hegarty, N. Kopteva, E. O'Riordan, M. Stynes (eds.), *BAIL* 2008 – *Boundary and Interior Layers.*
70. M. Bercovier, M.J. Gander, R. Kornhuber, O. Widlund (eds.), *Domain Decomposition Methods in Science and Engineering XVIII.*

71. B. Koren, C. Vuik (eds.), *Advanced Computational Methods in Science and Engineering.*
72. M. Peters (ed.), *Computational Fluid Dynamics for Sport Simulation.*
73. H.-J. Bungartz, M. Mehl, M. Schäfer (eds.), *Fluid Structure Interaction II – Modelling, Simulation, Optimization.*
74. D. Tromeur-Dervout, G. Brenner, D.R. Emerson, J. Erhel (eds.), *Parallel Computational Fluid Dynamics 2008.*
75. A.N. Gorban, D. Roose (eds.), *Coping with Complexity: Model Reduction and Data Analysis.*
76. J.S. Hesthaven, E.M. Rønquist (eds.), *Spectral and High Order Methods for Partial Differential Equations.*
77. M. Holtz, *Sparse Grid Quadrature in High Dimensions with Applications in Finance and Insurance.*
78. Y. Huang, R. Kornhuber, O. Widlund, J. Xu (eds.), *Domain Decomposition Methods in Science and Engineering XIX.*
79. M. Griebel, M.A. Schweitzer (eds.), *Meshfree Methods for Partial Differential Equations V.*
80. P.H. Lauritzen, C. Jablonowski, M.A. Taylor, R.D. Nair (eds.), *Numerical Techniques for Global Atmospheric Models.*
81. C. Clavero, J.L. Gracia, F. Lisbona (eds.), *BAIL 2010 – Boundary and Interior Layers, Computational and Asymptotic Methods.*
82. B. Engquist, O. Runborg, Y.R. Tsai (eds.), *Numerical Analysis and Multiscale Computations.*
83. I.G. Graham, T.Y. Hou, O. Lakkis, R. Scheichl (eds.), *Numerical Analysis of Multiscale Problems.*
84. A. Logg, K.-A. Mardal, G. Wells (eds.), *Automated Solution of Differential Equations by the Finite Element Method.*
85. J. Blowey, M. Jensen (eds.), *Frontiers in Numerical Analysis - Durham 2010.*
86. O. Kolditz, U.-J. Gorke, H. Shao, W. Wang (eds.), *Thermo-Hydro-Mechanical-Chemical Processes in Fractured Porous Media - Benchmarks and Examples.*
87. S. Forth, P. Hovland, E. Phipps, J. Utke, A. Walther (eds.), *Recent Advances in Algorithmic Differentiation.*
88. J. Garcke, M. Griebel (eds.), *Sparse Grids and Applications.*
89. M. Griebel, M. A. Schweitzer (eds.), *Meshfree Methods for Partial Differential Equations VI.*
90. C. Pechstein, *Finite and Boundary Element Tearing and Interconnecting Solvers for Multiscale Problems.*

For further information on these books please have a look at our mathematics catalogue at the following URL: www.springer.com/series/3527

Printed by Printforce, the Netherlands